The HOME THEATER Companion

The HOME THEATER Companion

Buying, Installing, and Using Today's Audio-Visual Equipment

Howard Ferstler

Schirmer Books
An Imprint of Simon & Schuster Macmillan
New York

Prentice Hall International
London • Mexico City • New Delhi • Singapore • Sydney • Toronto

Schirmer Books
An Imprint of Simon & Schuster Macmillan
1633 Broadway
New York, New York 10019

Library of Congress Catalog Number: 97-7337

Printed in the United States of America

Printing number
1 2 3 4 5 6 7 8 9 10

Library of Congress Cataloging-in-Publication Data

Ferstler, Howard, 1943–
The home theater companion : buying, installing, and using today's audio-visual equipment / Howard Ferstler.
p. cm.
Includes bibliographical references and index.
ISBN 0-02-864715-7
1. Home entertainment systems. 2. Home video systems. 3. Stereophonic sound systems. I. Title.
TK7881.3.F47 1997
621.389'7—dc21 97-7337
CIP

This paper meets the requirements ANSI/NISO Z39.48-1992 (Permanence of Paper).

Contents

Preface

The ideal TV screen should act as a perfectly transparent window, reproducing a movie or other program without introducing distortion of its own.

—Carleton Sarver, "Clearing the Picture," *High Fidelity* magazine (November 1988)

The purpose of high-fidelity equipment is to reproduce as closely as possible the experience of the concert hall, not to transcend or improve it.

—Edgar Villchur, *Reproduction of Sound* (1965)

IF YOU ARE CRAZY about watching TV—whether it involves cable, videotapes, discs, or satellite-TV programs—and/or you collect and listen to compact-disc recordings, then it is possible that you already own a decent audio or audio-video system. However, the fact that you are looking at this book indicates that you may be interested in obtaining a more potent "home-theater" and/or sound-reproduction system than you now own, or are at least interested in making what you presently have work better.

This book will appeal to the frugal reader, because one of its goals is to show how you can obtain maximum audio-visual bang for the buck, particularly when purchasing items from department or appliance stores, or by mail order. For the really parsimonious, it even explores the value of shopping for used equipment. The book will also be of great value to anyone with a "cost-is-no-object" mentality. These buyers may be surprised to find that less-expensive equipment might perform as well as—if not better than—the pricey competition.

Whether you are a big spender or just an old-fashioned tightfisted individual, you will find that this book is just the thing to keep you from being victimized by predatory equipment salespeople who would happily sell you gear that is inferior to other products that cost a lot less.

Those with a background in both audio and video or no formal training in electronics, optics, acoustics, or physics will be happy to know that the book is not beyond them by any means. While the book *can* certainly be used as a basic text on the subject of audio and video, it is first and foremost a shopping guide and manual for equipment use. Each chapter is structured so that the reader may, if the going gets tough, skip over the gory details and plow right into shopping techniques, set-up methods, and operational procedures. At the same time, it has been designed so that careful reading will advance the readers' knowledge enough for them to hold their own in future battles against even the most predatory sales clerk, or understand the proclamations of even the most jargon-spouting journalists.

This new volume is an effort of one serious home-theater and audio enthusiast to outline a number of ways that fairly well-heeled, level-headed, intelligent, but still essentially bargain-oriented, individuals can acquire systems of surprising quality. While primarily aimed at readers with budgetary constraints, the book still discusses "high-end" video and upscale audio in order to highlight the cutting edge of these fast-evolving areas of home-entertainment technology. Happily, the important, initially costly breakthroughs will eventually show up in more reasonably priced hardware. The book also includes a large number of "how-to" and "this-is-how" drawings and graphs, as well as many often-anecdotal sidebars.

The book generally bypasses blatantly "high-end" audio, except as it relates to surround sound and deep-bass reproduction (spending big bucks can pay off here). However, those wanting upscale, basic audio performance usually need only obtain more powerful amplifiers (even some self-contained audio receivers have near-earthmoving power; and THX amplifier certification is a good, although not necessarily essential, guarantee of amplifier design integrity), maybe a good subwoofer, and a pair of equally substantial main-channel speakers.

THX

Ultra-expensive amps and speaker systems may be nothing more than overkill, like putting a racing engine in a garden tiller. What's more, some high-end amplifiers are actually detrimental to good performance, because they exhibit substandard behavior compared with a number of reasonably priced models. In addition, some high-end speakers, particularly those in really small cabinets, perform no better than many economy models.

The material that follows builds upon topics that I first explored when writing guest essays for *Stereo Review, Digital Audio* (now called *CD Review*), *High Fidelity*, and the Boston Audio Society magazine, *Speaker*, as well as regular columns for *Fanfare, The Sensible Sound*, and *The American Record Guide*. It also contains ideas that I explored in my book of record reviews,

High Definition Compact Disc Recordings (McFarland, 1994), and in my earlier book on A/V, *High Fidelity Audio/Video Systems* (McFarland, 1991). While this new book covers a lot more area—particularly in the realms of home theater, video, and surround sound—its main function is to simplify, clarify, or elaborate (often extensively) upon the themes dealt with in my earlier books, my essays, and essays written by others—and make them more understandable. I believe this is the most comprehensive, single-volume piece of literature on home theater and audio-video systems available.

While most equipment manufacturers (and most electrical engineers and audio-video journalists) have a good grasp of basic electronics, a surprising number seem ignorant of home-theater performance requirements and even basic acoustics. Of course, there are exceptions to every rule, and I must thank Roy Allison, Fred Davis, Tom Nousaine, Harry Munz, and Tom Tyson for their assistance. If these accomplished audio and video experts had not diligently proofread (and reread) assorted draft chapters and spotted numerous technical and grammatical errors—and discussed audio and video theory with me at length, both by mail and in phone conversations—this volume would never have been taken seriously by any publisher. I should also thank my wife for putting up with my frequently odd, reclusive behavior during lengthy periods of literary toil.

To aid you in finding useful information as quickly as possible, I have created the following three icons that will alert you to different levels of help:

LINKS to helpful information related to this topic elsewhere in this book.

TIPS that are especially helpful to less-experienced or novice buyers.

DEEP WATER points out information that is appropriate for the "high divers." This material packs a bit more detail, but it's well worth the time and may save you money!

key terms

For rapid information searches, key terms appear in a distinctive typeface to the left of the passage where they first occur, and glossary items have been *italicized.*

1 Introduction

HOME ENTERTAINMENT—THAT IS, *ELECTRONIC* home entertainment—has come of age. For Americans who want to spend their surplus cash intelligently, exotic vacations, dog and/or cat shows, tractor pulls, swimming pools, bungee jumping, Jacuzzis, and kindred activities are passé. Replacing these diversions are three marvelous, rapidly evolving electronic pastimes: home computers, home theater, and home audio. The concept of the "electronic hearth" where the family gathers to socialize has now expanded to include entertainment and learning possibilities of astonishing quality. These days, it seems that everyone wants to have a PC and a decent audio-video installation.

home computers

Home computer systems will no doubt someday integrate most home-based, electromechanical systems: heating and cooling, security, cooking, planning, and, of course, home-theater and audio installations. However, this is still some way off in the future, although Sony's WebTV system can now link our TV sets to the Internet. For some of us, the home computer is becoming, at best, more of a home necessity and education tool than a home-entertainment item; at worst, the PC is nothing but a reminder, or for many workers even an extension, of the grind at the office. For now, let's analyze what many would deem the more rewarding aspects of the electronic revolution: straightforward home theater and home audio.

video

Video as a "serious" hobby is a fairly new phenomenon, but as it developed, it was destined to become the foundation of the home-theater revolution. Prior to serious video, the only way to even remotely simulate a movie-theater experience in the home involved building a small projection room in your house and obtaining 16-mm copies of commercial theater presentations. Unfortunately, this was nearly impossible for people not connected with the motion-picture establishment or with lots of cash to spare. What's more, while those film formats worked, in actuality they lacked flexibility, were awkward to work with, and had unimpressive sound. There is little doubt that current, video-based, home-theater systems are better in nearly every way than old-fashioned film projectors.

As one might expect, video and modern home theater have grown up together. Prior to the home-theater revolution, only a few years ago (particularly during the black-and-white era), we had "television." However, the lack of high-fidelity sound, the small size of the screen, and the not-always sharp, not-always color-accurate picture made the idea of theater in the home appear impossible. Basic TV performance was capable of giving us home entertainment, but it was not suitable for high-impact home theater.

home theater

As video pictures got better—colored, larger, and clearer—the idea of a *theater* in the home began to form among a few manufacturers and journalists. Until relatively high-quality, larger-screened color television sets were built and the video tape-rental revolution triggered the outpouring of reasonably good audio-video copies of motion-picture films, there was no way for mass-market consumers to have a theater-type motion-picture experience anywhere but in a theater.

While good video is mandatory for modern home theater, high-impact sound is nearly as important. For many long-time audio enthusiasts who have also embraced the home-theater revolution, good sound can often be more critical than video. Indeed, some high-end, home-theater journals, many of which are managed and staffed by serious audio enthusiasts, frequently put more emphasis on the sound of home-theater installations than they do on picture quality. High-fidelity audio has been a viable hobby far longer than high-fidelity video, and home-based sound systems, even before the appearance of the compact disc (CD), were more notable for quality performance than any home-video system. When video screen images began to get really big and—possibly even more important—technically better, it seemed only rational to couple improved video with high-definition audio.

The advent of hi-fi videotape recorders and matrixed Dolby Surround Sound were the final ingredients necessary to have decent, theater-like sound in the home. However, digital technology has had a great impact also, and has allowed cutting-edge home theater to improve upon the technically flawed matrix-surround concept. Digital audio not only makes it possible for those with the requisite cash to create impressive "theater-like" environments in their living, family, or media rooms, but also allows them to go far beyond what was possible with two-channel stereo systems.

WHAT IS HOME THEATER?

Video-oriented home theater is actually whatever you want it to be, provided it allows you to enjoy the "theater" presentations you want to watch. While

Figure 1.1 A Small Home-Theater System

(Photo courtesy Roy Allison)

Provided you use it in a less-than-cavernous room, a modest home-theater system gives up little to its more expensive counterparts, other than the ability to simulate a big, theater-like image, play really loud, and deliver super-deep bass. The small speakers are "satellite" models, and the black, cubical box on the floor is an 8-inch woofer system that handles the bass between 50 and 150 Hz; these were all designed by Roy Allison. The left and right speakers handle the usual stereo sound, and the one above the TV set handles the center, often mistakenly called "dialogue," channel. Two additional satellites out of the picture reproduce the "surround" sound. This economically priced group of speakers and 27-inch (or, better yet, larger) television monitor will provide your family with an exciting window to the world of home-theater entertainment.

this can simply involve a decent, fairly large TV set and basic antenna- or cable-delivered programs, home theater is usually at its best if it at least includes quality source material (even VHS videotape can be satisfactory, provided the tape and the VCR playing it are configured for "hi-fi" use), a TV set of reasonably large size (say at least 27 diagonal inches, but preferably more), and two-channel stereophonic sound.

While both digital and even nondigital surround sound are mandatory for some fastidious individuals, a passable home-theater system does not actually have to include this feature. Indeed, home theater does not really even require two-channel stereophonic sound, provided the user's main interests involve watching old, prestereo "classic" films. However, for the enjoyment of modern films, particularly those displaying lots of sonic pyrotechnics, "flyovers," and enveloping action, having a Dolby processor, left-, center-, and right-channel speakers, plus a pair, or even two pairs, of small surround speakers properly positioned—and maybe a subwoofer or two—certainly does not hurt.

Do You Need A "Killer" Home-Entertainment System?

You can have a satisfying home-theater experience with a surprising minimum of equipment. Many individuals own basic TV sets, simple VCRs, and no fancy audio gear at all; they get more enjoyment out of watching old and new movies on rented videotapes than many other people who watch the same programs on large-scale, elaborate systems, complete with projection TV, videodiscs, big amplifiers, towering speaker systems, and full-blown surround sound. For some people, the content of a motion picture transcends the delivery medium.

There are also people with diminutive stereo systems who enjoy music more than even the most "serious" audio buffs who own very elaborate and very expensive sound systems—some of which cost more than a good car. Indeed, many music lovers, particularly those who play an instrument, would rather invest spare cash in tickets to live concerts or genuine stage musicals than in flashy audio hardware. No matter how good the system, the home stereo will never replace a live musical experience.

Similarly, many film buffs, particularly those living in larger cities, would rather visit a first-rate theater than experience a video at home, no matter how elaborate the audio-video playback system. While a good home system can in many ways surpass most theaters in terms of wide-bandwidth sound quality, no home system, even one with digital-video inputs, can approach the visual clarity, color depth, and large-screen impact of a decent motion-picture emporium, particularly one that is THX certified.

However, for a substantial number of individuals, the home-theater and high-fidelity audio revolutions, with their built-in comfort, flexibility, and convenience, are an important addition to their other viewing or listening habits. You can now watch a movie when and how you want. No problems with snack, phone-call, or bathroom breaks: just put the video recorder or disc player on pause. Missed some tricky dialogue? Just back up the VCR or disc player a bit and listen again. Dull movie? No need to have a ruined evening: chuck the disc or tape and watch something else—or listen to a compact disc.

A home-based system is another way for families to conveniently enjoy being together. It may not be the spiritual equal of sitting around the piano singing songs as our ancestors did, but it sure beats a long drive through traffic to take in a film or concert. Rainy night? Pull out that Disney, LucasFilm, or Spielberg movie from your collection or switch on the satellite receiver and watch a show in a clean, healthy environment with the spouse and kids. Let's face it, from a practical standpoint, theater at home is superior to theater

Figure 1.2 Digital Video Disc Player

(Photo courtesy Toshiba America Consumer Products)

The Digital Video Disc (DVD) is indicative of the rapid improvements that are occurring in the world of electronic home-entertainment systems—and home theater in particular. Interfaced with a good TV monitor, a DVD player like this one from Toshiba (a pioneer in digital video) will complement both modest and the very best high-end video systems.

downtown—as anyone who has had to put up with noisy fellow audience members, uncomfortable seats, and sticky auditorium floors will attest.

Those not so interested in home theater but still interested in listening to high-quality music recordings will also benefit from this new technology. Home-based audio can in some ways go live sound one better. In some situations, notably when solo vocalists or solo instrumentalists are involved, the music may sound better on a recording than in a live situation. A recording engineer working in a studio may be able to highlight soloists effectively, allowing the recording to have a wonderful sense of clarity and balance. At a live performance such highlighting may result in the performer booming out over a public-address-quality system. If you're sitting in the "cheap seats" at a local concert hall with only a marginal PA system, your experience may be no better than hearing the concert over a transistor radio or boom box!

Quality differences are especially evident when popular music is involved, with the edge nearly always going to recorded sound. Rock and country lovers, at least those who have grown tired of often obnoxious crowds and sound systems of dubious quality, may actually prefer the more satisfying impact, clarity, and smoothness of recorded sound. Indeed, loudness capabilities notwithstanding, nearly any rock recording is technically superior to the overbearing sound installations at typical rock concerts. Plus, as rock fans age, they may be less interested in hobnobbing with the rowdy adolescents who frequent these events.

Home-entertainment systems are here to stay and the better ones, even those that are not particularly expensive, offer us more than the sum of their parts. However, unless you're going to buy a preassembled system—or hire an expensive consultant to do the work for you—you're going to have to put together your home theater by purchasing separately its various components. And that's where this book can be most helpful.

What's Inside?

Some readers will enjoy reading this book cover to cover, learning everything there is to know about audio and video! However, because some themes will appeal more to certain readers than others, I'll briefly outline what you can find in it, so you can quickly turn to the part that interests you the most. The main text is divided into three parts: first, *audio performance* (Chapters 2–4); second, *video performance* (Chapters 5 and 6); and third, *system performance*, which pulls the two earlier sections together for a complete picture (Chapters 7 and 8). The individual chapter rundown is as follows. Chapter 2 deals with speakers, perhaps the most important of all audio-video components. Speaker-system behavior is pivotal to both audio-only and home-theater performance, because the speakers are the last and most vital link between the audio input and the listener's ears. If we do not have good audio to back up the video, our enjoyment of the home-theater experience will be greatly limited; not to mention that simply listening to our CD collection will also be a (sonically) frustrating experience. And without good speakers there is no good audio. This is a long chapter, because speaker systems, in spite of their design simplicity, behave in a complex manner when located in typical home listening rooms. The chapter initially discusses the basic function and design of speaker systems. However, it goes on to analyze speaker-room interactions and the results of those interactions with the listener—an offshoot of a discipline called *psychoacoustics.* Most importantly for a lot of readers, detailed hints for shopping intelligently for speaker systems as well as how to set them up and keep them working the way they should for a long time are given.

psychoacoustics

Chapter 3 deals with audio signal preamplification (control functions), electrical amplification, and tuner (radio) performance. The primary stress is on what is required to deliver clean, subjectively undistorted electrical energy to typical loudspeaker systems. While preamplifiers, amplifiers, and radios are also available as single-chassis components (often called "separates"), we show why the receiver, which encloses these three basic components within one box, is a better buy for most enthusiasts—even serious ones. Finally, costs and their relation to performance are dealt with, as are shopping techniques

separates

Figure 1.3 NAD Receiver

(Photo courtesy NAD Corporation)

While not exactly a household word, the NAD corporation markets some very good, straightforward products, including this fine, medium-priced receiver. Clean looking, with a minimum of frills, this model still manages to produce a solid 40+ watts per channel into 8-ohm speakers and can deliver short-term peaks of 90 or more watts per channel, making it more than able to handle critical musical needs in modestly sized rooms. If you need more power, it has external amp/preamp connections that facilitate interfacing it with a still more powerful amplifier. These connections also allow for an outboard-mounted surround processor, including models made by NAD, to be added on when your finances permit. A music-loving enthusiast need not feel one bit embarrassed about owning this receiver.

Figure 1.4 Rotel CD Changer

(Photo courtesy Rotel of America)

In the audio community, Rotel has earned a reputation as a producer of high-end-quality goods at sane, mid-level prices. This compact-disc changer is an example of the company's no-nonsense design philosophy. Indeed, at first glance it is hard to tell that this is a CD changer at all and not a basic, no-frills single-disc player. Those with even less to spend, fortunately, will be happy to know that most cheaper models—even much cheaper ones—should perform subjectively as well as this unit or even the most expensive high-end designs.

and equipment care. Although most current receivers have surround-sound capabilities, this aspect of performance is only touched upon here—receiving in-depth treatment in Chapter 10.

Chapter 4 discusses audio components, both playback systems (LP and CD) and recording ones (analog and digital formats). In spite of what some analog-sound fanatics believe, the LP has been totally outclassed in terms of absolute performance by virtually all digital formats—and has been for years. However, many people still have large LP collections, and those need to be properly maintained and played back with decent fidelity. CD systems are evaluated, along with how to maintain your CDs—including unnecessary and even potentially damaging maintenance techniques that should be avoided.

The chapter next deals with audio-recording technology. All of the better-known analog (including hi-fi videotape) and digital systems are discussed and compared, and, after the pros and cons are dealt with, a surprising amount of real-world-use praise is lavished on the good-old, analog-cassette format. The book discusses why digital-recording systems have not been the sales successes that many company executives predicted, and why only Digital Audio Tape (DAT), of all the new audio-only recording formats, has done passably well. Because so many new, video-oriented systems use similar technology, special attention is paid to the data-reduction and signal-masking techniques used for the MiniDisc and Digital Compact Cassette (DCC) systems. The excellence of "digital" notwithstanding, this section investigates why the analog cassette system will probably remain top sales dog for some time to

analog cassette

Figure 1.5 Onkyo Cassette Recorder

(Photo courtesy Onkyo)

Although digital recorders offer better all-out performance, analog cassette players still rule the sales world. The DCC and MiniDisc formats introduced a few years ago have failed to sweep the analog-cassette medium aside, as some enthusiasts predicted. This is because the older format performs surprisingly well, is low in cost, and remains the medium of choice for those wanting to record music for playing in their car (or on portable players and boom boxes). Early dual-well units were not all that great, but current medium- and better-grade models, like this one from Onkyo, offer good performance as well as double-cassette convenience.

automotive audio

come, particularly in automotive and mass-merchandising markets. Although dealing mainly with record-and-playback systems for home use, there is also a short section on the limitations of automotive audio—even when digital technology is involved. The chapter ends with shopping techniques, self-test procedures for both hardware and software, and tape and disc preservation.

Chapter 5 discusses television sets and monitors. Despite their pervasiveness, TV sets and their performance are little understood by most individuals. This chapter explains what to look for in a good set, particularly when shopping for one in a less-than-ideal, appliance-store environment. The various types of sets are discussed (monitors, projection models, direct-view), as are other basics such as screen sizes, aspect ratios, satellite systems, hookups, and control functions. Techies will be pleased to find that esoteric subjects like horizontal and vertical resolution requirements, the Kell Factor, overscan, digital video, interlaced and progressive scan, line doubling, color performance, Motion Pictures Experts Group (MPEG) data reduction, and National Television System Committee (NTSC) vs. High Definition Television (HDTV) are also explored. These technical details may be more arcane than some readers feel comfortable with, but they can easily skip over the tough stuff (hopefully returning to it later to complete their training) and peruse the sections on shopping techniques, operational procedures, and long-term care. You do not have to have a strong technical background in video to be able to intelligently purchase a decent TV set or TV monitor.

Figure 1.6 ProScan Wide-Screen Set

(Photo courtesy Thomson Consumer Electronics)

While not immense enough to mimic an uptown theater screen, this tastefully styled, 34-inch, wide-screen set by ProScan is an example of the new breed of television monitors available. This model can work very effectively as a high-end picture-delivering system for those with modest space and relatively large budgets. One notable feature of this set, and one reason for its high-end status, is its built-in, "line-doubling" circuit, sometimes called Improved-Definition Television. IDTV, which is also available in outboard-mounted adapters from a few other manufacturers—unfortunately, at often very high cost—allows a properly configured, conventional set with interlaced scan to behave something like a data-grade, progressive-scan, computer monitor. Line doubling smooths out the picture and helps to minimize the kind of eye-fatiguing problems—such as "dot crawl"—that are often painfully apparent on many large-screen sets. This set is ready to deliver top performance with conventional, 4:3-ratio television programs (including tapes and discs), letterboxed programs, and true wide-screen programs delivered by both digital-video discs and digital satellite. For more on wide-screen TV, see Chapters 5 and 7.

Chapter 6 focuses on both traditional videotape recorders and the newer digital frontier. For traditional VHS systems, the reader is given a full technical discussion, as well as hookup techniques, copying (dubbing), the advantages of the different formats, tape types and use, programming techniques, programming technology, how to shop, and how to care for both recorders and recording tape. The balance of the chapter deals with laser-read videodiscs, both the older analog and the newer all-digital varieties. The history of both of these formats is dealt with to a small extent, as is the remarkable technology behind them. An analysis of the theory behind, operation of, and pros and cons of the 12-inch, analog-video format follows, as well as how the older format compares with the Digital Video (or Versatile) Disc (DVD) system that is replacing it. As with the other chapters, the section ends with shopping techniques for both hardware and software and also spends some time on practical hookups and caring for both hardware and software.

Chapter 7 summarizes the earlier chapters by discussing surround sound and its relationship to video, audio, and home theater. Although modern sur-

Figure 1.7 Yamaha DSP-A2090

(Photo courtesy Yamaha Electronics Corp., USA)

This Yamaha, Dolby AC-3 adaptable, DSP-A2090 audio-video receiver is an example of what is available to those wanting high-end surround-sound performance at a medium-end, or at least upper-medium-end, price. With the proper speakers, good input hardware, and decent software, this system, with its 100+ watts of power for each of the three front speaker systems, can deliver high-end amplifier performance even in large listening-viewing rooms. Yamaha, like Lexicon, Fosgate, and a few others, designs its top surround-sound equipment to power four, instead of the usual two, surround speakers, with this particular model capable of delivering 35 watts to each. While this may have only limited impact under some conditions, it can work wonders with DSP-synthesized effects from two-channel, audio-only source material, as well as with video program material.

round-sound system performance will, of course, be of major interest to home-theater buffs, those who are interested in music-only playback systems will find a wealth of information here as well. A comprehensive history of surround sound both in theater environments and in the home is given to help the reader understand how to shop for and effectively use surround hardware and software. The chapter also illuminates the operational theory behind Dolby Surround as well as the THX amendments to it (both the regular theater and home-theater versions, including Pro Logic), and also discusses the THX software program. The latest in surround-sound technology—including the competing digital 5.1 and 7.1 channel systems, in both the theater and home-theater manifestations—are also analyzed. The chapter also deals with a variety of popular and esoteric surround-sound extraction and synthesizing techniques designed to obtain the best, "concert-hall-like" effects from conventional, two-channel recordings. Finally, cost factors are discussed, because this is one of the areas where spending a bit more than the minimum required may actually pay off.

THX

Chapter 8 deals with good, bad, and useless accessories. Good ideas and useful accessories are applauded, while the near-racket-like behavior of certain snake-oil salespeople and manufacturers is exposed. While not all accessories and attitudes are analyzed, the chapter helps the reader develop the kind of critical thinking that careful shoppers should cultivate.

The two appendices include a list of top-quality audio and video recordings (items capable of revealing the positive and negative qualities of even the best audio and home-theater playback systems) and an analytical listing of some of the more notable (and for some enthusiasts, notorious) audio and video journals available at the time the book was written. An extensive glossary, followed by a comprehensive index and bibliography, will assist those needing brief definitions of terminology used throughout the book and those who wish to locate important individuals or themes and/or do additional research.

BUYING HARDWARE

Low-Cost Hardware

The equipment in any typically well-stocked audio-and-video store (that does not also deal in washing machines and vacuum cleaners) normally is divided into two sections: the "high-end" room and the rest of the place. The main showroom usually features both inexpensive and mid-level gear and even some mildly expensive hardware (especially video hardware), but the high-end area will nearly always contain only *really* expensive items. Often, this equipment will be displayed in specially adapted rooms (which may be strategically lined with acoustic padding to optimize performance) and will not contain any video componentry at all. Many establishments may keep these special areas locked up tight and accessible only by permission and/or appointment. High-end equipment demands respect.

high-end gear

High-end video gear usually includes "elite-grade" video-disc players, upscale Super VHS recorders, and large-screen projection TV monitors. Very large *rear*-projection sets may qualify for this category, but the elite of the elite will be the exotic *front*-projection designs, with ceiling- or floor-mounted projection modules, digital line-doubling circuitry, and wall-mounted, motorized screens. More-diminutive TV monitors, even fairly costly examples, will usually be berthed with the mid-level gear out in the main showroom.

High-end audio hardware usually includes high-powered, free-standing amplifiers and preamplifiers costing a thousand bucks apiece and up (often *way* up) and top-of-the-line CD players costing ten times (or more) what a discount house charges for more mundane models. This aristocratic category also includes LP record players resembling machine sculpture, pro-grade DAT recorders, FM tuners costing more than some premium-level TV sets, and, needless to say, some very formidable speaker systems. Many of the latter resemble huge, well-finished wardrobe cabinets with cloth fronts, while others are made up of multiple sections, with some parts resembling room dividers or even robots: and all costing megabucks. Buyers of these systems are obviously very serious about what they are doing (or a bit crazy) and are nearly always quite wealthy. While these systems can behave in uniquely spectacular ways, the less well-heeled enthusiast should know that it's not necessary to spend big bucks to get similar performance.

video installations

I have experienced a number of decent and not so decent *video* installations and have come to the conclusion that the paramount difference between a good expensive system and a good cheaper one (sound-system quality aside) is screen size and shape. While the new breed of wide-screen NTSC standard and advanced-digital HDTV sets are better at presenting wide-screen film images than conventional, 4:3-ratio sets, and premium-quality designs (which often have better high-voltage regulation than cheaper models) will have better pictures than economy-grade versions, expensive TV monitors of reasonably large size are not appreciably better in terms of "practical" picture quality than many smaller, much cheaper versions with identical width-to-height ratios. Even the largest projection sets (including the wide-screen models), in spite of their often overwhelming visual impact, are not significantly better in terms of everyday, *usable* sharpness, contrast, color noise, etc., than more diminutive, modestly priced projection sets or direct-view monitors.[1]

The trick is to adjust the set properly.

Modern, mid-priced, and even some surprisingly low-cost video hard-

[1]We will discuss this in greater detail in Chapter 5.

ware can deliver quality that will satisfy fairly critical individuals if it is properly chosen, installed, adjusted, and maintained. Each of these aspects of performance will be covered in the video chapters later in this book, along with a lengthy analysis, in Chapter 10, of surround sound: the "secret ingredient" of total-video satisfaction. For surprisingly few dollars, one can upgrade a modest TV installation (containing only a decent TV and hi-fi VCR) and end up with a small "home theater" that easily surpasses the fun quotient of any expensive, large-screen television set operating by itself.

audio installations

I have auditioned a number of *audio* installations over the years and have concluded that the key difference between good expensive and good cheap versions—most of the time—involves the former's low-distortion loudness potential and room shape and size. There are certainly other factors, like deep-bass performance, surround processing, and image-control features, where spending extra money may pay off. However, most cheaper systems—because of their smaller sizes and lower amplifier power—fall behind bigger, more expensive ones because they are unable to fill a large room with attention-getting sound levels. It is important to remember that large-scale sound is not the end-all of audio or home theater. In average-size rooms and at reasonable sound levels, particularly when surround sound is involved, good smaller and cheaper systems can do as well as, and in some cases better than, their larger brethren—at least in terms of sound staging and clarity.

Audio performance depends upon the listening-and-viewing room, the way a system is installed in that room, and, of course, the components chosen. Interestingly, many customized high-end installations stress decor and convenience at the expense of good performance; in some cases, *very* expensive components will be compromised to a startling degree by decor considerations simply because the owner did not want his or her elaborate and expensive system to visibly intrude. Tailoring an installation often includes speaker placement oriented toward visual rather than aural impact. This is the single biggest error so-called custom installers (even those who are "certified") make when trying to satisfy both the sonic and ornamental needs of their clientele. Evidence of this problem occasionally can be found in the "system-of-the-month" spreads in certain audio-and-video or home-theater monthlies or in the glitzy "A/V-decor" publications found at well-stocked newsstands.

Who makes one's components is important, of course, because some companies offer more bang for the buck than others. Most electronics companies, high-end or mainstream—Japanese, European, or American—assemble their products to a reasonable standard, with the main differences being reliability, operational ease, and flexibility. However, speaker companies are quite another matter. A world of difference exists between the better full-system manufacturers and the "box builders." I have seen $550 speaker systems made by one company handily outperform $2000 models made by another.[2] *Some* modestly priced (and even budget-oriented) equipment is OK. The trick is to separate the wheat from the chaff when shopping and then use the newly acquired hardware properly when you get it home. Hopefully, as you proceed through the rest of this book, you will gain the expertise necessary to do so.

Used Hardware

The biggest problem with used equipment is finding components in decent shape. There are some fine buys out there, but they must be hunted down. The most obvious way to shop for used A/V gear is to scan local want ads to see what is available.[3] Usually, private owners will quote prices that are sub-

Figure 1.8 AR Receiver

(Photo by the author)

The AR receiver was designed in the 1960s and delivered basic performance that eclipsed many competitive designs. This particular sample is over 25 years old and still works very well, although its owner had to replace a noisy volume control a few years back. Clean and uncluttered in appearance, it lacks the kind of input flexibility required for many modern setups, including any kind of video connections. However, for basic, stereo-system use, even the most modern designs will not offer better audible performance. Interestingly, this model did not have an AM tuner, because its designers felt that only FM was capable of delivering high-fidelity radio performance. Most modern AM tuners are still miserable performers.

[2]We will deal with loudspeaker performance standards in Chapter 2.

[3]For some time, a good national source for used audio and video gear, and also a way of assessing the prices of used equipment for sale locally, has been Audiomart, Rt. 3, Box 692, Crewe VA, 23930.

Figure 1.9 Allison CD-9

(Photo courtesy Roy Allison)

Certainly the most exotic-looking model Roy Allison ever produced, this discontinued CD-9 speaker system combined close-to-boundary woofer placement (for uniform midbass) with Allison's unique tweeter and midrange driver designs. This was a robust system, built to last for years, and was designed to work at its best in typical, living-room-like listening-and-viewing rooms. If you find a pair of them for sale used and they do not look abused, either buy them yourself or drop me a line so that I can buy them!

stantially lower than specialty stores, which obviously have to cover their overhead. Another advantage of buying from a private person is that you will have some idea of how the equipment has been treated. Evidence of sloppy maintenance should be fairly obvious if you take a careful look around the seller's setup. Meticulous, finicky individuals can be a real find for the used-product shopper, as are true audio-videophiles; they often upgrade to the latest models, unloading used equipment long before it has deteriorated.

specialty shops

I must sheepishly confess, however, that I am a sucker for the kind of deals available on used gear in specialty shops. A specialty operation will probably offer a warranty of some kind, a bonus rarely found when purchasing from a private party, even a meticulous one. Many big-city shops will have a good selection of used hardware: trade-ins or left-over demo displays that have been superseded by newer, differently styled (but not necessarily better) models. Some items may be discounted because of a slight blemish on the case but will still have a new-product warranty.

However, a lot of products should not be purchased used. I do not advise buying a used phonograph stylus (needle), for instance, unless the owner still has it sealed in its original box. If you buy a used turntable and cartridge, it would not be a bad idea to install a new stylus, unless you are certain that it isn't worn. Tube-type amplifiers are risky buys also, unless they are recent high-end models. I'm also leery of used loudspeaker systems unless I am very certain of how they were treated. The only way I can get excited about a used speaker system is if it is a real top-grade, "classic" model, built by a manufacturer known for doing a first-rate job and who is still in business and can therefore supply replacement parts. If this is not the case, the repair or replacement of key components could be a problem.

Of course, many people do not live in large cities, and most small-town purchases of new equipment will be made at chain stores that do not take trade-ins and rarely discount demo models. However, much of the new low-cost, and nearly all of the middle-priced, *electronic* gear available is sonically equal to the more expensive stuff, new or used—although spending a bit extra will pay off if purchasing certain kinds of surround-sound equipment. In addition, remember that it will be best to purchase "mechanical" hardware new, anyway. Discount stores, in small towns or large, will commonly have deals on new equipment that make the purchase of even top-grade used gear a waste of time.

Pros and Cons of Mail Order

There are many mail-order outlets for audio and video equipment, most of which advertise in the major audio and video magazines. They are successful because most provide fine products at low prices and give decent service to boot. Although local stores may promote their ability to repair your equipment, in most cases the work—especially warranty work—will be done at a centralized repair facility. An item that malfunctions is usually shipped off for several weeks, whether you bought it locally or by mail.

I have purchased quite a few items via mail order and, even counting the few returns I have had to make, have never had any serious problems. When I've occasionally encountered an incompetent or surly order clerk, I simply hung up the phone and tried another vendor. I have seen irate letters to the popular hobby journals that complain about shabby treatment, but those usually involve a misunderstanding on the part of the customer. Occasionally, for instance, an item will be out of stock, and the customer will be upset because of the delay. Often, a mail-order house will advertise an item that is scheduled for delivery in the near future and then be caught in a pinch if its production is delayed. Sometimes a popular item will sell out rapidly, and factory replacements will be back-ordered.

delivery delays

Delivery delays, unfortunately, are typical of many mail-order houses, even the first-rate ones. While their phone clerks are sometimes knowledgeable and attempt to be helpful, there is no way they can duplicate the intimate service of a good local store's sales clerk, particularly if the store is well stocked with a variety of good equipment. Another big problem with mail order is that the customer cannot actually see the goods. If you know what you want and that item is in stock, there's no problem. However, it is risky to select an alternate item by phone if what you want is unavailable. Worse yet, attempting to actually "shop" by looking through a mail-order advertisement or catalog may lead to unpleasant surprises. Following the suggestions of phone clerks can also lead to disappointment; they may suggest a "special" item that is either a high-profit piece or something that is overstocked or discontinued and in need of liquidation. Returning unsatisfactory equipment can be another hassle, because you will have to repack it (often in the original box) and drive to a local UPS facility to have it shipped.

A few American manufacturers actually sell their products *only* by mail order. Usually, those companies produce loudspeakers, some of which are of surprisingly high quality. What's more, my experience has been that their phone personnel are more knowledgeable than most of those who work for retail-only operations. It's sometimes possible to actually speak to a design engineer about a product's operational principles, features, and advantages before ordering it. Because these manufacturers depend entirely on mail or phone orders, their return policies are often very liberal.

As a mail-order customer, you need to *know* the products you want to purchase. Studying test reports can tell you which items work well, what their features are, and their weak points. However, tests reports cannot tell you if you'll like the way equipment looks or the way it handles. One solution is to pay a visit to any local stores that sell the item, with the intention of only studying it for a later mail-order purchase. You may find that the local price for your desired product is only marginally higher than that offered by a mail-order house and that the prospect of immediate gratification (and ease of delivery—and return, should the product have problems) more than offsets the small cash difference.

Perhaps the best feature of mail order is that of product availability for those of us who live in small towns that do not have specialty or discount stores with a large selection of good equipment. Some of the mail-order houses publish large catalogs full of fine merchandise that is simply not available in many local stores. If you live in a large metropolitan area, mail-order purchases may be a waste of time and energy. If you live in the sticks, the back pages of the hobby magazines may be your ticket to first-class performance.

And now, on with the show.

2 Speakers

WHETHER YOU'RE LISTENING TO A CD of Beethoven's Ninth or watching an enhanced videodisc of *Jurassic Park*, the quality of the speakers that you purchase will have a dramatic effect on your entire home theater. Allowing for differences in the room where your system is installed and, of course, the quality of the audio or video material you're playing, the speakers will have more impact on the sound and visceral impact of a system than anything else. Before purchasing a speaker system, it is important to understand some of the different designs that are available so that you can select the one best suited to your listening and viewing habits. In this chapter, we will discuss all types of speakers, as well as briefly discuss the least expensive—and least social—of all listening setups, headphones.

ANATOMY OF A SPEAKER

driver

crossovers

Speakers are available in assorted shapes, sizes, and operational configurations, but most are box-shaped and contain two or three noise-making elements or *drivers*. Commonly, separate drivers are used to cover the lowest frequencies (often called a "woofer"), the middle range, and the higher frequencies (the "tweeter"). The signals coming from an amplifier or receiver are routed to the appropriate drivers; the routing device is called a *crossover network*, a kind of internal traffic cop that determines which frequencies are best reproduced by which driver. Most conventional systems are either two- or three-way designs; they use either two or three drivers, although some double up drivers to cover specific frequency ranges. A fourth type of driver, a subwoofer, once an exotic add-on, is now fairly common in many systems.

These drivers may be designed in many different ways. The basic layout

Figure 2.1 Woofer Cutaway

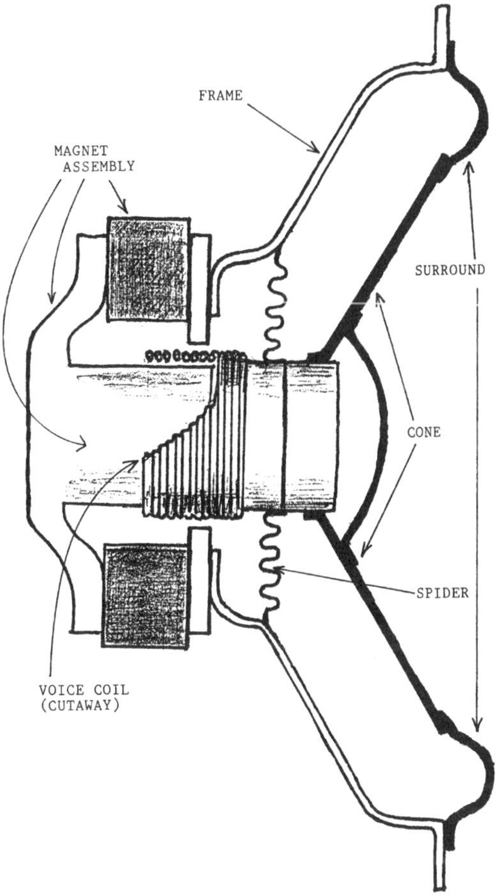

coil

spider

surround

uses either a cone- or a dome-shaped diaphragm that vibrates back and forth to project the sounds that we hear. As you can see from the cutaway drawing, the five basic parts of a cone driver are the magnet assembly, the voice coil, the centering spider, the cone/surround, and the frame. The coil carries the current from the amplifier and reacts as any motor coil would when immersed in a magnetic field. Instead of going round and round, however, the coil moves back and forth. It is attached to the cone which, therefore, also goes back and forth to generate the minus and plus variations in the sound wave air pressure cycle. The spider allows the cone to move easily in that direction and functions to keep the rear of the cone centered in the magnetic field. The surround, which is actually the outer part of the cone and also allows it to move freely, keeps the front of the cone aligned. Some manufacturers use die-cast

Systems

Driver Type	Typical Operating Range	Common Cabinet and Driver Designs	Less Common Designs
Woofer	Bass frequencies: 20–40 Hz to 200–500 Hz	Bass reflex; drone cone; bandpass; acoustic suspension	Horn; transmission line; ribbon; electrostatic; planar magnetic
Midrange	Middle frequencies: 200–500 Hz to 3–4 kHz	Dome; cone	Ribbon; electrostatic; horn; planar magnetic; EMIM
Tweeter	Upper frequencies: 3–4 kHz to 15+ kHz	Dome; cone	Horn; ribbon; electrostatic; Linaeum; EMIT; convex dome; planar magnetic

aluminum frames to hold the works together, and others use simple, stamped steel construction. While the former is cosmetically more impressive, the latter works fine for all but the largest (15+-inch) drivers.

Most stereo sets used to have two-speaker systems, each with two or three drivers. Nowadays, three-section systems are popular, employing three or more drivers; in these setups, the separate, larger woofer module is often placed somewhere between two other "satellite" boxes, each of which contains tweeters and midranges. And, of course, there are surround-sound systems that feature even more speaker enclosures.

woofer

The best-known speaker driver is the "woofer." In a three-way system, it handles the low or bass frequencies. Woofers normally have to be large and well made in order to handle the long movements involved in reproducing bass-frequency wavelengths. Most are about 8 inches in diameter, but smaller sizes are popular too; a number are 10 inches or larger. Generally, the bigger the woofer, the deeper the bass notes it can reproduce and the louder it can play at really low frequencies. However, some big woofers are inferior to a few of the better small ones because of poor design or cheap construction.

Most woofer systems use *bass reflex* (or a variation to this design that adds a *drone cone*), *bandpass*, or *acoustic suspension* design principles.[1] These designs are used in speakers in all price categories, and each has the potential to reproduce fairly deep, distortion-free bass sound. The major design differences, their advantages and disadvantages, and some past and present manufacturers who have used these designs are listed in the table on page 24.

Woofers are not well-suited for reproducing frequencies above about 500

[1]Horn and transmission-line systems are also used, but lack the popularity of the other designs. Folded-duct transmission lines are particularly tricky to design.

Figure 2.2 Bass Reflex Cutaway

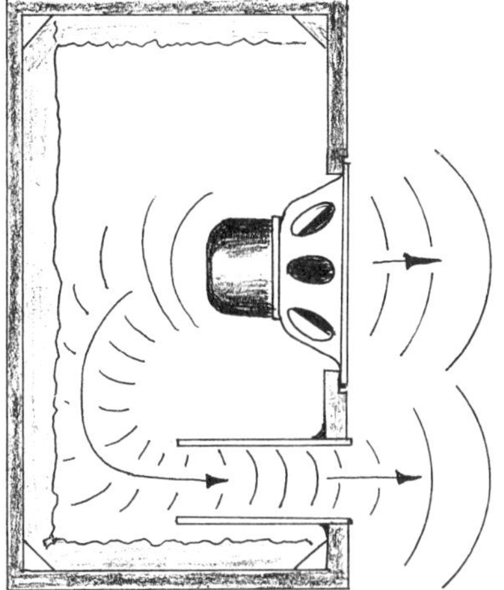

Figure 2.3 Drone Cone Cutaway

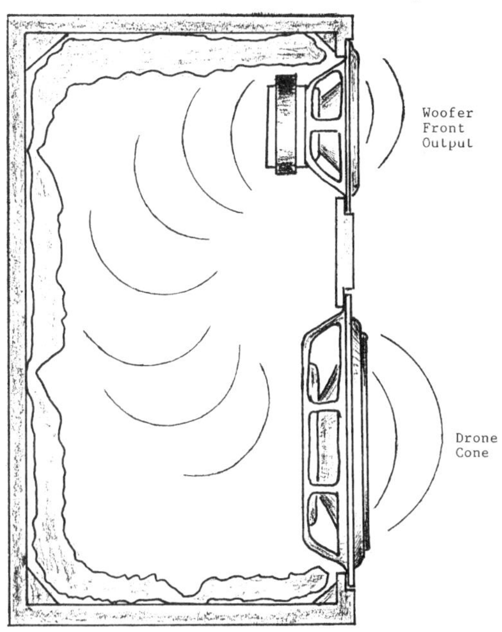

Figure 2.4 Bandpass Woofer Cutaway (one type)

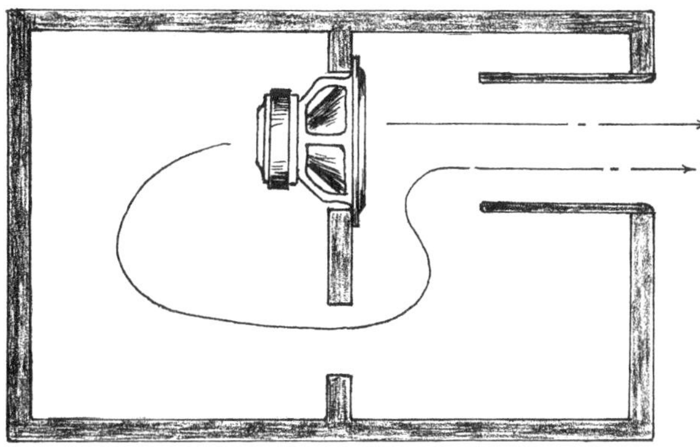

Figure 2.5 Acoustic-Suspension Woofer Cutaway

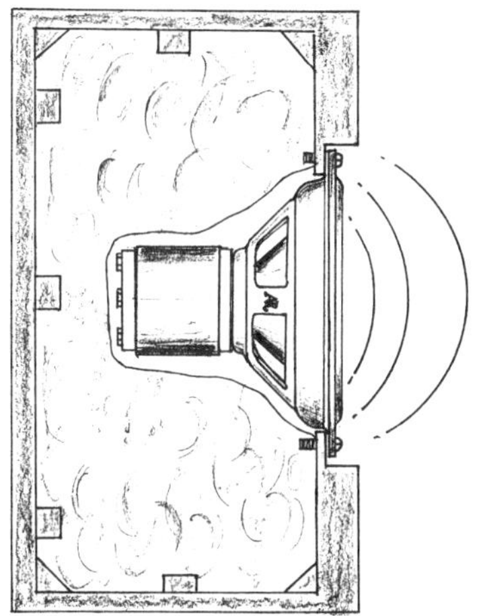

Figure 2.6 Dome Speaker Cutaway

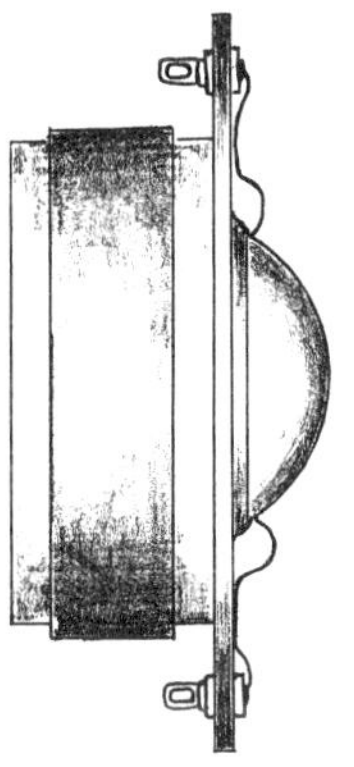

Most Common Tweeter/Midrange Designs				
Type	**Design**	**Advantages**	**Disadvantages**	**Some Users**
Cone	Cone-shaped diaphragm, magnet/coil motor; voice coil attached near center	Cheap and easy to build; many versions available from OEM suppliers; consistent performance	Cone breakup possible (particularly with tweeters); lower power handling than similar-sized dome	Ubiquitous as midrange; rarely used as tweeter
Dome	Dome-shaped diaphragm, magnet/coil motor; voice coil attached near rim	High power handling for size; many versions available from OEM suppliers; inherent structural rigidity insures low distortion; uniform dispersion	Usually more expensive to build than cone; exposed dome prone to damage	Ubiquitous as tweeter. Used as midrange by Allison Acoustics, Cello, Duntech, Dynaudio, Energy, Infinity, Jamo, Legacy, RA Labs, Shahinian
Horn	Diaphragm coupled to air through a horn; magnet/coil motor	Very high sound output possible; very robust; useful if directional control required	Large, complex and expensive to build; too directional for most home uses; potential resonance in horn wall	Rarely used as tweeter; used as midrange by Klipsch, Jamo, Radio Shack, Westlake

Hz, because their large and heavy cones can neither reproduce nor project these higher frequencies efficiently. The obvious solution is to route these middle frequencies to a "midrange" driver that is smaller in size than the woofer. Typical midrange drivers are surrounded with a sealed "sub-enclosure" within the speaker cabinet to avoid any interference with the operation of the woofers. The most popular midrange drivers are cone, dome, and horn designs.

midrange

As the frequency climbs into the treble range (about 3 or 4 kHz), midrange drivers will display the same problems that woofers do at the top of their operating ranges. The solution is to route these signals to a "tweeter" driver, which should be much smaller than the midrange one. Most good tweeters are domes; some inexpensive systems use cones, while horns are usually best for auditorium work. There are also a couple of more exotic designs that are usually reserved for more expensive systems, including the Linaeum driver, the excellent EMIT design used by Infinity, and assorted electrostatic, planar-magnetic, and ribbon models. While all of these designs are worth a thorough investigation, conventional cones and

tweeter

Sound and Soundwaves

The term *Hertz* (*Hz* or cycles per second) usually refers to the number of times per second a sound wave oscillates. Human hearing normally spans a range from 20 Hz at the low end to 18 or even 20 kHz ("k" denotes times 1,000, and *kHz* is pronounced kiloHertz) at the high end, if the listener is young and healthy. However, most adults, particularly middle-aged males, have a significant hearing loss above 13 to 15 kHz that often gets progressively worse with age (this effect is called *presbycusis*). However, this condition is not as serious as it appears, because there is a surprisingly limited amount of sound in the upper musical octave, whether recorded or performed live.

This drawing gives you an idea of what we are talking about when discussing cycles or Hz. (Note that a plus indicates air pressure above ambient [room] pressure and a minus indicates air pressure below ambient.) Curve A is a sine wave (a pure, clean, uncomplicated signal) spread out over a one-second period of time; in other words, it is a 1-Hz signal. Curve B is 2 Hz. Curve C is both of them combined over the same period of time. In the real world, musical signals spanning the audio bandwidth are combined in much more complex ways to create the sounds we hear. No matter how complex, however, it will still be a single wave that strikes our ears and causes us to hear.

Incidentally, to calculate the actual length of any given wave, divide the speed of sound (1,130 feet per second at sea level) by the frequency. A 1-Hz wave is 1,130 feet long; a 2-Hz wave is 565 feet long; a 100-Hz wave, 11.3 feet long; a 1,000-Hz wave, 1.13 feet long; a 10-kHz wave, 0.113 feet long (1.36 inches); and so on.

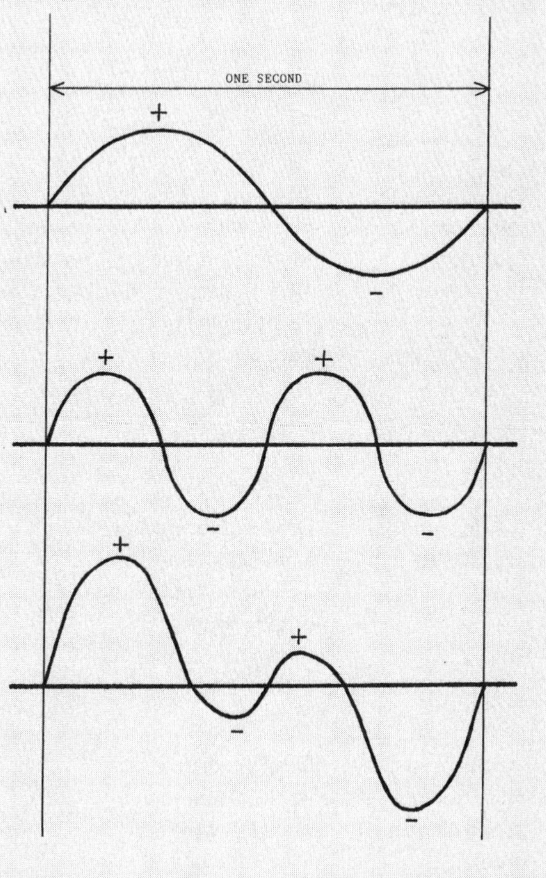

domes will do yeoman's work at reproducing sound (often equaling some of the more exotic designs) and often have a reliability edge over at least some of the exotics in the same way that a Toyota is more reliable than a Ferrari.

Most Common Woofer Cabinet Designs

Type	Design	Advantages	Disadvantages	Some Users
Bass reflex (also called tuned port or vented enclosure)	Enclosure has a hole (port) in the cabinet, which often includes a tubular duct, tuned to allow the low-frequency sound from the back of the driver to reinforce that of the front at specific frequencies	Higher output than direct radiator at frequencies above point where back and front signals begin to null each other	Possible port wind noise; 24-dB-per-octave nulling of output below port resonance may limit deep bass; driver not self-protected below resonance, because of loss of cabinet back pressure; port and duct size calculations tricky	B&O, B&W, Bose, Camber, Canton, Celestion, Cerwin Vega, Definitive Technology, Hsu Research, Infinity, JBL, KEF, Klipsch, Mirage, NHT, PSB, Paradigm, Waveform
Drone cone (also called passive radiator)	Variation on bass reflex, with a "dummy" driver used in place of port to augment low bass	Eliminates port wind noise; simplifies design compared with bass reflex	Steep, 24-dB-per-octave attenuation of output below resonance; woofer not self-protecting below resonance; performance can deteriorate as mechanical components age	Bozak, Infinity, Polk, Shahinian, Thiel, VMPS, Velodyne
Bandpass (includes coupled cavity)	Driver(s) located in a multichambered enclosure, with all output routed through one or more ports	Port(s) can be tuned to cover a specific frequency range, reducing unwanted signals above the woofer's operating range	Some versions may produce port wind noise; steep, 24-dB cutoff below resonance; construction complex	ADS, Bose, Boston Acoustics, Cambridge Soundworks, KEF, Mission, Polk, Tannoy, Warfedale
Acoustic suspension (sometimes called air suspension)	Direct radiating driver is mounted in a sealed, glass-wool-filled cabinet and uses the trapped air to dampen cone movement	Low distortion, due to linearity of air mass; woofer is self-protecting below resonance; reasonably easy to design and build; good, easily equalized, 12-dB-per-octave rolloff below resonance	Substantial electrical input may be required, particularly if the woofer is to deliver strong output at very low frequencies or is in a small cabinet	ADS, Acoustic Research, Allison Acoustics, Boston Acoustics, Cambridge Soundworks, Kenwood, M & K, NHT, Phase Technology, RA Labs, Snell, Velodyne

Distortion/Decibels Down

Distortion is one of the most nettling topics in audio. This chart gives us an idea of just how *loud* a specific harmonic-distortion percentage is in relation to the fundamental that produced it. Good speakers produce distortion levels below 1 or 2 percent (40 to 35 dB below the fundamental) in the midrange but may generate levels of 10 percent or more in the deep bass. (Excellent systems keep the distortion at near midrange levels in the lowest octave, between 20 and 40 Hz.) Good amplifiers and CD players can have levels below—often, way below—0.1 percent, which is 60 dB below the fundamental. Most listening rooms have a noise floor that is louder than that. Interestingly, if an amplifier's power output is 100 watts and a speaker is generating distortion levels of 10 percent, those distortions will only be the equivalent of 1 watt loud. A distortion level of about 0.1 percent, which is typical for decent amplifiers, would be 60 dB down, or only about 1/10,000 watt!

Obviously, under some conditions, distortion levels as high as 5 or 10 percent will be inaudible, as would be the case if the source material being played were complex (symphonies, busy jazz, runaway rock), because the cacophony of musical signals would mask the distortion. Under other conditions, distortion levels as low as 1 percent might be audible—as would be the case with test tones or solo instruments like a flute. Different kinds of distortion (such as even-order harmonic, odd-order harmonic, and intermodulation) may have different thresholds of audibility. Most modern audio systems have low distortion when played at reasonably loud levels in normal-sized rooms.

Distortion level	Decibels down
100%	0
60%	5
50%	6
30%	10
20%	15
10%	20
6%	25
5%	26
3%	30
2%	35
1%	40
0.6%	45
0.5%	46
0.3%	50
0.2%	55
0.1%	60
0.06%	65
0.05%	66
0.03%	70
0.02%	75
0.01%	80
0.006%	85
0.005%	86
0.001%	100

Two-Way Systems

The midrange drivers are the most expensive part of many three-way systems and require a rather complex crossover circuit. Because of this, the two-way system is popular among those interested in good sound at a reasonable cost.

While a few high-end enthusiasts feel that so-called top-grade two-way systems are actually better performers than most three-way models, the fact is that the laws of physics force the builders of two-way models to make some compromises.

The woofer in a two-way system must also double as a "most-of-the-midrange" driver. Consequently, two-way systems often have smaller and lighter woofers than three-way versions, with different electrical and magnetic configurations, restricting performance at very low frequencies. Some companies compromise by using two small woofers to do the job, but this pushes the cost of the systems upward, in some cases to nearly the same level as three-way designs. Plus, no matter how small they are, woofers will have trouble dispersing higher-frequency sounds.

The tweeters found in two-way systems have problems of their own. For every halving of the frequency being reproduced, the cone, dome, or diaphragm will have to quadruple its and back-and-forth movement and there will be an additional thermal load on the thin voice-coil wire. In dropping the crossover point from the 4 kHz of a hypothetical three-way system to the 2 kHz of a two-way design, a manufacturer puts an extra burden on the tweeter. In many cases the tweeter may not be adaptable at all, and the manufacturer will have to substitute a different one with heavier voice-coil wire and a larger cone or dome for adequate output in the middle range, resulting in reduced performance at higher frequencies. The audio spectrum practically demands at least a three-way design for optimum performance.

For more on subwoofers, see pages 63–70.

Some companies have gone to using four- or even five-way configurations for their top models. In theory, this should result in better sound. However, if good drivers are used and the system crossover points are optimally located, a three-way speaker system will satisfy most musical requirements. Going beyond a three-way design is only beneficial when *extremely* deep bass is required. In that case, properly adding an outboard subwoofer will turn a three-way into a four-way system.

With good drivers and a good crossover, a two-way system can work surprisingly well. Two-way system performance limitations include reduced maximum output and a bit more distortion in the middle range. The former problem should not be serious for those of us who listen at less-than-thunderous levels, and the latter may not be audibly significant, because complex music signals often mask these anomalies. In addition, a lack of super-deep bass performance will ordinarily not be missed by most listeners unless they are serious synthesizer, pipe-organ, or *Jurassic Park*–style, action-movie buffs. The only audible glitch in two-way system performance involves dispersion characteristics in the middle range (see the "technical digression" section ahead).

Three-Way/Two-Way Basic Curves

In these curves and many of those to be follow, a "perfect" signal will overlap the "0" line exactly. Signals that are plotted above the line will be perceived as being louder, and those below it will sound softer. In most cases, a gradual downward tilt above 1 or 2 kHz will correct some of the harshness found on certain recordings or movie soundtracks.

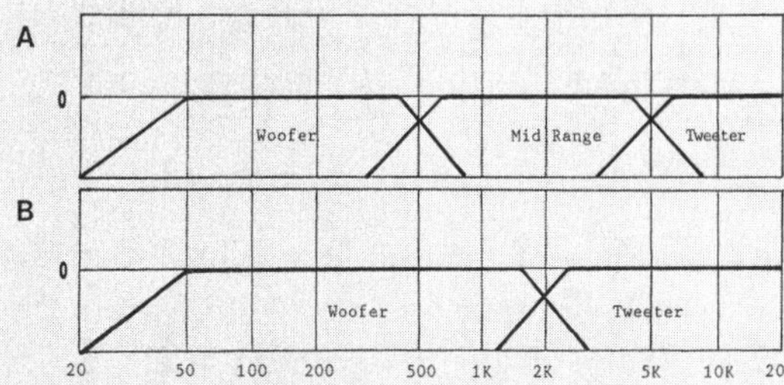

Diagrams A and B give you an idea of the frequency ranges covered by typical speaker-system drivers, representing the *on-axis*, directly-in-front-of-the-center, output of each driver. Curve A shows the response of a three-way design, with the woofer covering the range from 20 Hz to 500 Hz (the fall-off below 50 Hz is typical for many systems), a midrange covering from 500 to 5 kHz, and a tweeter covering from 5 kHz on up. The notches at the overlap points are intentionally set by each system's crossover network to offset the increase in output that would result if both drivers were operating at full power at those points.

Curve B shows the response of a two-way system. It can be seen that the woofer and tweeter in this design must share the duties of the midrange in a three-way system. In the real audio world, speaker curves would differ considerably from these due to mechanical aberrations and off-center, *off-axis* losses. Full "system" curves, measuring everything together, would also show the effects of phase interference between drivers that would be insignificant with most good systems in typical listening rooms. In addition, most real-world tweeters will not have such a flat output above 10 kHz, especially at angles well off to the sides.

Note that these diagrams have a "logarithmic" scale. That is, the range from 20 Hz to 500 Hz has almost the same span as the range from 500 Hz clear out to 20 kHz. There is a reason for this.

First, if the chart were made linear, the left half would be compressed to a very small width if the right side were expanded to take up its full share. Because a *lot* of recorded music fills up that 20- to 500-Hz space, we would be desperately short of accurate information about the behavior of any speaker. By the same token, if the whole chart were expanded linearly relative to the 20- to 500-Hz section, it would run fill up several additional pages.

Second, the ear hears logarithmically. That is, a shift in frequencies from 50 Hz to 100 Hz is clearly audible. If a similar linear 50-Hz shift from, say 5,000 to 5,050 Hz takes place, it would take an astute listener to hear the difference; a shift of similar subjective impact would actually be from 5 kHz to 10 kHz. In each case, the change would be *one octave* wide. Because the ear hears in such bites, the impact of the latter change would be no more dramatic than the 50- to 100-Hz jump.

To show how this works in the "real"

(continued)

Three-Way/Two-Way Basic Curves *(continued)*

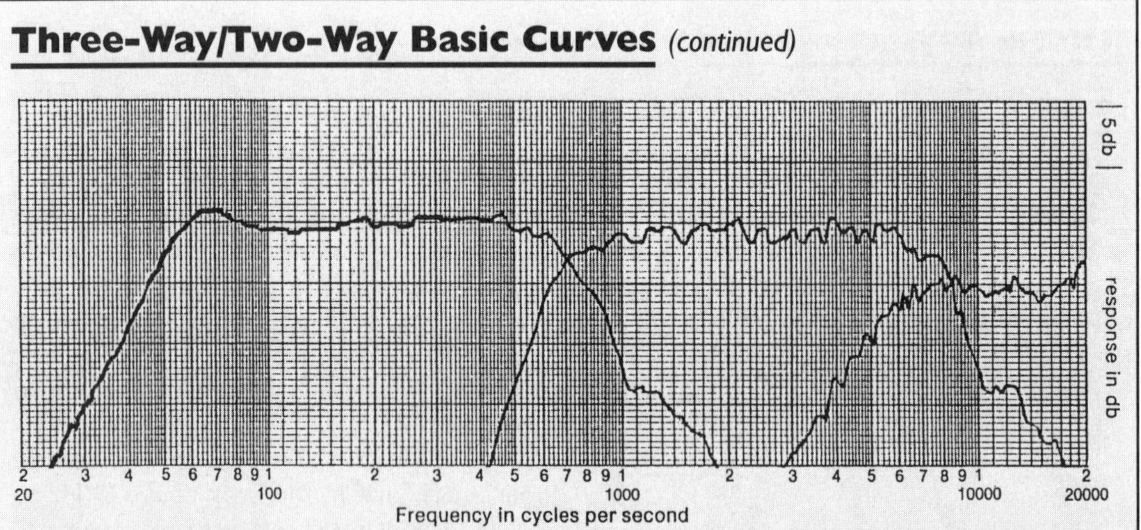

world, here's an unretouched, on-axis response of each of the three drivers used in the Acoustic Research AR-5, a speaker system that was popular in the 1970s. As can be seen, it is rougher looking than the "perfect" curves shown earlier, but the overall shape is similar. Actually, on-axis curves like these tell us little about the overall performance of a system in a typical room. For a working knowledge of how this design behaves under real-world conditions, additional curves would have to be run at a number of off-axis locations. Note that the tweeter output was somewhat attenuated to offset the stridency common in LP recordings and certain phonograph cartridges of the era. (More than a few modern recordings, particularly CD reissues of LP originals, may still benefit from such an adjustment.) The system contained level controls on its rear panel to allow individuals to contour the sound to suit tastes and room conditions.

Three-Piece Systems

Probably the strongest competitor to the two-way design in the low-to-moderate cost hi-fi speaker sweepstakes is the three-piece, or woofer/satellite, system. This configuration has separate mid-/high-range boxes ("satellites"), just like any stereo pair of speakers, and a single woofer module to handle the blended left- and right-channel bass.[2] A few designs come in four parts, with two bass modules to enhance bass capabilities, aid positioning flexibility, and more effectively localize some of the pink-noise-like sound effects that are intrinsic to certain action-adventure film soundtracks.

[2]The woofer module may still contain two woofer drivers, however.

A woofer/satellite system can provide decent service if the woofer-to-upper-range crossover point is low enough (optimally below 100 Hz), because of a phenomenon called the *Franssen effect.* Here's how it works. Imagine a tone of several seconds' duration with very abrupt start and stop points that are substantially steeper in rise and fall times than the frequency of the tone itself. The tone is sent to a divider network that routes its abrupt beginning and end to one speaker system and dispatches the steady-state middle part to another. A listener sitting at roughly equal distance from both speakers will perceive that the *entire* tone is coming from the one handling only the abrupt beginning and ending segments.

In a woofer/satellite system, the woofer handles the steady-state part of any bass signals, and the satellites handle their initial transients and endings (if any), as well as all midrange and treble signals. With the exception of exotic test tones, the listener will "hear" the sound as emanating from the small satellites, not from the woofer module(s). Even if a bass instrument generates a sharp, directional transient like the "thwack" of a hard mallet on a bass drum, the initial attack portion of that sound will be reproduced by the upper-range drivers, and the reverberating "bass" sound will be a nondirectional, supporting tone. (It is also nondirectional in a standard system, which is why a woofer can be placed at the bottom of the cabinet.) This allows for great flexibility in placing the bass modules.

Most three-piece systems use the three-way speaker design, with the tweeters and midrange drivers mounted in the small, easily positioned satellites. These can be aimed independently for more flexible imaging control and better sound dispersion. The woofer section (often a "bandpass" design) is usually in a somewhat larger, floor-placed enclosure that is easily hidden away in a corner, under an end table, or behind a couch.

A few years ago three-piece systems were unavailable, but they are now common for a number of reasons. As already noted, their components are easily placed in nonintrusive locations. Second, although they are usually cost competitive with two-way systems, most are three-way and therefore share the latter design's advantages of good midrange dispersion. Finally, they may actually produce even better musical sound than many three-way models because of the flexibility the separate woofer module provides. In some rooms, positioning standard, two-piece speaker cabinets for optimum bass performance may result in less than ideal music reproduction from the upper-range drivers; conversely, locating a pair of systems for optimum upper-range performance can result in less-than-enchanting bass performance. Three-piece systems may overcome these deficiencies.

multiple drivers

There are a number of speaker systems on the market with multiple woofers, midranges, and/or tweeters. The use of double woofers, assuming they are decently designed, will improve power handling and reduce distor-

Five-Piece System

(Photo courtesy Roy Allison)

Satellite/woofer systems, like this Roy Allison–designed combination of five satellites and a woofer module, make use of the Franssen effect. Two of the satellites can be placed for the best front-channel, midrange, and high-frequency imaging, while the cube-shaped woofer module, which in this case contains a dual-voice-coil 8-inch driver within its small cabinet, can be placed for the best bass propagation. In most rooms, it is best to have the woofer section at the front part of the listening area, with the optimum position being within the area *between* the front main speakers.

While musical bass can be subjectively directionless, certain kinds of "waterfall-like" noises that might appear during film soundtracks (jet-engine noises, rocket whooshes, and, of course, waterfalls) can call attention to the woofer's location if it is too far off center and the crossover is between 200 and 150 Hz. This is usually more apparent in smaller rooms, at closer listening distances, than in larger ones. Set up correctly, a system like this will often surpass the sound of similarly priced, standard stereo speakers, although the latter will usually have better deep bass.

Single-woofer-box systems do have potential drawbacks, even when working only with music. Some individuals believe that perceived spaciousness may be adversely affected, and the really low bass response can be limited compared with a dual-woofer system. Nevertheless, as a convenient and attractive alternative to a large, two-box conventional system, satellite-speaker systems will be hard to top, particularly for video use—although it is a good idea for the satellites to be magnetically shielded (as these are) to protect direct-view TV picture tubes from interference. This particular system is designed specifically for audio and/or video use, with the extra three satellites normally assigned to center- and surround-channel duty (see Chapter 7).

Figure 2.7. Advent Laureate

This moderately priced two-way model makes use of two modest-sized woofers to increase power handling in the lower octaves and to reduce distortion. Together, they have a low-range capacity similar to that of a single larger driver but without the midrange horizontal "beaming" that would plague a large-woofer two-way speaker design.

tion. In two-way systems, stacked smaller woofers will disperse sound more uniformly in the horizontal plane than will a single, larger-diameter woofer with similar bass potential; stacked tweeters have similar dispersion qualities but at higher frequencies. Laterally placed, multiple tweeters can offer superior horizontal dispersion, but only if they are mounted on panels that are strongly angled outward from each other.[3]

multiple tweeters

A number of systems have *lots* of drivers. Usually, such combinations serve to increase power handling, although in some cases the drivers are con-

[3]For more on the behavior of double woofers, see the subwoofer section; we'll discuss bass-woofer, "coherent" bass reinforcement later in the chapter (see pages 64 and 66).

Figure 2.8. Definitive Technology Speakers

(Photo courtesy Definitive Technology)

These three Definitive Technology speakers are "bipolar" models, with drivers facing both front and rear. Assuming that you place the main speakers out some distance from the front wall, the big advantage of having a second set of drivers facing away from the listening area (aside from increased power handling) is the heightened sense of space and openness that results from having half of the system's full-bandwidth output reflected from the front wall.

The two smaller systems shown are four-driver, two-way models with the front and rear panels each containing a woofer and tweeter. The larger system is also a two-way design, but the front and rear panels each have a single tweeter and pair of woofers (for a total of three drivers facing the front and three facing the rear), with each pair of woofers mounted above and below each tweeter, D'Appolito style. The latter limits vertical dispersion in the midrange and improves both imaging and clarity for listeners positioning themselves so that their ears are close to the same vertical plane as the tweeter.

figured to control imaging, as with some NHT models, or add ambiance. For many enthusiasts, these systems are too expensive to contemplate owning, but smart shoppers would do well to keep their eyes peeled for used, larger Allison or early dbx models. Perhaps the most extreme example of multiple-driver use is the highly regarded Bose 901, which has *nine* identical mid-range-sized speaker drivers to cover the audible bandwidth in a *one*-way design.[4] In any case, do not confuse multiway designs with multidriver designs; the latter is used to enhance the performance of the former.

OEM suppliers

Most manufacturers assemble their own crossover networks, but surprisingly few build the drivers used in their systems. Instead, they shop from independent OEM (Original Equipment Manufacturer) suppliers for drivers that satisfy the rather broad parameters they have selected. In some cases, design parameters are compromised because the speaker-system builder must accept the limitations of the available drivers. (Of course, if the expected sales volume is high enough, OEM suppliers may be more than happy to custom-build drivers.) Also, if a driver becomes unavailable, a different model may have to be substituted later in the production run.

I believe that in many cases speaker manufacturers who make their own drivers rate higher than "box builders" who install off-the-shelf components made by others. A manufacturer's ability to produce its own drivers indicates that it is both serious and competent and that it uses very rigid guidelines that must be satisfied by in-house, customized work. Do not hesi-tate to query store personnel or even manufacturers themselves about who builds the individual parts in the speaker systems you are interested in purchasing.

NARROW VS. WIDE DISPERSION

In spite of their outward simplicity, speakers are easily the most controversial part of the audio-video world. Whereas decent electronic audio (and video) components tend to behave with exacting consistency, speaker systems, even different models made by the same company, usually sound surprisingly *unlike* each other. While most amplifiers, for instance, differ mainly in maximum power output, speakers can vary greatly in quality because of their mechanical operation and the three-dimensional environment (i.e., the room) in which they operate.

Many experts agree that flat (or at least smooth) power input to a room and proper high- and middle-range dispersion are the most critical aspects of

[4]One driver faces forward, to "lock in" the first-arrival image, and the other eight are grouped on a divided rear panel, facing the front wall at moderate outward and inward angles to impart a spacious effect.

Flat-Panels, Dipoles, and Line Sources

Visit a high-end audio-video store and you may discover that some of the speakers on display do not look much like speakers at all. These usually quite- expensive models often resemble Chinese room dividers or flat pieces of modern sculpture. Many are over six feet tall and a foot or two wide and only a few inches thick, although one, the Quad, is much shorter and nearly square.

Unlike "conventional" speaker systems—which use cones, domes, and boxed enclosures—these systems often employ a thin, electrostatically driven, large-surface diaphragm or a thin sheet of similar material (usually Mylar™) laced with inductive wiring that interacts with many small magnets fixed in place over a grid that covers and protects the radiating surface; the former are called *electrostatic speakers* and the latter, *planar-magnetic speakers*.

These panel-system drivers are not mounted in an enclosure but are simply attached within a frame that is open front and rear: hence the room-divider look. As a result, the sound is radiated from both the side facing the listener and the side facing the front wall; these systems are sometimes called *dipole* radiators. This characteristic necessarily results in front sound waves being out of phase with the rear and has two consequences. First, a cancellation "null" exists to the sides of the speakers at all frequencies, limiting side-wall reflections. Panel-speaker enthusiasts often claim that this improves clarity and soundstage realism. Second, below certain frequencies the cancellation effect attenuates the bass at 6 dB per octave, resulting in the characteristically anemic deep bass of such systems.[1] A third characteristic of flat-panel performance is the strong front-wall reflection. This is said by partisans to impart a realistic sense of depth to some program material.

While a few panel-type speakers are one-way models and use a single large radiating surface, most are two- or three-way designs and employ multiple drivers to handle the bass, mid-, and high ranges. (Some designs employ one or more cone woofers in a cabinet for the bass.) Often, the "ribbon-like" treble drivers are only a half-inch wide and may extend from the top of the system to the bottom, with the larger bass-range panels arrayed to the side. Like the larger drivers, the ribbons are dipoles, but being narrow gives them better horizontal dispersion over a ±45-degree angle—front and rear—than larger, single-panel, one-way versions, although they still exhibit a null directly to the sides. While two-way panel designs will display horizontal comb-filtering artifacts at and near crossover points, because of the side-by-side layout of the high- and low-range drivers, their superior dispersion partially compensates for this by producing more reflected sound from the walls to the side, smoothing out some response irregularities.

In addition to the dipole effect, the vertical length of most panel-system drivers results in them behaving as *line sources*.[2] While the use of

[1]A few models, notably some dynamic-woofered ones produced by Carver, are electro-mechanically configured to compensate for this effect.

[2]See the section on THX speakers, particularly Figure 7.5, to get a handle on line-source behavior.

(continued)

Flat-Panels, Dipoles, and Line Sources *(continued)*

a long line source will result in substantial comb filtering, thereby affecting frequency response in both the direct and reverberant fields, many panel-speaker enthusiasts believe that the attenuated vertical radiation ensures greater clarity and better imaging. The nature of these systems makes it usually impossible to get fully into the reverberant sound field in most listening rooms (see "Near/Far, Direct/Reverb Explanation," page 42). Consequently, they often have a headphone-like clarity that many people consider the height of audio realism. Live-music enthusiasts may disagree with this conclusion, but few doubt that panel systems often have an uncanny ability to reproduce sounds on a recording that would not be audible even from the front row at a live concert.

A system does not necessarily have to use electrostatic, planar-magnetic or continuous-length ribbon drivers to deliver panel or line-source performance. It is quite possible to simulate the behavior of a large one-way panel system by employing an array of conventional drivers spaced over a large open-backed surface (the Carver system's woofers, for example) or clone the sound of a long ribbon tweeter by employing a vertical line of open-back conventional drivers. The sound characteristic of most flat-panel systems is mainly the result of driver layout and not driver design. It is also possible to mimic the line or large-radiating-surface characteristics of such systems without the dipole effect by mounting an array or row of drivers in a conventional, sealed box. Indeed, a "short" vertical line-source array is a marvelous way to control unwanted vertical dispersion without adding too many comb-filtering effects.[3]

Flat-panel systems are mysterious performers, and even their most fervent advocates will admit that they are not always suitable in every kind of listening environment. In addition, few panel-type systems would satisfy a die-hard "videophile" without the addition of a subwoofer to fill out the bottom end. While there are a handful of models available that do not cost a king's ransom, most are expensive enough to justify careful shopping by even well-heeled enthusiasts.

[3]Configurations of this kind can be found in the Allison IC-20 and the large Polk Audio Theatre system. The D'Appolito arrangement, employed with the larger Definitive Technology speaker systems (as well as models from a number of other manufacturers, including many that are THX-certified), also takes advantage of the phenomenon.

speaker performance. Low distortion and wide frequency response are important also, but these other characteristics are what separate bona fide high-fidelity speaker systems from the bush-league competition. However, while audio enthusiasts in general agree on the need for smooth power input, there is significant disagreement about desired dispersion characteristics for speaker systems.

One school of thought considers very wide dispersion at all frequencies to be not only unnecessary but actually detrimental to good performance. They believe that the first-arrival signal from the speaker should be flat in

Anechoic vs. Reverberant

The terms "anechoic" and "reverberant" are bantered about by audio buffs all the time. Those who are not well-versed in this terminology may be confused by the assorted definitions.

Anechoic literally means "without echo." An anechoic situation exists when acoustic signals produced by a source (speaker or other noise maker) are not reflected back to it or anywhere else. If you were outdoors on the Great Plains, you would be in an almost anechoic environment, but the ground would still reflect some sound. To be fully anechoic, you would have to parachute out of a plane; on the way down, you would be in an anechoic environment. Back on earth, partial anechoic environments can be made by treating rooms with exotic sound-absorbing materials, but even those will be not be anechoic at low frequencies.

A *reverberant* environment is the opposite of an anechoic one: sounds are reflected all over the place. A large reverberant area would be the Grand Canyon or the Vehicle Assembly Building at Cape Kennedy, where an echo will return from great distances. A smaller one would be a typical bathroom when the shower curtain and towels are removed. Reverberant rooms have hard walls that reflect a lot of energy. Theoretically, a perfect example would allow the sound to bounce around indefinitely. However, even stiff walls flex some, and air absorbs energy, so the sound is converted to heat and dissipated.

Acoustics experts will talk about the reverberation time of a room, which deals with the time it takes a signal of a given intensity to fall to a much smaller intensity. Proper listening rooms have a mix of anechoic and reverberant characteristics at assorted frequencies. Fortunately, most well-furnished living rooms have a good anechoic/reverberant mix.

amplitude and that all signals be phase coherent.[5] This clean, first-arrival signal should not have its purity muddied by multiple-driver interactions and by many strong, nearby wall and cabinet reflections arriving a very short time afterwards. These buffs prefer a single, two- or three-way driver system for all frequency ranges, often of moderately large size at least in the midrange and treble, and usually with the drivers mounted in some kind of "phase coherent" alignment. However, these systems will probably perform at their best under *direct field* (close-up) listening conditions, with each of a stereo pair angled inward somewhat and carefully "aimed" at the solo listener. Such systems are said to have a narrow-field radiation pattern.

direct field

Another school of thought considers extremely wide dispersion at all frequencies to be a positive thing. These individuals believe that first-arrival signals mainly allow the listener to accurately position various instruments and

[5] This relates to their arrival times from the various drivers in the cabinet and those reflected from its front surface and nearby walls.

How Off-Axis Behavior Affects Speaker Sound

Those who insist that *extreme* off-axis behavior is not important may point out that the kinds of systems they admire have significantly reduced output off to the sides anyway, making the irregularities that result from such behavior inconsequential. While there is no doubt that a speaker system that radiates a substantial amount of energy 45 to 90 degrees off axis will sound different from one that does not, there is also no doubt that a system that has a nonlinear frequency response that far off center will not be as satisfactory as one that has a smooth output over the same angular range, no matter how strong or weak their relative outputs at those angles. These diagrams show why even a strongly attenuated off-axis signal, irregular or not, can be very audible.

The two hemispheres indicate the total frontal radiation of a typical speaker system. The shaded area of hemisphere A is the angle

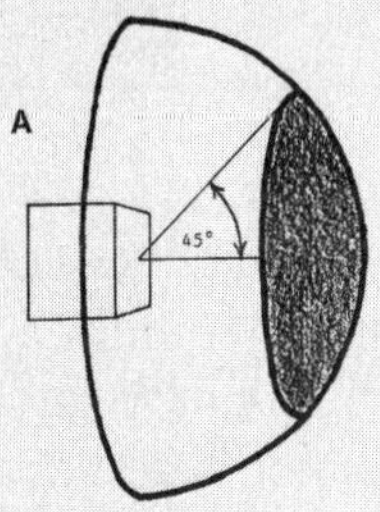

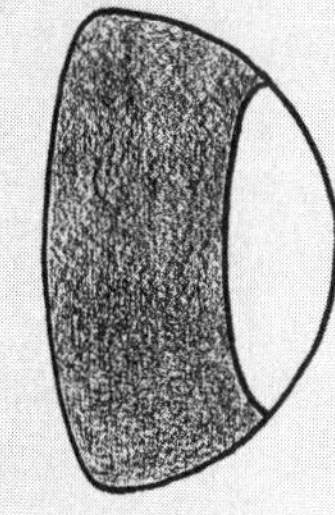

covered by the 0- to 45-degree off-axis radiation of a system. Most speaker testers measure only within this area or even less (more than a few near-field-oriented individuals stop making measurements at a rather narrow 30 degrees off axis), and many manufacturers consider it to be by far the most significant part of a speaker's output into any kind of room.

The shaded area surrounding the bulk of hemisphere B indicates the level of coverage from 45 to 90 degrees off-axis. While the total angular coverage is the same as in hemisphere A, the total *area* covered is almost 2.5 times as large. (Comparing area covered within the 0- to 30-degree angle favored by the many near-field-oriented speaker testers with the remaining coverage over the 30- to 90-degree angle results in an even larger area difference.) Even if the off-axis signals radiating into this segment are substantially lower in level than those covering the forward angle, the larger area being affected amplifies their impact. Because off-axis irregularities are common, the result is that many of the speaker systems available, even when clean and smooth over a narrow dispersion angle up front, produce inappropriate sound when located in a typical home-listening environment.

Uniform radiation in the vertical direction should not be as critical as that in the horizontal, although it may be desirable to limit it for analytic imaging and precise, close-up articulation. (The THX people at LucasFilm—see Chapter 7—are acutely aware of this, although some knowledgeable enthusiasts prefer the more diffuse, spacious soundfield presented by systems with wide vertical, as well as horizontal, dispersion.) However, over the full horizontal angle in front of a speaker, the effects of nonuniform, broad-band polar response, particularly as it reflects the crossover-controlled transition points between different-sized drivers, will be considerable. Designers who ignore extreme-angle off-axis behavior do so

(continued)

How Off-Axis Behavior Affects Speaker Sound *(continued)*

at the risk of producing very clear-sounding, one-listener-at-a-time-oriented loudspeakers that, instead of sounding like live music, sound like, well . . . loudspeakers.

Although it is difficult for any two-way system to produce uniform sound both on- and off-axis while maintaining decent frequency response, it is possible to create compromise solutions that work quite well, at least over the horizontal plane. Of course, a good three-way system can do a better job of delivering uniform sound over the front hemisphere than any two-way model. We will discover the reasons for this up ahead.

When discussing polar behavior, there are other variables to consider. For example, a speaker will also radiate sound into the hemisphere behind it. If the response over the 45- to 90-degree angle is smooth, the attenuated, rearward response that results from cabinet shadowing should also be smooth. However, if the system's response at extreme angles is rough, imaging will be erratic, and the reflections coming from the wall behind the speaker will be rough and will color the overall sound. "Omnidirectional" or partially omnidirectional speakers[1] actively strengthen and control radiation to the rear, and successful designs will have a rearward radiation pattern as uniform as that delivered to the forward hemisphere.

[1]As exemplified by the dbx Soundfield One, certain Definitive Technology, Snell, and DCM models, the Ohm Walsh design, the old AMT driver and some Mirage models, as well as the Bose 901 and even many electrostatic-panel and magnetic-planar models—although the latter two do not have particularly strong radiation to the sides.

performers; the initial signal "captures" the ear but does not dictate the ultimate acoustic balance. Because the majority of sound any speaker system generates is going to be reflected from room boundary surfaces before being heard, it is vital that *all* off-axis signals be as smooth and clean as those heading straight out towards the front.

Listeners who believe that the first-arrival signal is critical tend to sit in the direct field, fairly close to their speakers, usually with those speakers pulled out away from nearby reflective walls and carefully aimed at the prime listening position. However, wide-dispersion advocates point out that most home listeners sit in the *reverberant sound field,* where the reflected sound will be considerably stronger than the direct sound.

reverberant sound field

Depending on the dispersion pattern of the system, the frequencies involved, and the room's reflecting and absorbing characteristics, the transition from the direct to the reverberant field (called the "critical distance" of a room) can take place anywhere from 3 to 20 feet from the loudspeakers. With speaker systems having erratic mid- and/or high-frequency dispersion, the transition point will shift as the frequency of the musical program changes.

critical distance

Narrow vs. Wide Dispersion

Narrow Dispersion Advocates	Wide Dispersion Advocates
Sit close to and directly in front of speakers for best sound reproduction	Sit in the "reverberant field," where the reflected sound will dominate
Believe first-arrival signal should be flat in amplitude and not be muddied by multiple-driver interactions or strong reverberations from walls or other reflecting surfaces; the only reverberation should come from the rear of the room ("live end/dead end" [LEDE] room)	Believe first-arrival signal allows listener to "place" a sound in the sonic field, but that reverberant sound is more important to the overall experience and allows for room reverberation as natural part of listening
Control the listening environment as much as possible; move speakers out from walls; add sound-deadening materials to eliminate off-axis signals	View extreme room modifications as costly and ultimately aesthetically unacceptable and unnecessary; if enough power is supplied to the mid and treble ranges, room can be saturated with sound in a natural way
Prefer analytical, flat, tighter, and more focused sound	Prefer warmer, open, more natural sound

While direct-field listening may result in extreme clarity (the ultimate example would be the clarity produced by a pair of headphones), reverberant-field listening can open things up, while having no important negative effects on clarity. Most live music is enjoyed in the reverberant field.

radiation pattern

Wide and uniform dispersion champions point out that it is very difficult to obtain a *controlled*, narrow radiation pattern from speaker systems. If a driver is small in relation to the audio wavelengths it must reproduce, it will radiate them uniformly over a wide area. Woofers, for example, radiate the low bass omnidirectionally because they are small in relation to the long bass-frequency wavelengths. As the signal handled by the woofer climbs in frequency, those wavelengths become shorter, and at some point the off-axis output will begin to narrow. However, as the frequency continues to climb, the crossover network will route the signal to another, smaller driver (midrange in a three-way system or a tweeter in a two-way system), and the *total* system off-axis response increases in strength. The result is that while the on-axis response of many highly regarded two- and three-way systems may be flat over much of the audio bandwidth, their off-axis response has a sawtoothed appearance, particularly at extreme angles off to the sides.

power response

This sawtoothed off-axis characteristic affects the behavior of a sound system in two ways. First, because the ear perceives the complete wavefront as unified, the *total* output of a speaker system—its *power response*—will be irregular if the off-axis output is not smooth. Compensating for this by increasing the output of the woofer and/or midrange drivers as they reach the

Near/Far, Direct/Reverb Explanation

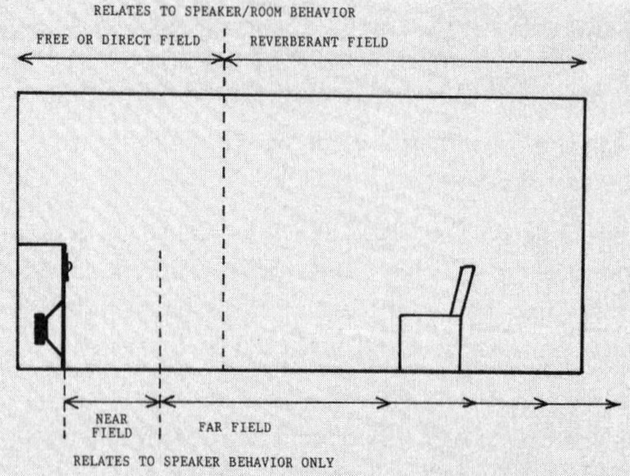

The *free field* is the condition whereby a signal reaches the listener without having been reflected from any surfaces. It correlates with the *direct field*, which is the area in a room where the direct sound is louder than the accumulated sound reflected from wall, ceiling, floor, and other surfaces. The *reverberant field* is the sound that exists when the reflected sound in a listening room predominates over the direct sound. The point between those areas is often called the "critical distance."

The *near field* is technically the region where particle velocity is mostly out of phase with the sound pressure and relates to speaker behavior apart from any room considerations. It may exist as close as the distance between the longest dimensions of the sound source (which means that it will change size as frequency varies and assorted drivers cut in or interact) or as close as one or two diameters of the wavelength being reproduced. The *far field* exists when each doubling of the listening distance results in a 6-dB reduction in the direct sound level due to the inverse-square law. It is typically 1 to 3 times the distance between the most separated points of a speaker system that are radiating at the same frequency. It is obvious that the distance from the speaker to the transition point between the near and far fields will vary somewhat, depending on which drivers of a system are operating or whether the drivers are very large, as would be the case with flat-panel systems or systems employing long ribbon-like drivers.

The important factors for people listening to recorded music in real rooms involves the direct- or reverberant-field distances. These will dictate how most non-exotic speaker systems sound under typical conditions. In the reverberant field, the peaks and dips that dominate the direct field are submerged in the mix of off-axis reflections formed over each 30-ms (millisecond) period of time following each direct signal. The mass of reflections contained within this multitude of 3/100-second intervals will themselves run together to form a continuum. What the listener will hear, in other words, will depend upon the uniformity of the total output of the speaker system, or its *power response*, in combination with its radiation pattern. Even easily measurable peaks and dips in the direct field may be of little consequence if the systems are wide-dispersion models and are listened to at normal distances.[1] The most important characteristic of a speaker's performance will relate to its behavior in the reverberant field.

[1]This statement assumes three things, however. *First*, that those peaks are less than 1/3 octave wide (which relates to the signal filtering behavior of the human ear); *second*, that they are not related to system box or driver resonances (which can cause peaks in the reverberant field); and *third*, that they are symmetrical in each speaker's radiation pattern (if not, aural images will shift about and be poorly defined).

Figure 2.9 AR-LST

Photo Courtesy of Acoustic Research

In the 1960s and 1970s, the company most identified with good reverberant-field performance was Acoustic Research (AR). One of the most famous of their systems was the AR-3a, which used midrange and tweeter drivers that were very small for the frequency ranges they handled, giving the system very wide and uniform dispersion in the front hemisphere. Lower-cost versions, the AR-5 and AR-2ax, were only marginally inferior to the 3a (with the AR-5 matching it in everything but deep-bass capacity) and were proof that excellent sound could come from a low-cost package. However, perhaps the best speaker of that era was their original AR-LST shown here, which used four of the 3a tweeters and four of its midrange drivers (in combination with its robust woofer) on three angled front panels, to present a nearly perfect hemispherical radiation pattern.

Like the very different Bose 901, the LST, designed by Roy Allison, has long been considered a "classic" and is still able to hold its own against many contemporary systems. Although the original version has been discontinued for some time, a number of these systems are still in use as studio monitors, lab units, and home systems. The first LST was an expensive system for its time ($1,200 a pair, in 1975), but the cheaper AR models could give it quite a battle in a musical-accuracy contest. Over the years, the 3a and LST have gone into and out of production, with the former often showing up as a high-end export to Japan and the latter sometimes marketed as an even higher-end item produced under the Cello name for domestic sales.

tops of their frequency ranges may result in close-up, on-axis performance that is erratic and/or overly bright. Second, the ear is also aware of the *difference* in broad-band performance between a system's on-axis and extreme off-axis output. The sounds being reflected from nearby boundaries should be similar in character to those of the first-arrival signal. If they are not, the speaker system may sound isolated, constricted, and "box like." No amount of driver-response contouring, or equalizer or tone-control manipulation, will correct this kind of imbalance.

There are arguments to be made for both narrow- and wide-response groups. Defenders of wide-dispersion speakers point out that most people do not want to turn their homes into studios, and that most live music is heard in

Figure 2.10 Ohm Walsh Speaker

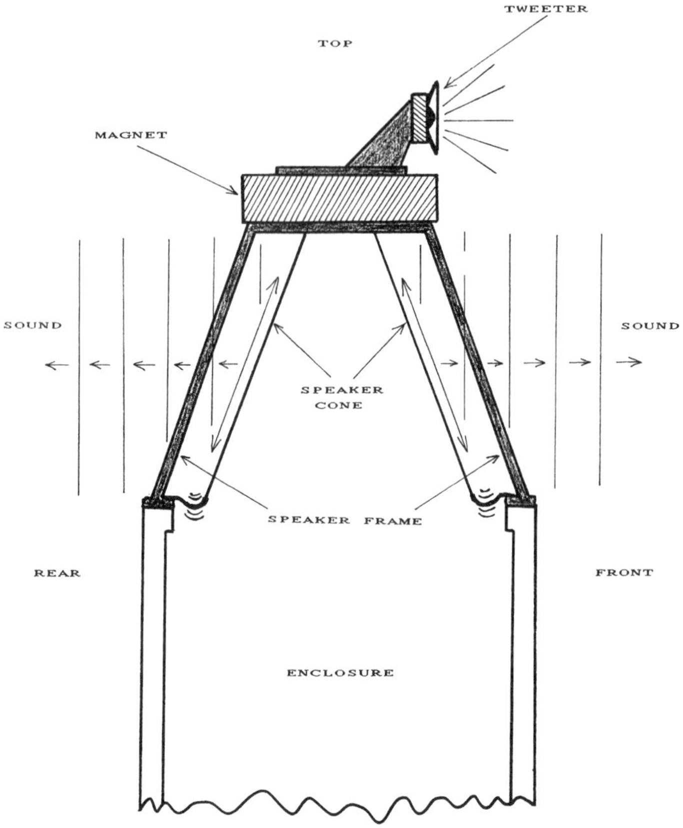

Most of the Ohm Corporation's "Walsh" speakers are pretty expensive. However, a few of them are not outrageously so. The very interesting Walsh driver illustrated here is an elongated and inverted conventional speaker cone that radiates sound from its backside in a 360° arc around the system. (We are viewing it here in cross-section, from the side.) Because of its design, the horizontal wave front is roughly cylindrical in shape and is quite phase coherent. The single driver is not adept at producing higher frequencies, so a separate, top-mounted tweeter is included to fill out the high end and shape the radiation pattern for good imaging over a wide listening area. To protect the drivers, a screen "cage" encloses the assembly, and a cloth-covered frame fits over the entire top to make the system compatible with a typical living room's decor.

Three-Way Flat On-Axis Curves

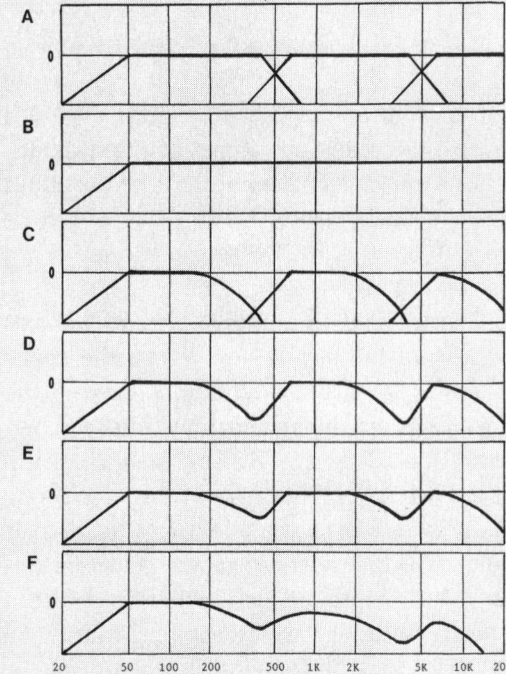

This series of response curves is indicative of the performance of a three-way system optimized for a flat first-arrival signal. The required compromise involves giving up a flat overall power response. The curves here are "idealized," in that they do not reflect the behavior of an actual system, which would picture other visible aberrations in a real-world analysis. The speaker drivers employed in a system of this kind might include a single 1-inch-dome tweeter, 4-inch-cone midrange, and 10-inch woofer in a short, tower-type cabinet having dimensions of, say, 30 × 12 × 11 inches.

Curve A indicates the on-axis response of the three individual speaker drivers. The woofer-to-midrange crossover point is at 500 Hz and the mid-to-tweeter point is at 5 kHz. Curve B is the on-axis, *system* response of the three drivers; the notches of curve A are eliminated due to signal overlap. This is what would be heard if the system were auditioned from very close up, especially in a fairly nonreflective, acoustically "dead" room.

Curve C indicates the 75°- to 90°-off-axis response of the three drivers. The steeper upper-end slopes are the result of their being too large for the crossover frequencies chosen. Curve D is the off-axis, *system* response of the three drivers. Because of the steep rolloffs at the tops of their operating ranges, the crossover overlaps are not great enough to offset signal losses. The curve takes on a sawtoothed look. This is the sound that would be reflected off nearby walls in a typical listening room.

Curve E is the front-hemisphere response of the system, with curves B and D, together with all the curves in-between, added together to form a composite. The sawtoothed characteristic is lessened but still remains—and still remains audible. This kind of curve is important for pinpointing design flaws in forward-facing systems.

Curve F is the *power response* of this system, with the signal losses that result from the additional but suppressed higher frequencies being radiated behind the system and bounced from the front wall added into the picture. Because the ear senses this combined output under typical listening conditions, it is obvious that this system will lack balance when heard in most rooms. However, the sound will be detailed and clean when the system is auditioned from close up and/or in acoustically treated (padded) rooms. When two-way systems are designed for flat on-axis behavior, the resulting power response may be even more irregular.

Two-Way Flat Power Curves

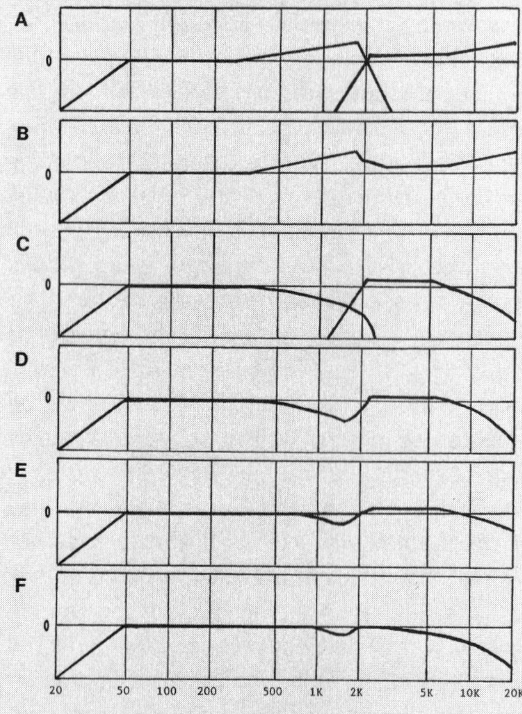

This series of response curves shows the performance of a well-designed, economical two-way system that has been optimized for *flat power response*. The required compromise involves giving up a flat first-arrival signal. As in "Three-Way Flat On-Axis Curves," page 45, the curves here are "idealized." A system of this kind might have a single 1-inch dome tweeter and an 8-inch woofer in a cabinet roughly 20 × 11 × 10 inches.

Curve A indicates the on-axis response of the two drivers. The woofer-to-tweeter crossover point is at 2 kHz. The outputs of both drivers have been purposely increased at the tops of their respective ranges. Curve B is the on-axis, *system* response of the two drivers. The curve is uneven because of the above response contouring. This is what would be heard if the system were auditioned from close up in a fairly dead room.

Curve C indicates the 75°- to 90°-off-axis response of the two drivers. While the upper end slopes are still steeper than in curve A, the contouring has minimized their effect somewhat. Curve D is the off-axis, *system* response of the two drivers. Because of the contouring applied, the sawtooth effect is less than it would be if the system had flat response on-axis. This is the sound that would be reflected off nearby walls in a typical listening room.

Curve E is the front-hemisphere response of the system, with curves B and D, together with all the curves in between, added together to form a composite. The contouring has mostly corrected the sawtooth effect. Curve F is the *power response* of this system with the signal losses that result from the additional but suppressed higher frequencies being radiated behind the system and reflected off the front wall further compensated for by the contouring revealed in curves A and B. Because the ear senses this total output when listening under typical conditions, it is obvious that this system will sound balanced in most situations. However, it will sound a bit rough under direct-field listening conditions or in rooms with heavy acoustic treatment on adjacent walls. (So-called "near-field" studio monitors are used this way.) When such contouring is applied to three-way systems, even better performance results.

the reverberant sound field, not from 6 to 8 feet away. Direct-field proponents, who often *do* sit pretty close to their speakers, counterargue that any room colorations affect the listening experience. They insist that room effects should be reduced by sound-deadening treatments, speakers be placed away from floor and wall surfaces, and those speakers have a clean, flat first-arrival signal.

wide dispersion

Those who champion wide-dispersion, wide-bandwidth systems counter that it is impossible to control room acoustics with any reasonable precision (especially in the real world of the typical homeowner) and that any system with erratic off-axis behavior will have imperfect power response—something that multiple surround speakers cannot correct. They also note that a reduction in high- and midfrequency reflections from near a speaker will make it sound weaker at those frequencies unless the midrange and tweeter drivers are given increased power inputs, reducing peak subjective volume potential and increasing the possibility of audible distortion and/or voice-coil burnout.

The dispersion controversy has raged in hi-fi circles for years. It is a fact that some recordings sound great on narrow-dispersion systems in specially

Figure 2.11 LEDE Power/Direct Curves

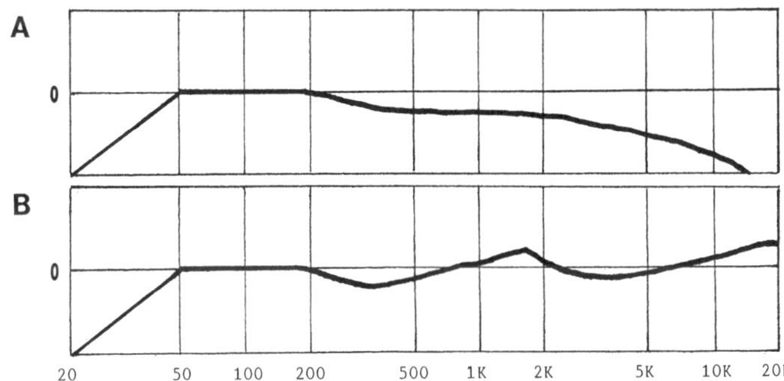

If the two-way, "flat-power" system were placed in an LEDE-type room, the off-axis absorption would produce effects similar to these. Curve A is the room response in this environment and indicates what would happen to the power response. The treble part of it would be attenuated by the absorbing qualities of the padding up front. Curve B is the on-axis, direct signal that would dominate under such conditions. A pair of stereo speakers might have remarkable detail and soundstage "imaging," but would lack the sense of space and integrated feel of live music because of the differences between the two curves. The pair would sound somewhat isolated and "speaker-like." Surprisingly, many buffs (even knowledgeable ones) prefer these attributes above all others and spend great amounts of time, money, and energy creating these environments in their homes. However, only systems designed for a flat first-arrival signal will sound balanced in an LEDE-type room, and then only within a small area.

LEDE-Type Room

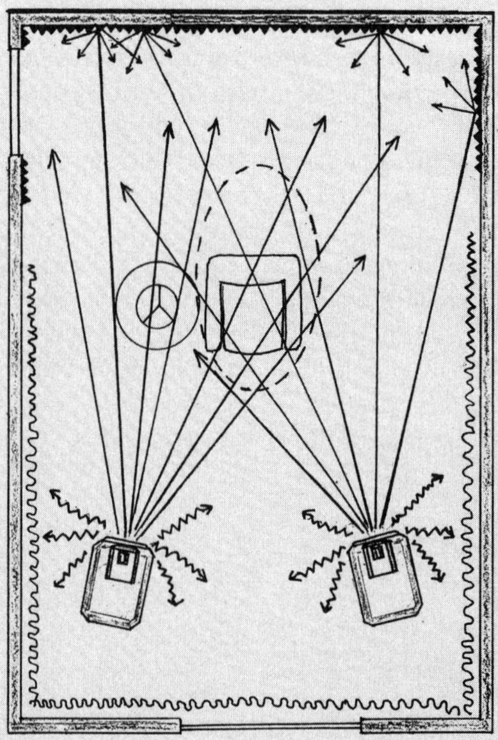

In a normal room, speakers designed for direct-field use may have poor spectral balance. This drawing gives an idea of how a pair would behave in a "dedicated," LEDE-type room. "Toed in," they project a strong, direct first-arrival signal to the listening position, while their somewhat erratic, but subdued, off-axis middle- and upper-range sounds are rendered even more ineffectual by the adjacent side- and front-wall padding. A significant amount of clean, "aimed" sound also reaches the rear wall behind the listener, where it is reflected but also scattered by hard, specially designed diffusion surfaces back to the listening area—appreciably delayed in time. The diffusion surfaces eliminate potential hot-spot reflections, and the delay makes it less likely for them to negatively affect the ambiance on the recording itself.[1] These setups may provide almost "headphone-like" detail and present a very precise sound stage. However, they normally cannot dodge the reality of their lack of frontal spaciousness and warmth and tend to sound like "hi-fi systems." The behavior of such a room/speaker configuration precludes having a good seat anywhere but near the central "sweet spot."

Note that adding four or more "surround" speakers to augment the main ones, at least if the surround processor is an elaborate DSP-type device (controlling two speakers placed to the sides, or even up towards the front, as well as two additional ones further toward the rear), can often successfully fill out the deficiencies in spaciousness and combine pinpoint imaging with concert-hall ambience. Of course, these processors can be very expensive, and additional, good-quality speakers must be used—and the room must still be dedicated to the system and employ sound-deadening materials. In addition, unless the processor is an ambience-extraction type (as opposed to one that synthesizes ambience), the additional reverb may do more to "mask" the recorded reverberation than any pair of wide-dispersion speakers operating alone. (See Chapter 7 for more on the use of surround sound.)

[1]Sound waves that are larger than a reflecting surface are diffused rather than solidly reflected at a fixed angle equal to but opposite to the angle of incidence—and a properly designed diffusion array contains a multitude of small, hard surfaces at different angles.

treated rooms, that some recordings sound equally good on wide-dispersion models in more typically furnished rooms, and that some recordings sound excellent (or bad) in either. It is also a fact that some listeners like the open sound of the wide-dispersion designs and that others prefer the tighter, more analytical characteristics of the more focused competition. Tastes vary, and either configuration should satisfy the definition of high fidelity provided that each delivers the requisite *flat power to the listener's ears.* While it is considerably easier (and usually cheaper) for a pair of wide-dispersion systems to

flat power

"Normal" Room and Speakers

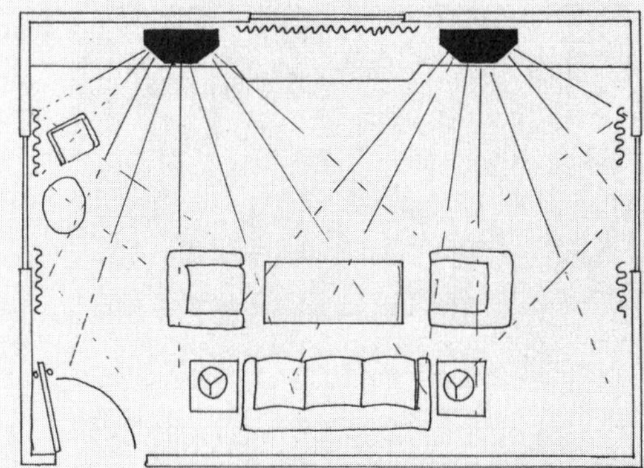

Speakers with very wide and uniform dispersion tend to have good spectral balance in "normal" rooms. This sketch gives you an idea of how they would behave in a room that has a balanced mixture of reflective and normal surfaces. The first-arrival signals, although not strong compared with the reverberant-field sound, are still adequately loud to localize the recorded images, because of the Haas or "precedence" effect. A reasonable amount of padding on the floors (carpeting), along the walls (furniture, drapes, bookcases, etc.), and out in the room itself (more furniture and people) will further smooth things out.

Most good rooms tend to straddle the line between extreme reflectivity and heavy absorption. When a decent balance is found, the frontal spaciousness and lack of "speaker-like" sound can make for a superior listening experience, particularly when only two speakers are involved. Note that the lack of directionality with systems of this kind insures good listening positions over a fairly broad area.

In spite of what some critics say, wide-dispersion "main" speakers can work effectively with surround processors, especially the more basic models that have only two surround outputs. This is because most such devices require that the surround speakers be placed straight out to the sides or somewhat to the rear. When this is done, the wide-dispersion main speakers may actually enhance the overall surround-sound effect, because of the intensified spaciousness up front. That spaciousness will not negatively affect the imaging of video-oriented systems that employ a "steered" or discrete center channel, because the center speaker (which, like the main speakers, will also be more effective if it has wide and smooth dispersion) will keep dialogue and centralized sound effects and music focused in the middle where they should be. (See Chapter 7 for more on the use of center-channel and surround speakers.)

Figure 2.12

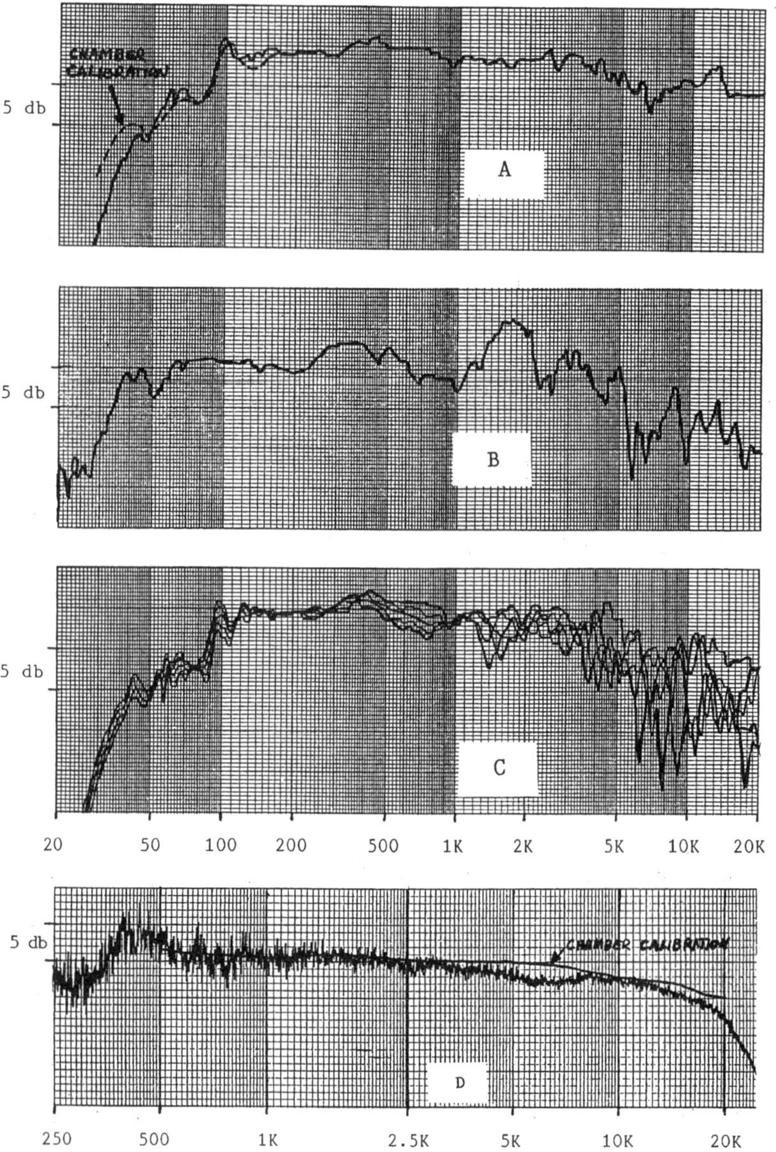

These curves—produced by an AR-3a loudspeaker under very controlled conditions—give you an idea of the performance of a real system under assorted conditions. Curve A is its on-axis performance in an anechoic chamber. This is the "first-arrival" signal and is what localizes the image. Curve B is the 60°-off-axis performance of this same system. The roughness is the result of molding-induced diffractions and driver interference effects that cause comb-filtering peaks and dips. At any other angles off-axis, different peaks and dips would appear at different places. Curve C is an anechoic curve of one of many possible 0°- to 90°- angular spreads for this system, including diffraction and interference effects. Note the random overlap of the signals. Curve D is the power response of this same system from 250 Hz to 20 kHz, combining all the curves around the entire 360° perimeter. The power response indicates that this design is very smooth under real-world listening conditions where cabinet diffraction and driver interference effects are blended together to form a totality. It is this totality which we mostly hear when listening to a speaker system.

achieve this goal in most rooms, there is no reason why narrow-dispersion models, particularly if they are part of a surround-sound installation, cannot deliver the goods with a bit of effort on the part of the owner. That effort plus the cost factor are obviously going to be important to most people, however.

Most dealer showrooms have problems demonstrating wide-dispersion speakers.

In my opinion, audio enthusiasts and music lovers on any kind of reasonable budget are going to have to go with wide-dispersion speakers by default, no matter how they may feel about the wide-vs.-narrow controversy. Otherwise, besides possibly paying for elaborate modifications to the room, the listener will be required to place him- or herself in a position optimally suited to the speakers (perhaps not the most comfortable spot for listening); groups of listeners will have to sit closely together, and some members of the group will still probably suffer. Given the fact that most recordings sound at least as realistic and usually considerably better on wide-dispersion systems in typical home listening rooms, the latter appears to be a better choice for almost everyone.

CAN A CHEAP SPEAKER DO THE JOB?

While two-way speakers may have problems with producing both a smooth power response *and* a uniform radiation pattern, there is no reason for budget-minded enthusiasts to despair. Because most listening is done in the reverberant field, power response will be the most important performance criteria. If you sit back more than 8 or 9 feet from your speakers, the effects of on- and off-axis differences will not be as apparent as an irregular power response; indeed, poor power response will create even worse off-axis behavior. Good power response is *the* prime factor to consider when selecting a system, and many reasonably priced models meet this criterion.

radiation pattern

If a system has good power response, the added benefit of a uniform radiation pattern is the icing on the audio cake. You should not get too worked up if the radiation pattern of a cheaper system is not world class. The usual effect of such an inadequacy is a bit more "speaker-like" sound and less-than-exact imaging, something that is easier to live with than you might think. By the way, just about all current, low-cost systems have this problem, meaning that if you want better performance, an extra outlay for a more expensive (i.e., three-way) system may be your only alternative. The path to world-class reproduction is going to require a few more bucks up front! Whether the audible difference between a good, economically priced two-way system and a somewhat better, but more expensive, three-way model is economically significant is up to you. With most people, even some very astute ones, I would guess that the extra cost will not be justified. I know some dedicated musicians who are very pleased with fairly modest sound systems.

Figure 2.13 Three-Way Flat Power Curves

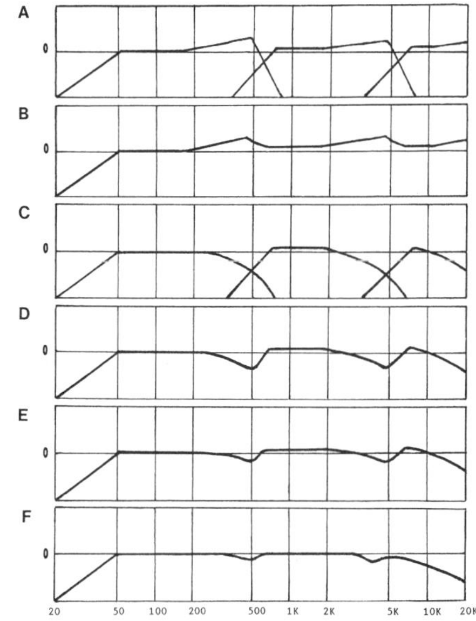

This series of curves reflects the performance of a three-way system making use of the same drivers as those employed in the flat-on-axis, three-way series ("Three-Way Flat On-Axis Curves," page 45) but, in this case, with them optimized for flat power response. Flat-on-axis behavior has been sacrificed, as was also the case with the two-way curves ("Two-Way Flat Power Curves," page 46), in order to deliver more effective real-world performance. However, the direct-field response (often called near-field response) will be a bit uneven. Note that this design surpasses the two-way version in terms of midrange smoothness and should also have better power handling. With a proper woofer, it would also have deeper bass than the two-way system without sacrificing good midrange performance.

low-cost systems

As noted in Chapter 1, the main difference between a good, low-cost system and a good, expensive one will be maximum loudness potential, at least if listening is done in the reverberant field. However, *most* recordings heard by *most* people in *most* rooms will not overtax the volume capability of decent-quality low-cost systems. In addition, most people do not require potent low-bass performance from their speakers. Any decent moderately priced model should be able to deliver fairly flat response (within 3 dB or so) to 50 Hz; even in this compact-disc age, very little recorded music has significant bass content below 40 or 50 Hz. The extra bass potential that is gained by going to a higher-priced, three-way system is easily recognizable on a number of recordings (particularly action movies on videodisc), so if you make a steady diet of viewing Rambo films, you might benefit from the extra

Figure 2.14 Three-Way Flat-Power/Flat-on-Axis Curves

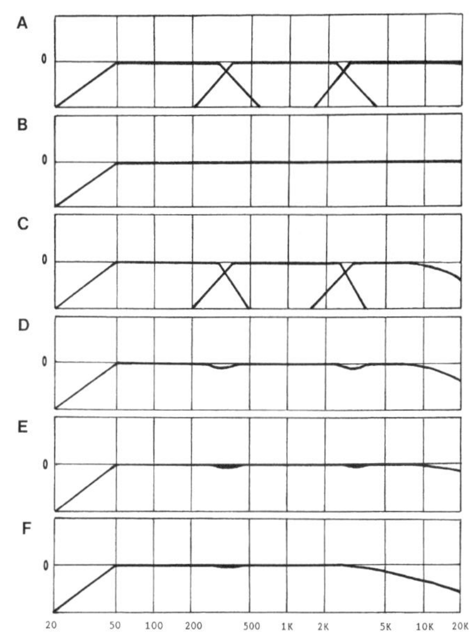

This series of curves reflects the performance of a three-way system having superior drivers, working in combination with the kind of low crossover frequencies that take advantage of their potential. The result is response that is flat on- *and* off-axis. This technique limits the negative effects of off-axis attenuation and allows the manufacturer to maintain flat on-axis contouring as well. The overall result is a smooth front-hemisphere response and a power response shaped for the best performance with most recordings. While this system sounds smooth in both the direct and reverberant fields, only a handful of companies successfully employ this kind of driver and crossover combination.

power. Then too, if you get a smaller pair of speakers at first, you can always add a potent subwoofer later and move into the big leagues that way.

Of course, the serendipitous byproduct of getting a three-way system for better bass sound will be an improved midrange and a more uniform radiation pattern. An improved treble range may also result, but if you are dealing with a good company and choosing between its two- and three-way models, the improvement will often be minor. I know of several companies who produce two-way models so good that they surpass the bulk of their competition's more expensive three-way models in both flat power output *and* radiation pattern uniformity—and even give their own three-way models a run for the money. Indeed, in some cases, the strongest competition their mid-priced, three-way models face will be their own two-way, "econobox" stablemates—and some of these two-way speakers are surprisingly cheap.

The Decibel

Technically, the *decibel (dB)* is one-tenth of a "Bel." The number of Bels is the common logarithm of the ratio of two powers; if they differ by 1 Bel, the greater will be 10 times the lesser. A 100-watt amplifier is 1 Bel, or 10 dB, higher in maximum output than a 10-watt model. Thus, decibels are fixed ratios, not fixed quantities. Under most conditions, wide-band musical signal changes of as little as 1 dB can be heard by careful listeners, and critical A/B listening with test tones or certain kinds of music may reveal broadband differences even smaller. (During A/B comparisons of such exacting items as amplifiers and CD players, such level contrasts are often perceived as quality, rather than quantity, differences.)

A 3-dB change should be clearly audible over even a fairly narrow frequency band, and a 10-dB change will be perceived by most people as being "twice" or "half" as loud, depending on whether the sound is made louder or softer. Interestingly, very narrow but prominent (high-Q) peaks are usually less audible than broader but less intense (low-Q) peaks, but the latter will usually be easier to manage with tone controls or an equalizer. Narrow dips are usually less audible than similarly shaped peaks, but broadband dips are just as audible as broadband peaks (a broad midrange dip, for instance, has the same subjective effect as a bass and treble combination of broadband peaks). Good speakers, of course, have few peaks or dips of any kind.

THE SPEAKER AND THE ROOM

There is more to speaker sound than speakers, of course. The listening room will influence the sound of a pair of speakers (or group of them, as would be the case with a surround-sound installation) to an amazing degree. This interaction can have some surprising results.

The size of a woofer driver (or drivers), plus the shape of typical speaker cabinets and the location of woofers in those cabinets, makes it nearly impossible for most systems to deliver flat bass response in typical listening rooms. Under most conditions, a woofer's cone will be a foot or two from the nearest boundary surface such as a wall or floor. At very low frequencies (below 80 Hz or so), the cone moves back and forth over a substantial distance. Because of the speed of sound and the woofer-to-wall distances involved, the reflected wave will reinforce and solidify the direct sound from the woofer—increasing deep bass loudness. On the other hand, from the upper bass on up (above 400 to 500 Hz), the typical cabinet front is large enough to function as a "wall" that is acoustically close to the driver or drivers producing the direct sound. Thus, over the low bass range, as well as throughout the upper-bass, midrange, and treble frequencies, the power response of a speaker system can be flat—that is, over those ranges it has the potential to *input* flat power to a listening room.

However, with a surprisingly large number of systems, there will be trouble in the middle bass. At those frequencies, the woofer cone will be called upon to reverse its direction more quickly than it did at low frequencies. Because of this, it will often be working in opposition to and out of phase with signals being reflected off nearby room boundaries. This will result in a null or "suckout" at some point in the midbass, usually in the 150- to 250-Hz range. This phenomenon is sometimes called the "Allison effect," in honor of Roy Allison, the man who discovered it.

Note that suckout is not the same as the mid- and high-frequency boundary reflections sometimes mentioned in hi-fi magazine test reports, nor can it be controlled by any kind of LEDE room manipulations. The subjective effects of high- and midrange room reflections will be dependent on listener location and should have little effect on the total power input to a typical listening room. However, midbass suckout is not listener-location dependent and will impact directly on power response and therefore be very audible.

Only a handful of companies, such as AR, Allison Acoustics, Boston Acoustics, NHT, RA Labs, and RDL, have purposely built systems to combat the Allison effect. (In-wall speakers handle this almost automatically or at least simplify the solution.) The best way you can overcome the problem is to not let it be exacerbated by a cumulative doubling or tripling of similar boundary reflections. Make certain that the distance of the woofer to each wall or floor surface is *substantially* different, as measured from a point directly in front of the *center* of the cone to the floor or side wall, or from the center to the edge of the box and then straight back to the wall behind. (A point exactly between each woofer in a stereo pair located close together in a small A/V layout will acoustically simulate a solid wall, so consider that measurement also if it approximates the other woofer-to-boundary distances.) Setting them at 20, 22, and 24 inches will not do much to level out the compounding effect; for best results, something on the order of 20, 30, and 40 inches would be better. (Note that typical speaker stands, because of the way they position even a low-mounted woofer in relation to the floor and wall behind, may cause serious suckout problems if the user is not careful.) This placement puts the woofer section into what speaker-tester Dave Moran calls a "least-cubes" situation, spreading the effect of suckout over several center frequencies and limiting its negative impact.[6]

Consider the midpoint between a pair of stereo speakers also as an "acoustic" boundary.

While mid-bass suckout involves room interactions at close distances,

[6]Incidentally, the least-cubes correction also applies to one's listening position. Make certain that your head is also positioned at reasonably staggered distances from nearby walls and floor to insure smooth midbass performance. It is important to remember that driver-induced suckout and listening-position suckout can work together to generate very large irregularities in the middle bass.

additional interference within the bass range will involve the interaction of a woofer or woofers with room walls at greater distances. Even if a system delivers flat bass to a room and has overcome the suckout phenomenon, it will almost certainly have problems with standing waves, particularly in smaller rooms with similar height, width, and depth dimensions. When a room dimension is such that it is half the wavelength of a signal, the reflected sound at that

The Boundary Effect

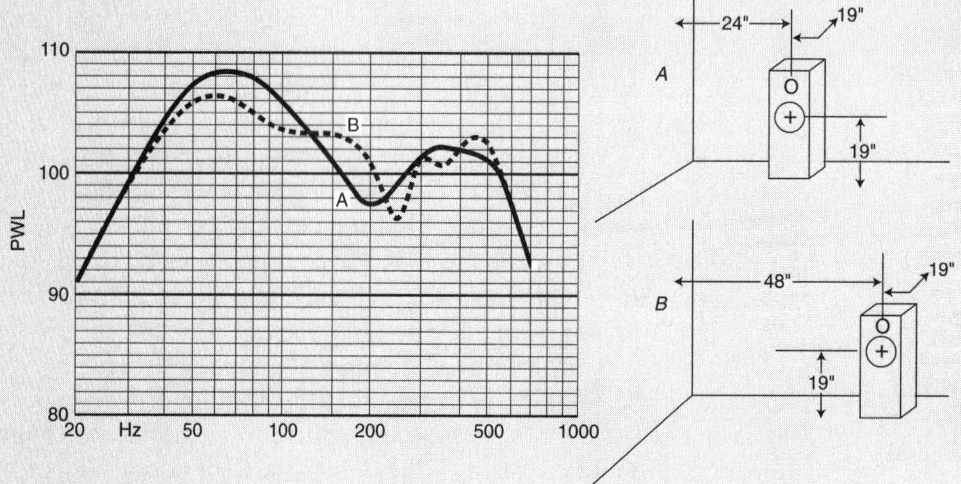

Suckout problems plague more speaker systems than audio writers care to discuss. Aftermarket speaker stands and the mounting of woofers fairly high up on larger floor-standing systems can both have negative impacts on mid-bass smoothness (see "Speakers and Stands," page 58). Here we see the theoretical response of a 10-inch woofer that, when measured in an anechoic environment, has admirably flat response from 50 to 500 Hz. Curves A and B both exhibit the same woofer driver in a 14-inch-wide, 34-inch-tall, 11-inch deep, floor-standing system with the *center* of the cone 19 inches off the floor. In one case, the woofer center is 24 inches from the side wall (boosting the output from 50 to 100 Hz) and in the other it is 48 inches away, resulting in a loss of power over that range. However, note that in both locations there is a dip ("suckout") within the 150- to 300-Hz octave, because the distance from the woofer's center to the wall behind the cabinet is also about 19 inches (the box is 1 inch from the front wall).

The solution to this predicament is to reposition the system so that *all three dimensions* are substantially different. Pulling the cabinet out somewhat is about the only solution, but the system in A should definitely not be pulled forward 6 inches from the front wall, because that would place the cone center 24 inches out, equal to the distance to the side wall. Boundary-distance asymmetry should be our goal.

frequency will be in phase with the direct signal at both ends of that dimension, and the sound at those locations will be measurably louder; conversely, in the middle of that room dimension it will be measurably weaker. These peaks and nulls can greatly complicate both speaker and listener positioning.

In addition to the initial waves, resonances for each dimension will also

Figure 2.15 AL-125

(Photo by the author)

The midpriced, three-way Allison AL-125 delivers a lot of bang for the buck. Note that the dual 6-inch woofers are located at the bottom of the cabinet, close to the floor, making it easier to correct for boundary-derived "suckout" effects. In addition, the midrange driver is functionally identical to those used in the much more expensive, but unfortunately discontinued, IC-10 and IC-20 models, as well as the still earlier Allison One and Two (although those four larger systems used more of them), and the same goes for the tweeter. In terms of bass performance, this system has surprisingly low distortion because of the "push-pull" woofer mounting system (which nulls even-order harmonic distortion), although it is not fair to expect such small woofers to deliver crushing bass at the very bottom of the audible range. However, in terms of reverberant-sound-field smoothness and normal-room listenability, there are few other speakers at any price that are markedly better than this one.

Speakers and Stands

Most two-way speaker models are so small that when they are used as floor systems the sound appears to be coming from the ground. Also, the midrange may be colored by the floor cancellation dip that results from having the woofer/midrange driver so close to the boundary. Many enthusiasts, including a few equipment reviewers, feel that raising these systems off the floor will solve these problems. A number of expensive two- and even three-way models come with dedicated stands for just this reason.

However, the solution may be worse than the disease because it creates two other, possibly worse problems. *First*, if the woofer-to-boundary distances are not all staggered, the suckout effect will be amplified. If staggering is properly done, the abruptness of any boundary cancellations will be smoothed, and the result will be a gently depressed, but still acceptable, response. *Second*, if the woofer is small, the midbass dip will possibly be lowered in frequency enough to have it drop below the woofer's inherent bass rolloff point. This will weaken its already limited deep-bass potential.

Oddly enough, this loss of deep bass is one reason that stands often appear to improve midrange performance. The ear-brain coalition tends to let bass signals that are simultaneous with middle-range signals partially mask the detail of the latter, even with *live* music. Thus, systems with weak bass often sound cleaner than those with equal midrange performance but better deep-bass output, at least when deep bass is present on a recording.

If you must use a stand, it is not necessary to buy an expensive, massively heavy, exotic design to hold your speakers securely. Research by Tom Nousaine has shown that the minuscule rocking motion or vibrations that lighter-weight stands allow are inaudible. You could set your speakers on stands made of wicker and still get good sound, provided you followed the above placement guidelines.

standing waves

happen at every multiple of that wave's frequency. A room that is 20 feet long will have a standing wave resonance of about 28 Hz and will have succeeding resonances at about 56, 84, 113, 140 Hz, and on up the frequency ladder. At higher frequencies the resonances move progressively closer together, and the peaks and nulls eventually become so narrow that they give uniform support to all upper frequencies because of the filtering action of the human hearing mechanism. Unfortunately, the wider spread of those peaks and nulls throughout the bass range is a serious problem.

In addition, when the three dimensions for most rooms are similar, the various waves and nulls in these "axial" modes reinforce each other at certain locations.[7] The worst case would be that of a cube-shaped room. A close sec-

[7]There are also three more "tangential" series, as well as an "oblique" series, that are more difficult to calculate and are thought by some experts to be as negative in impact as the axial ones.

Figure 2.16 Real-Time Analyzer (RTA)

(Photo by the author)

If intelligently used, sophisticated and expensive one-third-octave "real-time analyzers" (RTAs) such as the AudioControl SA-3050A and R-130, Ivie IE-30, and dbx RTA-1 can do a marvelous job of measuring audio-system behavior. However, a one-octave model (like this hand-held unit by AudioSource) can work near wonders at solving placement problems if your speaker systems are top quality. While this unit is not something that an average consumer would buy to do a one-time measurement of a sound system, a lot of dealers will lend out similar or even more elaborate gear to assist customers in positioning their speakers for best results.

The Audiosource RTA measures discrete, one-octave intervals centered at 31.5, 63, 120, 250, and 500 Hz, as well as 1, 2, 4, 8, and 16 kHz. An LED-panel array presents the readout, with the source signal coming from a supplied, outboard "pink-noise" sound generator or a test disc feeding a sound system's amplifier or receiver. While this model can be configured to read electrical inputs directly, its 2.5 dB of resolution makes it unsuitable for making effective use of this ability, and its primary use will be as a speaker measurement and equalizing tool—with the speaker output measured by a microphone built into the case. Although I would never recommend it as a testing tool for actually rating or comparing speakers, it can be helpful when setting up systems of known high quality and in determining or fine-tuning a good listening position. Basic measurement and adjustment tools like this will also be valuable to anyone trying to accurately calibrate the multiple channels of a surround-sound installation.

ond would be a room with the width and length nearly the same but with a smaller height dimension. Fortunately, most listening-room measurements are not so closely matched, but it is still possible for some of the secondary resonances to add to each other. For example, a 15-foot measurement has resonance peaks at about 38, 76, 113, and 151 Hz, and still higher. A $7\frac{1}{2}$-foot ceiling height has peaks at about 76 and 151 Hz. Therefore, incorporating the 20-foot calculations in the paragraph above, a room with dimensions of $20 \times 15 \times 7\frac{1}{2}$ feet would have multiple axial mode peaks at 76, 113, and 151 Hz at numerous locations throughout the room.

In small rooms (and cars) at very low frequencies, a "pressure pot" forms, and peaks and nulls disappear.

The best way to control such anomalies is to see to it that the listening room you use has asymmetrical dimensions and that those dimensions do not let secondary resonances develop. First, measure the room and divide 565 (half the speed of sound at sea level; those of you living in Denver will have to use the speed of sound at one mile up) by each dimension to see what the primary resonance frequencies are. Calculate the multiples, and compare them for common resonances. The fewer the common peaks, the better (the number of nulls for any resonance axial mode will always be one less than the number of peaks). This procedure should limit the most serious problems.

If a room is L-shaped or has irregular boundaries, like a small alcove off to the side or a sloped or vaulted ceiling, measurements become more complex, but the very nature of such room shapes should automatically reduce detrimental effects in most cases. For such situations, make your measurements interact with the boundaries having the largest surface areas. If you are shopping for a house and make the above checks, remember to take into consideration how furnishings might be arranged. Often, a well-shaped or irregular-surfaced room that handles standing waves with aplomb will leave no place for either you or your speakers to be properly positioned without doing violence to the decor.

room modifications

If you are pretty much stuck with a room, one recourse is to modify it to break up standing waves. Lining a wall with bookcases (filled with books, of course), even short ones, can break up standing waves fairly well. If you are really ambitious, you can alter the dimensions of a room by constructing a heavily braced divider. However, I do not recommend this solution unless you have other reasons to make a change, or a spouse who is as committed to audio as you are. Whatever you decide to do, remember that creating asymmetrical dimensions is tricky work that may not give the results you desired. Be sure to carefully plan, including getting some advice from a room acoustics professional, before making major renovations to your home!

experimentation

Experimentation is the key if you are going to do it yourself. Often, merely repositioning the speakers a bit (particularly in relation to their height) can counteract a standing-wave listening-position peak or null with an opposite speaker-generated null or peak. If a speaker is located at a null position for a specific frequency, the effect will be to diminish any peak at that same frequency at locations throughout the room, one of which might

Figure 2.17 Mirror Image Speaker/Room

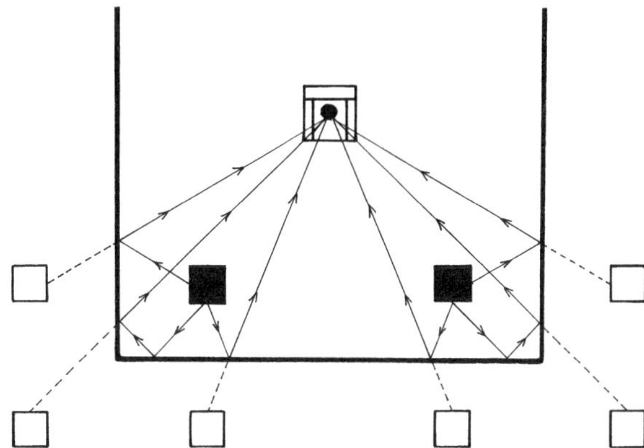

Advocates of LEDE room design often point out that when speakers are positioned in typical rooms, they will be plagued with early wall (and floor/ceiling) reflections that not only smear the "image" but also affect the direct-field frequency response because of cancellation effects. These anomalies result in "nulls" and "peaks" at assorted middle and high frequencies. This diagram shows that early boundary reflections (in this case only depicted in the horizontal plane) are actually weak, often fuzzy or distorted "mirror images" of the speakers themselves and behave as if the room walls, floor, and ceiling were missing and ersatz speaker systems were located at appropriate positions outside of the room area. This may or may not adversely color the sound in the reverberant field. (The dark squares are the speakers, and the three mirror images caused by the reflective room walls subjectively form outside of the room area.) The most important result of this phenomenon is a broadening of the sound stage and an apparent increase in stage depth. If only two speakers are involved and they are well placed in a normal room, this can produce a very realistic effect with some recordings.

be just where you want to sit. Also, oftentimes the suckout phenomenon can be used creatively to smooth things out. Obviously, careful listening and judicious experimentation over a period of time will help to solve bass standing-wave problems.

For more on tone controls, see page 92.

A fair number of home listening areas are simply not suited to complement any A/V system. Thin walls can weaken bass; uncarpeted floors, expansive window areas, and large open entries near or even behind speakers can foul up the sound of even the best designs. Tone controls and equalizers can only partially compensate for these anomalies. However, many rooms are more than adequate to team up with properly designed and positioned loudspeakers, and common sense dictates much of what is needed to correctly utilize these systems. For instance, you should avoid placing speaker

Hearing the Speakers

Let's temporarily ignore the reality that two ears cannot occupy the same place at the same time. When identical signals are reproduced by more than one driver in a speaker system—as would be the case at a crossover point or if several drivers were handling identical frequencies within some kind of "array"—there will be interference effects from the multiple sources unless the listener is exactly the same distance from each of them. At any other location, some frequencies will be "in phase" and others will be "out of phase" from similar ones produced by the other drivers, causing both cancellation nulls and reinforcement peaks. If this frequency response were drawn on paper, the result would have a choppy, comb-like look. (This is one source of the term "comb filter.") However, it is important to remember that this plot reflects the sound of the system *in the direct field.*

When auditioned from close up, these interactions will cause audible problems with nearly any speaker, especially with those employing multiple-driver arrays. While there is disagreement over whether anomalies less than ⅓-octave wide can be heard under such conditions (the peaks and dips will narrow in width as the frequency climbs and will eventually become extremely close), there is little doubt that peaks and dips wider than ⅓ octave will be audible during direct-field listening.

In typical rooms those peaks and valleys merge into the much more audible *reverberant field*, because the spate of direct-field responses are different in all directions and tend to cancel each other out when reflected from room walls and combined over the first 30-thousandths of a second. When listening to complex musical program material, rather than test tones, this "room response" is what the ear mostly senses and the hair-splitting debate about the audibility of less-than-⅓-octave-wide dips and peaks in the direct field, as well as driver phase effects and group delay, becomes academic.

boxes in front of large windows (that reach to the floor), and avoid placing them behind large couches, chairs, or other monolithic items. Speakers designed for along-the-wall placement should not ordinarily be placed too close to a corner or a large, open doorway and should, of course, be placed close to a wall.

Most competent manufacturers will provide general and sometimes very specific recommendations about how their various models should be located for best performance. If you have real placement problems compounded by serious room deficiencies, a letter or phone call to the manufacturer may offer some relief (the really good ones have a customer service 800 number). However, given speaker systems with flat power input to your room and a reasonably uniform radiation pattern, exotic, expensive, and aesthetically unpleasing solutions to the dilemma of room/speaker interactions should not be required.

SUBWOOFERS

I once considered outboard-mounted subwoofers to be "accessories"; after all, a good full-range system can handle the bass extension of most recorded material. However, these days many audio-video diehards find a high-quality subwoofer is essential—whether it is coupled with small, satellite-type main speakers, is configured to enhance rather substantial ones, or is part of an AC-3 surround-sound system (the latter has its own special subwoofer requirements; see Chapter 7).

Under certain conditions, a subwoofer can be very important—even essential—because few full-range speakers of any size can produce strong, flat bass in the octave below 40 Hz. Combining a pair of otherwise potent systems with a *true* subwoofer—one that produces solid bass down to 20 Hz and even below—will mightily impress anyone who enjoys synthesizer or organ music. Fans of up-to-date, "dynamite-grade-sound" action-adventure movies will also appreciate a decent subwoofer, which will often have substantial bass right down to 30 or even 25 Hz. An A/V system, no matter how advanced it may be otherwise, cannot really do justice to blockbuster films unless it has an authoritative subwoofer or very powerful main-system woofers, such as in upscale models from ADS, Allison Acoustics, B&W, Definitive Technology, Infinity, KEF, NHT, Ohm, Polk Audio, and Waveform.

However, before we head off on a frenzied hunt for the ultimate in deep-bass reproduction, it is important to remember that most good speaker systems, even modestly priced ones, can deliver decent bass down to below 50 Hz. Because a lot of program material—particularly pop, country, and easy listening—does not require earthshaking bass reproduction in the bottom octave, these systems can be very satisfactory. Indeed, any system that delivers flat response down to, say, 50 Hz will be able to accurately reproduce the full bass range of most compact discs (or DCC tapes and MiniDiscs) and of 99+ percent of the music available on LP records and nondigital audio cassettes.

decent bass

In typical listening rooms, most moderate- to large-sized systems should also be able to handle the bass on video-disc (and of course VHS-tape) soundtracks that are not in the *Terminator 2*, *Gettysburg*, *Apocalypse Now*, *Forrest Gump*, or *Jurassic Park* category.[8] Even these films will not lose that

[8]The digital-sound version of *Jurassic Park*, a well-known bass demonstrator, has strong response to 26 Hz at intervals, although much of the loudest material is at about 30 Hz.

much visceral impact when played back on speakers that "only" reach down to 40 Hz, particularly if they are part of a good surround-sound installation. Deep bass is often subtle and/or fleeting and will only be missed when an un-subwoofered system is directly compared with an installation that has flat, robust bass response to below 30 Hz.[9]

While many pop music recordings and motion-picture soundtracks may not demand exemplary reproduction below 40 or 50 Hz, they may still require forceful performance in the 50- to 100-Hz range—particularly if your system is placed in a large room. Top-grade subwoofers can assist smaller-scaled, otherwise competent main systems that might be overdriven by very potent bass in the next higher octave.

Subwoofers are usually configured as single-enclosure, monophonic systems. Because the bass below about 150 (and especially below 100) Hz is nondirectional, a single unit should not call attention to itself. The main-speaker "satellites" provide the required mid- and high-range directional cues and ambience, and the subwoofer fills in the low end. Most of the dedicated crossovers that come with expensive subwoofers electronically "sum" the bass from both main channels and send it to a single, very large driver. Other designs may have two acoustically isolated drivers close together within one box, each handling one channel, while a number of configurations use a single woofer driver that has two overlaid voice coils, one for each channel. These three summing-to-mono techniques have equivalent results in terms of the sound you hear.[10]

two-piece subwoofers

Two-piece subwoofer systems are now increasingly popular, and surround processors with dual subwoofer outputs are even showing up. Some listeners feel these work better than a mono configuration, because the two boxes can be stationed asymmetrically to better control low-frequency standing waves; others feel that, because standing waves are related to room size and shape, positioning tricks will have little impact and may produce even rougher sound.[11] However, all else being equal, two separate units will at least deliver 3 to 6 dB more deep-bass output than one unit of similar size, depending on the frequencies involved and the amount of separation between the boxes (see "Coherent Bass Reinforcement,"page 66). Many indi-

[9]Note that I did say usually: *Jurassic Park* is so full of deep-bass shake and thump that I imagine that only those with good subwoofers will fully experience what Spielberg had in mind, and one cannot appreciate the cannonades in *Gettysburg*, the distant B-52 attack in *Apocalypse Now*, the "chopper" scene in *Gump*, or the truck crash in *Terminator II* without a passable subwoofer.

[10]Review the Franssen effect on page 31 to better understand why a mono subwoofer that works well for music must be carefully positioned if it is to also be used to reproduce some motion-picture sound effects—at least if the crossover point is above about 100 Hz.

[11]Research by Tom Nousaine indicates that the best deep-bass results are produced by a single, corner-located subwoofer, crossed over to the satellites at about 80 Hz.

Figure 2.18 Definitive Technology Subwoofers

(Photo courtesy Definitive Technology)

Even the least expensive of these Definitive Technology subwoofers can upgrade nearly any A/V system. With one 18-inch and two 15-inch models available, each with a very flexible built-in crossover network and substantial amplifier, these systems can deliver impressive bass to below 20 Hz, with the main differences between them being output loudness capabilities at such low frequencies.[1]

There are a number of other fine subwoofer designs being produced, and new ones are showing up all the time. Velodyne is famous for its outstanding, although somewhat expensive, self-powered models (which usually incorporate motional-feedback circuitry to reduce distortion), and M & K, Allison Acoustics, and RA Labs have produced reasonably priced and impressive low-bass speakers that feature "push-pull" double woofers to keep distortion low. The Sunfire subwoofer is noted for both its performance and its small size. Even Cambridge Soundworks has a good single-enclosure subwoofer system, made even more impressive, at least down to a bit below 30 Hz, if their unpowered "slave" system is added in.

A strong contender for the "best buy in subwoofers company" award is Hsu Research, which has a variety of intelligently designed models that are not only among the lowest-priced available but also among the most potent in terms of powerful performance at extremely low frequencies. Poh Ser Hsu's most affordable super-grade models are unusual in that they are vented designs that are adjusted for a very low frequency cutoff point. Some of the models he has produced have been unpowered. However, they have the advantage of being highly efficient, meaning that you can get away with using a moderately powerful, reasonably priced power amp.

[1]As a point of reference, a single 18-inch, acoustic-suspension subwoofer with a linear movement of 0.75 inch (a typical specification) can move three times as much air as a pair of acoustic-suspension 10-inch woofers, each with linear movements of 0.5 inch. A 15-inch subwoofer with the same "overhang" can still double the air-moving capacity of those two 10-inchers and can move over three and a half times as much air as a pair of 8-inch woofers with a 0.5-inch overhang.

Coherent Bass Reinforcement

If a full-range speaker system is operating and a second, identical model, playing the same monophonic material at the same level, is positioned next to it, the sound level throughout most of the audible range will increase by 3 dB. However, throughout the bass range, at least at frequencies with wavelengths that are longer than the distance between the woofers in the two systems, the sound level will rise 6 dB, increasing the low-frequency output by four times. Why is this so?

Picture a single floor-standing system located at a solid floor/wall intersection and well away from any room corners. Assume that its input to the room is reasonably flat over the full audible range and that it radiates uniformly over a 180-degree forward angle. Now, move that same system into a corner. While the sound throughout the mid- and treble range is reflected from the walls and floor, those reflections are in random phase relationship with the direct outputs of the midrange and tweeter drivers. Consequently, the average power input of the system into the room at those frequencies is not changed.

However, with long-wavelength bass signals (those half the length or longer than the distance to a floor or wall boundary), the reflected energy arrives back at the woofer location in a short enough period of time to coherently augment its power input to the room beyond what was originally provided by only one wall and the floor. The reflected energy remains essentially in phase with the direct energy radiated by the woofer, and the system response is no longer flat, due to the increase in bass power input to the room. A system designed for corner use can have its bass output electrically attenuated by 3 dB, while maintaining flat bass power, to take advantage of this coherent-reinforcement phenomenon. This reduces the strain on the woofer compared with what would exist with an along-the-wall placement of the same system.

Let's move the system out of the corner and back to its original location. When an identical system, playing exactly the same material at exactly the same level, is placed next to it, the middle and high frequencies are doubled in output, because there are now two systems radiating equal sound; the bass response output is also doubled for the same reason. However, throughout the bass range the second system also simulates a reflecting boundary (a corner wall) between the systems. Because of this, the bass range is further augmented by an additional 3 dB.

If two sound systems are placed in a typical stereo-playback location and are playing monophonic signals, the treble, midrange, and most of the upper and middle bass range will add together in a noncoherent manner, increasing the sound level by 3 dB. However, at really long, low-bass wavelengths the two systems will be less than one wavelength apart, resulting in a 6-dB augmentation to monophonic deep-bass signals. (With speakers placed 11 feet apart, this additional augmentation will begin at about 100 Hz.)

Bass—particularly low bass—will be monophonic most of the time in almost any recording, so there is little doubt that a typically placed stereo pair of systems playing at a given level will exhibit richer, fuller *deep* bass than a single identical system playing at a mid-treble level equal to that of the pair together. This also means that a mono subwoofer, whether reproducing the bass of the entire system or just that of the center channel, has to be manifestly robust—and preferably corner located—to supply the same calibre of flat, low-distortion bass that really formidable woofers in a conventionally spaced stereo pair of full-range speakers can deliver.

viduals also believe that a double-subwoofer system will impart a better sense of hall depth and ambience to recordings made with multiple microphones—such as typical classical-music CDs—than a single-box system.

single subwoofer

Other researchers feel that a single subwoofer will actually do a better job of projecting low bass, because two units will introduce phase interference effects that detract from the coherent sound that good bass should exhibit. While the upper and middle bass (above 80 to 100 Hz) may benefit from "stereo" bass reproduction, truly deep bass sounds better if handled by a monolithic subwoofer, preferably mounted in a corner for flatter, multiple-boundary-augmented propagation. The argument against dual subwoofers is directly related to the Allison effect, with the midpoint between the two woofers simulating a distant boundary.

powered and unpowered subwoofers

Subwoofers are either *powered* or *unpowered*: that is, they either have an amplifier built in (or at least come with a "dedicated," outboard-mounted amp) or they have no self-contained amplifier at all. Powered subwoofers nearly always include a built-in, AC-powered, "electronic" crossover, although those blueprinted exclusively for THX installations and/or AC-3 (which has the low bass isolated from the upper bass, midrange, and treble during the recording process) do not. All accept line-level (RCA-plug) inputs from a preamplifier, and many will also accept speaker-wire inputs from a power amplifier.[12]

The crossover circuitry extracts the low bass from the two full-range channels and, if only a single subwoofer is being used, converts it to a monophonic signal. The low-bass signal or signals will then be forwarded to the usually built-in woofer amplifier (or sometimes two amplifiers, if two subs are used), and from there to the appropriate subwoofer driver or drivers. The crossover forwards the remaining stereophonic, full-range or high-passed signals either directly to the two main, "satellite" speakers (if speaker-wire inputs were used) or to a stereo power amplifier (if line-level inputs were used), where they are amplified and sent to the two main speakers. If line-level inputs are involved, the signal destined for the amp/satellite combination will nearly always be high-pass filtered.

active crossover

The better unpowered subwoofers are usually designed to be used with a separate, outboard-mounted crossover, either a dedicated model produced by the company that made it or a universal one built by an unaffiliated manufacturer. Most outboard-mounted crossovers will accept line-level inputs from preamps and route the appropriate line outputs to user-supplied, separate, main-satellite amps and subwoofer amp or amps. The universal models usually have selectable crossover points, and some dedicated models also offer this flexibility.

[12]Speaker-wire hookups are more complex and problematic than RCA lines, if you attempt to high-pass filter the speaker-level signal returned to the main speakers. Impedance-induced response irregularities may result unless the satellite speakers are designed to work with the crossover in the subwoofer unit.

As you will see in Chapter 7, THX and some other surround processors (as well as some A/V receivers, A/V preamplifiers, and integrated A/V amplifiers) come with electronic crossovers built right in to them. This allows a subwoofer-amplifier combination to be directly connected to the main systems. Unfortunately, some of those non-THX, "built-in" electronic crossovers have transition points in the 150- to 200-Hz range, which is high enough to cause some lower midrange bleedthrough to the subwoofer—which is not a good thing, particularly if it is a mono unit located off to the side.

Powered or unpowered subwoofers with frequency-dividing networks reduce the amount of work for the main-speaker woofers and amplifiers. The subwoofer amplifier handles the power-consuming bass, and the "satellites" only receive the high-passed, upper-range frequencies. Without the sub, the small woofers may reach their physical limits, or the main-channel amplifiers can run out of steam, when reproducing a full spectrum of sound at reasonable listening levels. With a good subwoofer crossover, dividing the highs and lows appropriately, a system delivers potentially higher playback levels (in the upper ranges *and* the bass), lower distortion, and, of course, deeper bass.

passive crossover

Some unpowered subwoofers come with a "passive" crossover, which requires no power source, being analogous to the built-in crossover circuitry found in full-range speaker systems. This low-cost alternative to the electronic version may either be built into the subwoofer enclosure or mounted outboard. Unlike electronic crossovers, passive designs can only be connected to the speaker outputs of the main-system amplifiers. With some outboard models, the "picked-off" bass is routed to a separate amplifier and sent on to the subwoofer(s), while the main speakers continue receiving a full-range signal. Other designs (usually built right into the woofer enclosure) route the split-off bass signals directly to the internal woofer driver(s) and, after suitable high-pass filtering, forward the upper frequencies to the satellites, with no additional bass amplification required.

While a passive crossover can protect the satellites from potentially damaging bass frequencies, it will not reduce the power demands on the main amplifier. No additional amplification is added to aid in bass reproduction. Indeed, it will actually increase power demands somewhat, due to power loss in the passive components.

Passive designs that low-pass-filter signals directly to the woofer(s) and *also* high-pass signals going to the satellites are limited by practical considerations to a crossover in the 150-Hz range.[13] With a crossover point this high, you must be careful to position the subwoofer *between* the main speakers, particularly if you plan to view movies that contain any random-noise-like sounds (waterfalls, jet engines, whooshes, etc.). Placed well off to one side, a

[13]Those that "pick off" bass signals for use with an outboard amplifier/subwoofer combination can have a crossover set at about 100 Hz or below.

Allison ESW

For a few years Allison Acoustics offered an "electronic" subwoofer, which was actually a low-bass equalizer designed to compensate for the inevitable rolloff below an acoustic-suspension woofer's resonance point. A front-panel control allowed the ESW to be dedicated to any number of Allison speakers, although it could also work well with models produced by a few other companies. However, because it produced 11 to 14 dB of boost at 20 Hz, it was not suitable for use with anything but robust acoustic-suspension woofers that, unlike bass-reflex models, are self-protecting below resonance. At least 100 watts per channel might be necessary to insure enough headroom in typical listening rooms when listening to organ and synthesizer music. Because the ESW provided little or no equalization above 40 Hz, its boost functions could not be duplicated with a conventional, ⅓-octave equalizer, although a good parametric model might do nearly as well. (See Chapter 8 for more on equalizers.)

The ESW also contained some sophisticated filtering circuitry that protected speakers from high-frequency hash above 20 kHz and also protected them from sub-20-Hz noise artifacts. The high-pass-filtering feature was a godsend in the rumble-prone, vinyl-record days; it comes in handy even today, because many otherwise fine, modern recordings have very-low-frequency studio or hall rumble, or even auto traffic noise leaking in from the outside. Switching the ESW to its "filter only" mode would eliminate most such noise. In addition, the high-pass filter was capable of protecting otherwise robust but still potentially damageable woofers from the subsonic, low-end pyrotechnics found on some recordings (such as the 12.5-Hz thumps on the woofer-killing Telarc *1812 Overture*), even, believe it or not, when the selector control was set to deliver house-shaking, flat-to-20-Hz bass.

Interestingly, the ESW did not sell well, mainly because when it was introduced the CD was still five years in the future and the number of LP recordings that could highlight the device's bass-boost abilities were limited, particularly when the already fine, unequalized performance of the Allison speaker line produced was taken into consideration. Some of the early versions of this unit had teething problems, but those with serial number higher than 1,000 have proven to be quite durable, and I wouldn't hesitate to purchase one of them used.

(Photo courtesy Roy Allison)

single subwoofer probably will adequately reproduce music, but may display image shifts with motion-picture sound effects.

For a passive system without a high-pass filter to work its best, the main-system satellites should already have decent low-end performance, and their low-end rolloff should dovetail with the low-pass contouring of the subwoofer. Passive-crossover subwoofers work best when interfaced with acoustic-suspension-woofered satellites, because of the latter's controlled 12-dB-per-octave attenuation below resonance and resistance to deep-bass overload. Some of the most potent subwoofers ever built incorporate a passive-crossover and outboard-amplifier combination.

A good electronic crossover will have controls that allow a close match between the subwoofer level (volume) and the level of the main speakers. Passive crossovers may not have these controls, because they are usually designed to integrate with a dedicated subwoofer-satellite system. If you connect an undedicated subwoofer to a passive network, make sure that the latter includes level controls. Passive subwoofers that route the bass to a separate amplifier may be best served if the amp has level controls of its own.

subwoofer level matching

Level matching can be a tricky proposition, and it requires great care if done by ear; flat, smooth frequency response from a subwoofer-equipped system will almost always be more easily obtained if measuring equipment is used. The better subwoofer companies will include detailed instructions on how to dovetail their woofers into your system.[14]

Remember that there are a number of so-called "subwoofers" out there that are misnamed. Most are sold as part of three- or four-piece satellite systems and reproduce the range between approximately 150 and 50 Hz, with little output for lower frequencies. These woofer modules are often quite small, offer placement ease and imaging accuracy, and can be driven by a single stereo amplifier, along with the rest of the speakers. Many of these systems deliver decent bass, but the three-piece versions, even those with dual drivers within the woofer enclosure, are rarely a bass-output match for a pair of moderately sized full-range systems. It would be better if such speakers were simply called remote, or maybe "hideaway," woofers.

HEADPHONES

Besides hearing aids and cell phones, headphones are the smallest common speakers. They range in price from a few bucks to many hundreds of dollars and range in size from pocket-fitting (when folded up) to bulbous; their sound quality ranges from stinky to utterly revealing. Beyond convenience and portability, headphones allow you to have a fine sound system without purchasing any kind of room-dominating speaker systems at all, or renovating your house to accomodate your system. Still, their biggest advantage over speakers of comparable tonal quality remains their low cost.

[14]Perhaps the best test disc for adjusting subwoofer levels is *Surround Spectacular* (Delos 3179), a two-disc set with both test signals and musical program material.

Shopping for headphones has an obvious advantage over shopping for speakers: you carry your listening room with you at all times. Speakers can sound very different in different locations, so a pair that sounded good in a dealer showroom may not sound so good once you get them home, particularly if those speakers are designed for direct field use and the showroom was dedicated to their idiosyncrasies. (Conversely, speakers that do well in living rooms often do not sound so hot in large, or acoustically treated, unhomelike showrooms.) The listening room for a pair of headphones combines the cavities of the phones themselves with the cavities of your ears. If they sound good in the showroom, they will sound equally good in your home, provided the associated hardware is of similar quality.

Always compare the same music selections.

Shopping for headphones also has drawbacks. Speaker system A/B comparison tests, although not particularly rigorous when performed in typical showrooms, at least allow for rapid switching; the very nature of headphone behavior makes rapid changes impossible. Not only is it awkward to quickly compare two sets of phones, but it is also difficult to accurately set level balances—a critical factor. If one pair of phones sounds inferior but is a bit louder than another because of greater sensitivity, it will be perceived as having more clarity, impact, and superior bass.

Headphones are mostly one-way in operation, placing great demands on their designers. One driver per ear will have serious trouble reproducing the full range of sounds handled by the more specialized drivers found in loudspeaker systems. This is not as great a limitation as might appear, because headphones do not have to contend with the power inputs or require the kind of uniform high-frequency radiation that speakers do. Indeed, a more directional, focused high-frequency output helps the phones to better control their response at such close listening distances. Headphones are the ultimate "near field" speakers.

The table on page 72 shows the three most common headphone styles; comfort factors aside, the on-the-ear design has the advantage.

Headphones will allow any listener to hear nuances that would never be experienced at a live concert. Shuffling feet, coughs from the players, pages turning, and conductor grunts are not unusual at any live concert, but they ordinarily go unnoticed, because the hall and its occupants absorb these sounds. No such masking will happen with headphones, particularly on recordings that were made with multiple microphones that were placed very close to the performers. Some aficionados like the immediacy of this kind of reproduction, and feel that many great composers, were they alive today, would approve of this micro-detail (discounting the coughs and grunts, of course). Other listeners feel that such revelations, while fine for some aspects of program monitoring during the recording process, are out of place in the final result.

Headphone style	Advantages	Disadvantages
On the ear, making use of acoustically transparent contact pads for comfort and lightness	Allows outside sounds to intrude, often adding a sense of spaciousness that advocates believe enhances realism	Audible to those who are in proximity to the user, as anyone who has been close to someone with a Walkman™ can confirm; unwanted extraneous noise can be a problem
Over the ear, surrounding the entire ear with an acoustically opaque shell	Enclosed space enhances deep-bass response; isolation blocks out unwanted extraneous noise	Much larger, hotter, and heavier than the on-the-ear style; many cannot enjoy music if their ears feel encapsulated
In-the-ear drivers	Allows maximum isolation from outside noise, controlled and very exact frequency response possible; notable bass performance	Usually more expensive; rarer than other two styles; awkward to use

The biggest problem with headphone use may be the lack of realism that results from having performers within the listener's head instead of on a stage up front. Many reverberant-field speaker advocates are turned off by the overactive nuance-gathering potential of headphones. Pop music aficionados, however, may love the effect, because recorded rock music often strives for a surrealistic effect that may be enhanced by headphone listening.

binaural recording

"Binaural" recording takes advantage of the way headphones work. Instead of the microphone placement typically used with most recordings, binaural techniques require two microphones located at ear position in a "dummy" head. A properly done recording will present a true "stage front" illusion when headphones are used during playback, because of the time-delay and intensity clues provided by the recording's specialized engineering. Unfortunately, binaural recordings do not work well when played back through speakers, so few of them have been made. Digital processing systems, however, can easily simulate binaural recording effects with some standard recordings. Expect to see such features (switchable, of course) on some future electronic components such as receivers, CD players, and surround-sound processors.

If you buy a pair of headphones, remember that it is easy to play them too loud—that is, too loud for your ears. Friends and neighbors may complain if you crank up the stereo too high, or your amplifier may go on the fritz powering your speakers to such brain-damaging volume. However, headphones are easily driven to very loud, essentially distortion-free, levels that people around you may not notice at all. Extended listening of this kind will probably result in the need for another miniature speaker: a hearing aid.

SHOPPING: NEW VS. USED

Buying new or used speakers is a demanding business. There are a wealth of models available, with new styles and designs hitting the market constantly. Price is not always a guide to quality; there are inexpensive speakers that might perform better in *your* home to meet *your* needs than the most expensive, high-style models made. Buying used is another way to get good quality for less bucks, although be forewarned that speakers are more easily damaged than other components, and you must be able to assess accurately the condition of any used equipment before purchasing it.

magazines

The best way to begin shopping for a good pair of speakers is to look over some test reports. Popular mainstream audio magazines like *Stereo Review* and *Audio* often feature speaker tests, sometimes several in an issue, as will video magazines like *Video*. However, these articles are often too involved for typical readers to easily understand and, because they rarely "rank" the systems under test in relation to other models, are not helpful to individuals who want simple answers to complex questions. Fortunately, these magazines rarely test truly bad equipment, so just seeing a speaker reviewed in print can be a recommendation of sorts.

Small-circulation "audiophile" or "high-end" magazines are usually staffed by members of the direct-field school of speaker evaluation; their subjective interactions with speakers can differ profoundly from what is experienced by typical listeners in standard, home listening rooms. The latter are interested in good sound at reasonable cost as well as a decent listening-room decor. High-end testers may value aspects of sound reproduction (such as microscopic-grade imaging and almost surrealistic instrumental detail) that typical listeners (or musicians!) couldn't care less about. Certainly such test procedures can be useful, but I would caution ordinary listeners to take them with a grain of salt.

Consumer Reports

Probably the best testing organization for most neophyte shoppers is Consumers Union (CU), the people who publish *Consumer Reports.* They specialize in power-response measurements, a key quality determinant for most listeners in most rooms. Even more importantly, they usually test a fairly large number of reasonably priced speakers at one time, which makes it possible to intelligently compare various models. While they certainly cannot test every system available, they do more comprehensive multispeaker comparisons than the hobby magazines do, allowing readers to discover which companies produce consistently good designs.

CU test results can also be useful when shopping for items that they have *not* tested. While some companies occasionally build speakers that rank near

Figure 2.19 RDL S-1

(Photo courtesy Roy Allison)

It is just about impossible to fault the diminutive, cube-shaped, S-1 system that was available for some time from RDL Acoustics. One of the latest incarnations incorporating the remarkable Roy Allison–designed, convex-dome tweeter (working in combination with an Allison-designed 8-inch woofer), this was an excellent full-range package for small- to medium-sized room use. The system had a particularly fine midrange sound for a two-way model, due to the combination of the very-wide-dispersion tweeter and the upward-facing woofer. The unusual placement of the latter delivered uniform horizontal radiation over the whole 180-degree-wide area in front of the system, while also maintaining smooth power response. This design was an 8-ohm version of the discontinued Allison Acoustics 4-ohm-rated CD-6 and still earlier Model 6, both of which received rave reviews from both audio-hobby and consumer-testing magazines for many years. Although decently capable in terms of bass response, if combined with a authoritative subwoofer, any of these speakers can easily hold its own against some top high-end designs. For those who could not afford a good subwoofer, taller, more bass-potent floor-standing versions—the F-1 from RDL and the CD-7 and Model 7 from Allison Acoustics—were also available, as was a bigger bookshelf system, the Allison 5, from the latter company. Lots of these systems were produced and, if found in good condition, any of them would be good buys used.

the top in a CU test, others have repeatedly put models at or near the top of their ratings for years. Back issues of the magazine can help individuals shopping for outstanding used-speaker buys and can also serve as guides to manufacturer quality in general.

The companies themselves, through advertisements and promotional brochures, provide a good deal of information about their products. However, it is important to separate objective information from hype. Often, a new "breathrough" is touted as if it were the proprietary "discovery" of a particular manufacturer, when in fact it is known—and used—throughout the industry. Also, mysterious new methods of manufacture, exotic construction materials, or miraculous design enhancements should be looked on with a skeptical eye.

voice-coil coolant

For example, Ferrofluid™ was introduced in 1977 as a liquid voice-coil coolant for mid- and high-range speaker drivers, and somewhat later, in modified form, as a coolant for woofers. Liquid cooling allows a few of the more conscientious speaker builders to make drivers having superior dispersion without sacrificing power-handling capabilities. However, it has mainly been used to protect rather-conventional drivers from damage, reducing warranty costs.[15] A number of older systems with air-cooled drivers are sonically equal or even superior to many of the newer, liquid-cooled models, at least at reasonable, non-voice-coil-frying, listening levels.

exotic materials

Exotic construction materials, like polypropylene, titanium, graphite, and even diamond film, are also advertised by some manufacturers. While these innovations may have theoretical advantages, their products are not appreciably more robust or audibly more exact under normal conditions than high-grade paper-material versions properly treated with resonance- and moisture-controlling compounds. Indeed, the exotics can be more prone to failure than some of the more conventional material designs because they are often difficult to properly construct.

Many companies, particularly those who do not make their own drivers, emphasize their advanced crossover designs. These are said to be the "key" to good performance, rather than driver design. However, crossover designs have changed little in the last 10 to 15 years, and those changes are mostly related to improvements in power handling. The so-called "minimum phase" crossovers noted in some advertising copy (actually, simple, traditional designs) have little significant effect on reverberant-field performance.

[15]Nonmagnetic silicone grease is actually superior to Ferrofluid™ for small tweeter cooling, provided the driver does not have to dip too far into the midrange so that the resultant excessive excursions splatter the coolant out of the magnetic gap.

cabinets

Cabinet designs are probably the most heavily promoted feature in speaker ads. Boxes have been well-designed and constructed for years, but some current builders discuss their new models as if previous speaker engineers were doing their math with Roman numerals. Honeycomb construction, plastic and metal materials, and "diffraction free" designs are heavily advertised and are said to offer improvements over simple, well-braced wood or particle-board assembly and standard, reasonably uncluttered enclosure front panels.

A number of bass performance refinements, such as the Bose Acoustimass design, the KEF, M&K, and Allison "push-pull" woofer feature, and Polk's drone-cone and power-port concepts, as well as some of the motional impedance devices for reducing distortion, work very well. However, all are refinements of the basic acoustic-suspension, bandpass, or bass-reflex configurations that have been in use for a long time. And many other cabinet-design features are more the result of an ad writer's imagination than legitimate improvements.

Certainly the most convenient way to shop for speakers is to simply "go shopping" for them. This will involve going to a dealer and making compar-

Shopping-Comparison Speaker Curves

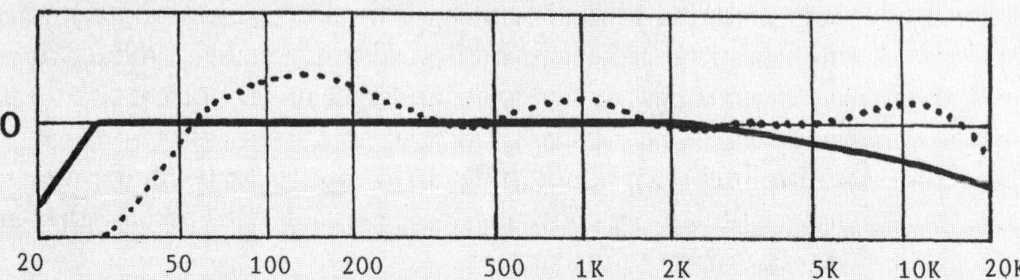

Any system capable of generating a power-response specification like that indicated by the solid-line curve is certainly a world-class design. Yet in many comparison tests, the dotted-line system would be preferred. The midbass "hump" would accentuate the bass punch of some pop recordings, the midrange rise would impose a sometimes-impressive forward quality on many others, and the treble elevation would add "sizzle" to all of them. However, over the long haul the instant-demo-standout quality of the lesser system would become oppressive, particularly on good recordings. When auditioning speakers, be careful to listen for qualities of smoothness, transparency, and neutrality. In short, shop for *realistic*-sounding speakers and not flash-in-the-pan models. If possible, compare what you are shopping for with top-level designs that you cannot afford, in order to get a point of reference for good sound. Make the top model your linchpin for comparisons.

Shopping Problems and Their Solutions

Problem	Solution
It is difficult (if not impossible) to compare speakers from one store to another; there is just no way that you will accurately remember how the speaker you heard an hour ago sounds in comparison with the one being auditioned right now. And even if you could accurately remember how a speaker sounded, the listening environments from dealer to dealer will vary immensely and will drastically alter the sounds of even identical speakers.	Bring along a recording that you are familiar with already; make sure it is a high-quality digital recording. Do your best to adjust volume levels similarly.
The louder-sounding of any items will nearly always sound better. A system that is more volume sensitive may sound superior in a short comparison even if it is markedly inferior in quality to a less sensitive one.	Make certain that the speakers you compare are located in similar positions in the store, placed similar distances apart, and above all are played at similar volume levels; bring along a cool-headed friend to help evaluate and adjust volume levels. And use recorded material that you are familiar with.
Most dealer showrooms are unlike home listening rooms, and therefore a system that works well in the store may be a flop once it is installed in your living room. Showrooms may enhance the performance of a system with a narrow, erratic radiation pattern and undermine the performance of one with superior dispersion that might sound excellent in a typical living room.	See if the dealer will lend you a demo pair. (A deposit may be required.) Read test reports to see if what you hear jibes with what the testers said.
Many dealer showrooms are centers of extraneous noise, making it impossible to fully concentrate on what any component is doing.	See above.
Some dealers are known for bad business practices.	Discuss with friends where they have purchased their equipment; ask about their experience with service. Be aware that some dealers may set up "straw men" systems of known high quality, adjust them so that they sound substandard, and then face them off against house brand or high-profit models of dubious quality.

isons. Although this can be a lot of fun as well as educational, there are a number of problems to consider before starting off.

Shopping for used gear in showroom situations obviously requires the same kind of care.

These comparisons may be all but impossible when you're buying a used piece of equipment from an individual. The best you might hope for would be an A/B standoff against the owner's new (and hopefully for his or her sake) superior replacement models. The key factor will be to make certain that any used speakers being considered are in decent shape.

There are two additional considerations in the evaluation of used speakers. First, the most drastic and (oddly enough) most common fault with speaker systems will be outright driver failure or a disconnected driver. Such defects are rarely subtle, although they could be, in a multidriver design. The best way to insure that all drivers are functioning is to closely listen to each one while playing music. A tweeter should be "tweeting" along in a very subdued manner if it is a three-way system, but will be much more audible if the speaker is a two-way model. Midrange drivers should be producing a solid amount of sound with no buzzy-ness (listen closely for this). Woofers should be clean sounding with no mechanical noise (their most common defect). On a two-way system, a woofer should also be producing a great deal of clean midrange sound.

Second, in the search for *defects* in used speakers, specialized test recordings will often work better than even the best CD or high-quality videodisc. While most of us don't have a library of test recordings, you may be able to use a dealer-provided one in checking out systems they have for sale. One of your audio-buff friends may also have a test disc of some kind.[16]

USE, CARE, AND DURABILITY

Speakers are pretty tough items. If a stereo pair are played at sane levels and located in a decent environment, they can easily be expected to operate up to specification for a dozen years or even more. However, I have seen systems that were located in a poor environment, such as a garage or a non-dehumidified room in a humid part of the country, go to pot in half that time. Obviously, environmental factors are critical here.[17]

About the only parts in a crossover network that will deteriorate over the long run are certain kinds of capacitors that will age whether they are in use or not. The resistors within the network should last a very long time, and the

[16]The frequency sweep segments of the Denon Audio Check CD (#33C39-7441), the Delos Surround Spectacular disc (#3179), and the Hi-Fi News test disc (#015), as well as some of the fixed test tones on the latter, are excellent for detecting serious mechanical or electrical aberrations in speakers, particularly woofers. The pink and/or white noise tests on those discs and the Elektra "Digital Domain" disc (#9 60303-2) are helpful tools for pinpointing more subtle problems.

[17]Foam-rubber woofer surrounds are the weak point in a lot of models. A number of companies offer after-market replacements that require only minimal skills to install. Most audio magazines will have ads from such companies.

inductors should last as long as the wires. Most capacitor and speaker-driver deterioration results from a molecular breakdown of the assorted glues, resins, plastics, and papers that are used to make them, and most deterioration is unrelated to the speaker's operation. They *will*, however, be related to such factors as humidity and heat. Over a period of time, any friction-held connectors within the speaker box (speaker pin connectors, in particular) may oxidize enough to cause an open-circuit situation, so even the wire can act up if the environment is not optimum. Connector problems can usually be corrected by plugging and unplugging the offender a few times, but if this kind of problem continues to remanifest itself, take a serious look at the surrounding environment.

Liquid contact enhancers may help control oxidation.

Surprisingly, the minuscule amount of back-and-forth diaphragm movement that happens when a system is played at even fairly high volume levels should result in negligible wear and tear. There is just not that much stress involved. Unless the speakers involved are "mini" models, playing a pair loud enough to actually damage them will submit the listener (and probably the neighbors) to sound levels that are genuinely uncomfortable.

There are exceptions, however. If you place bookshelf-sized speakers in a large room, you're asking for trouble if you try to simulate live concert-like volume levels. If you like to listen to rock music at "you-are-there" intensities, you risk burning out the voice coils of some of your drivers (particularly the tweeters), due to the sustained, high levels of heat-generating current flow. If you like to immerse yourself in front-row-center impact when experiencing demonstration-grade digital-sound software,[18] you risk damage to your woofers and even midrange drivers as they are pummeled back and forth by the recorded pyrotechnics. Speakers should be sized for the job they will be called upon to perform.

In the "old" days, one way to protect speakers was to equip them with fuses. The speaker is protected from abuse by the fuse that shuts it down if proper power levels are exceeded (just as your fusebox protects the electric service in your house). Today, many of the better speaker builders include driver protection devices within their systems. These are usually circuit breakers (often self-resetting) or bistable resistors; the latter are preferable because they have no mechanical parts and should never wear out. If a system is subjected to excessive drive levels, the bistable resistor increases circuit resistance dramatically, which limits current flow, protecting the speakers.

fuses

breakers

[18]Like the Telarc *Thriller,* the storm sequence in the *Grand Canyon Suite,* the sustained organ pedals found in a good recording of Dupré's *Symphony in G Minor,* the Argo recording of Mendelssohn's *Organ Works,* the jet takeoff at the beginning of the *Digital Domain* demonstration/test compact disc, the pyrotechnics of *True Lies,* the helicopter and firefight sequences in *Forrest Gump,* or the Brachiosaurus, Raptor, and Tyrannosaurus displays in *Jurassic Park,* among many others.

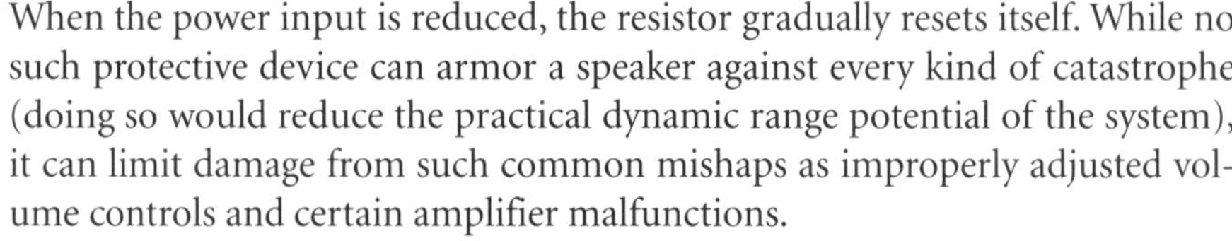

When the power input is reduced, the resistor gradually resets itself. While no such protective device can armor a speaker against every kind of catastrophe (doing so would reduce the practical dynamic range potential of the system), it can limit damage from such common mishaps as improperly adjusted volume controls and certain amplifier malfunctions.

While protected speakers are very durable, it is possible for individual drivers to have or gradually develop subtle problems, unrelated to excessive electrical inputs or even environmental factors like high humidity and heat. In other words, it is possible for both new and used speakers to have factory defects.

For more on random noise testing, see pages 144-45.

To check for these defects, use test recordings incorporating pink or white noise, or even the random noise generated by FM tuners when they are adjusted to a between-stations location on the dial (modern digital tuners can't be manipulated to produce old-fashioned static, unfortunately). This random noise is uniform in nature and has a consistency that recorded music cannot match. In a sense, it is the audio equivalent of a video or photographic test pattern.

When the speakers to be checked out are the same make and model, each speaker can be used as a reference against which to evaluate the other. First, set the amplifier or receiver mode switch to "mono." Then play the random noise or "hiss." With the volume set at a safe level (remember, tweeters can be damaged by this kind of sustained input if the volume is set too high), switch from one speaker to the other using the balance control, and compare the overall tonal similarities and differences.

It is almost a given that the speakers will sound surprisingly unlike each other. Room effects will be a major culprit, so try to move the speakers close to each other and well out into the room away from nearby walls. Then from a decent listening distance (probably over 8 feet), perform the hiss test once again. As before, the two speakers will probably sound quite different from each other, although not as much as in the first test. Even if the speakers are in fine shape, however, there will be audible differences because this test is so demanding.[19]

These differences may be subtle (such as one speaker emphasizing the highs or midrange a bit more than the other) but will usually be quite obvious. Small differences may be partially corrected with midrange and/or tweeter controls, if they are included with your speaker model. These controls may be set slightly differently because of wear or were always improperly calibrated. Adjusting them may fix the problem.

[19]To get an idea of how exacting random noise can be as a test instrument, rerun this test using recorded music and see just how surprisingly alike the speakers will probably sound.

Banana Plug

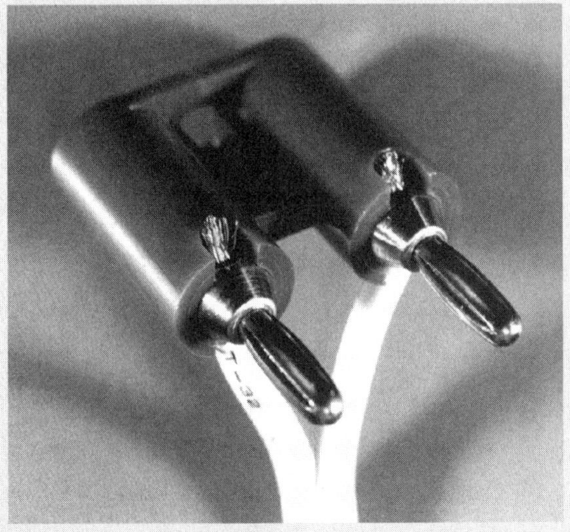

There are lots of ways to attach an amplifier to speakers, but the double banana plug is the best—if the connectors on the rear of the speaker cabinet will accommodate them. (Some excellent mini models, designed to be hung flush against the wall, will not work with banana-plug hookups.) The better amplifiers and receivers also have binding posts that will accommodate this kind of plug. While bare wires inserted into pressure connectors work fine if installed with care, there is always a chance that a short circuit between those leads will occur while the amplifier is running. (If banana plugs will not work with your component, I suggest you at least "tin" the twisted-wire leads with solder to prevent stray wire strands from causing mischief.) Also, if bare wires are pulled and reinstalled (say because of a cleaning job or a component relocation), there is the possibility that one of the leads will be reconnected backwards, resulting in speakers wired out of phase. This cannot happen with double bananas unless they are deliberately plugged back in upside down, an unlikely move given their design. On some budget-grade speakers, even the pressure connectors will handle double bananas.

(Photo by the author)

There may be a *gross* difference in the sound reproduction of the speakers resulting from a defective level control, crossover, or driver—or an internal disconnection. To determine the cause, first make certain that all drivers are producing sound by closely listening to each in turn. If they all pass that test, make a note of the *kind* of differences you are hearing, turn off the system momentarily to keep from accidentally shorting speaker leads together while the amp is running, and switch the speaker leads between the channels.[20] Turn the system back on and listen again. If the sound character shifted with the lead switching, your problem is probably not in the speakers, but somewhere else in the system. Possibly it involves the amplifier tone controls, so

[20]You can safely do this with the system left on if you have double "banana plug"–type connectors that prevent short circuits.

fool with them a bit to see. If exchanging the speaker leads did not affect the problem, you probably have a defective speaker component.

If you suspect speaker defects but are unsure as to what is exactly happening, you can isolate problems by creatively using the level controls on the back of the speaker cabinets (if available), or the tone controls on the preamplifier or receiver. In the old days, many speaker systems had continuously variable controls on the cabinet back panels that would allow you to almost shut off the various upper-range drivers. Most modern systems, if they have any controls at all, have switches that are more limited in function but more reliable; some systems have removable jumpers to isolate the bass section from the mid/tweeter section.

If all you have available are the amplifier's tone controls, you can selectively cut or boost the frequencies that coincide with the range of each driver in your speakers. Cutting the bass and treble on both channels while doing the random noise test will allow you to evaluate the sound of the midrange drivers more effectively. Cutting just the treble while boosting the bass will aid you in pinpointing woofer problems; doing the opposite will aid in checking tweeters (be careful to not overdrive them). If your preamplifier or receiver control panel has a midrange control or a graphic equalizer, even better results may be possible, but be sure that the equalizer is not contributing to the problem before employing it as a testing tool.

phase

Note that this test will be useless if the polarity of your main speakers is not correct. Both should be working "in phase" for the sound to be correctly reproduced, whether you are doing a test or listening to music. Match the color-coded input connections on the speakers to the similarly coded connections on your amplifier or receiver; that is, red (+ or hot) to red, and black (– or ground) to black. If you use typical 16-gauge lamp cord for the hookup (a good size for any run up to 20 feet), there will be a ridge molded into one side of the wire to aid with proper orientation.

If your speakers are properly phased, the sound from a central source (as would be the case with a mono recording or with the center performer in a stereo recording) will be centered and focused when heard from a position equidistant from and somewhat out in front of both speakers. If they are wired out of phase, the same mono signal will have a diffuse and directionless quality often with a loss in bass. Note that a phasing test must be done with any kind of center-channel steering—Pro Logic, for example—turned off. (If any individual drivers within either speaker system are wired improperly, it may be impossible to get a correctly focused image; contact the speaker manufacturer for help.)

There is more to speaker care than simply playing them at sane levels. Speakers are surprisingly tactile items, and typical owners will be tempted at

Absolute Polarity

A recurrent audio-extremist fixation is absolute polarity, or absolute phase. If *both* speakers of a stereo system are wired in reverse, the "phasy" quality discussed in the text will not materialize because the reversal is equally felt by both systems. Some enthusiasts, however, feel that this will still cause problems because it results in the speakers "pushing" when they should be "pulling" and vice versa. Drum "booms" will sound like "wooms" (or smoobs?), rocket takeoffs will sound like landings, and so forth.

Research by level-headed people like Richard Greiner and Stanley Lipshitz has shown that with test tones and some musical signals, extreme direct-field producers like headphones and near-field-oriented speaker systems may reveal polarity-reversal artifacts (assuming that the recording itself is fully phase coherent, with every microphone input having had the same polarity at the control console), while good speakers operating in the reverberant field probably will not. I say "probably" because speakers with high levels of asymmetric distortion (unwanted harmonic by-products that are even-order multiples of the fundamental) may highlight the problem; the asymmetry becomes reversed, and the ear may hear this. Speakers with respectably low asymmetrical distortion should have no problems.

The entire polarity controversy is genuinely bizarre because, although trained recording engineers are careful to wire all their microphones in the same relative electrical phase, most recordings are made with several, widely spaced microphones, making it impossible to phase-adjust all signals for absolute polarity for anything other than the bass.

times to remove the grille covers to show off the working parts. I always advise against this, not only to prevent excessive wear on the grille assembly, but mainly to limit the potential for damage to the exposed and often fragile drivers. (Back in the good old days, Acoustic Research and KLH *glued* the grille covers on their systems to keep curious fingers away from delicate parts.) You may want to pull the covers off once in a while to check things over, because driver mounting screws occasionally work loose. However, if guests wish to see how the innards work, hand them a copy of the owner's manual or a sales brochure.

Speakers are also not happy in wet environments unless they are designed for the outdoors. Walkman™ users please note that the same goes for headphones; avoid placing them in locations with high humidity. Consider a dehumidifier to be your latest A/V component if your home tends to get humid in the fall or spring, and remember that it will also protect other home furnishings. (A dehumidifier is mandatory in the deep South, y'all.)

dehumidifier

When cleaning or waxing speaker cabinets, take particular care not to spill liquids on the working parts. Also protect your speakers from inquisitive pets; remember, speaker grills are not scratching posts!

Finally, make certain any party guests know that your conveniently located speakers are off limits as drink holders. Do not make it a practice of transporting the speakers to other locations to provide party music unless you are ready to write them off. Party with friends who have good speakers of their own, or else purchase some fur-covered truck speakers with handles!

3 Audio Components

IN BUILDING A HOME-theater system, you need to have a power source to make your speakers work, as well as some way of receiving either audio or video signals to play through your system. While there are a multitude of electronic components available for A/V use, in this chapter we will deal only with the more basic ones: the *amplifier*, the *preamplifier*, the *tuner*, and the audio or audio-video *receiver*.

An audio receiver combines an amplifier, a preamplifier/control section, and a tuner (radio) in one chassis. Because this combination is convenient and economical, receivers far outsell separate amplifiers, preamplifiers, and tuners. We will not get too involved with analyzing the electrical operations of these components, but instead will discuss what each does—and does not—need to do to work well. We will also show why the supposedly lowly receiver is a more sensible investment for most enthusiasts than more expensive separates, unless extreme flexibility or really advanced surround-sound performance is required.

THE RECEIVER AND ITS COMPONENTS

Amplifiers

watts

The business end of any receiver is the amplifier. It has only one job: to increase the strength of an incoming signal to a level high enough to drive the speakers (or headphones) properly. The popular measurement for amplifier output is *watts*. However, what an amplifier actually does is amplify current and, more importantly, voltage (electrical pressure). When that voltage ("amplified" from the incoming voltage source of a CD player, tuner, DVD player audio output, etc.) is applied to an electrical *resistance*, the end result is electrical *current* flow.

Loudspeakers have a rather complex inductive *reactance* as well as *resistance*, and the combination is known as *impedance*. Although this impedance varies with frequency, a single value (i.e., 4, 6, or 8 ohms) is assumed for purposes of amplifier power calculation. The product of the current and the voltage (voltage times current, or the voltage squared, divided by the impedance) is the amplifier's *power* output in watts. You can see that an amplifier's capabilities are dependent upon the amount of voltage it can produce and its capacity to sustain a given amount of current flow without overheating.

power

A good amplifier must perform these functions without audibly changing or distorting the output. The distortion produced may assume any number of forms, from harmonic overtones added to the main signal to various kinds of intermodulation products. These aberrations will be subjectively inconsequential if kept below specific thresholds. All well-designed amplifiers, including those found in most fairly low-priced receivers, should have inaudible distortion under normal conditions.

clipping

The most obnoxious form of amplifier distortion is *clipping*, which occurs when an amplifier is overloaded; that is, it has to produce power above its capabilities. Very large quantities of harmonic distortion may be produced; what's more, the amplifier's built-in protection circuits can also contribute to the distortion when this overloading occurs. Therefore, the single most important aspect of a well-designed amplifier's performance will be power output below the clipping point.

The clipping point is measured in many ways by both equipment testers and manufacturers. Some will stress continuous-output capabilities, while others may favor peak or reserve-power potential (the ability of some designs to produce momentary outputs well above what they can generate at long-term sustained levels). Because musical and film soundtrack programs fluctuate in level, peak output performance is obviously important. However, if a program's dynamics are contained within the power limits of an amplifier with adequate continuous power output (even if it has no peak headroom at all), there should be no problem. The bottom line is that your amplifier must be able to reproduce the entire musical or film sound-source waveform without audible distress.

sensitivity

While speaker manufacturers disagree about many aspects of performance, the one thing that most do agree upon is *sensitivity*: the measurement of a speaker's ability to produce a given sound level with a set voltage input.

Figure 3.1 Decibel Levels

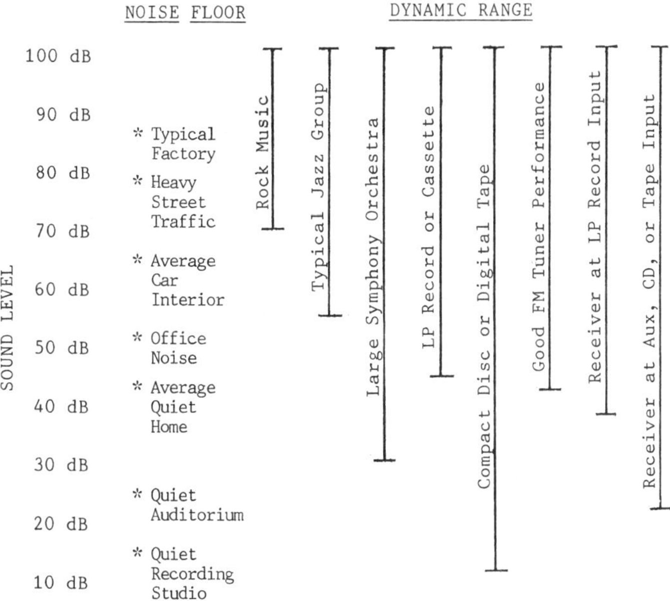

This chart shows the dynamic range, noise floor, and relative loudness levels of several different environments and types of music. The upper limit is 100 dB, although this is often exceeded by both live and recorded music. The CD, digital tape, and receiver high-level inputs easily span the required range of any source. However, it is easy to see why less-than-ideal, nondigital FM radio and LP record-player performance may produce distracting background noise. In addition, vinyl LP surface noise is so unlike the background ambience found in most home environments that it may be even more audible than this chart indicates. Finally, one glance at the indicated noise floor of an automobile will show why a car CD, MiniDisc, or DCC player offers no substantial advantage over a decent analog-cassette player unless absolute speed accuracy is required. No audio playback device can have a meaningful signal-to-noise ratio in a moving automobile.

The industry standard is popularly stated as the output in sound pressure level (dB SPL) with 1-watt input to a speaker of 8 ohms impedance at a measuring distance of 1 meter in front of the speaker cabinet. The more exact specification is voltage dependent and refers to an input of 2.83 AC volts. When applied to a speaker of 8 ohms impedance (a common value), this results in 1 watt of dissipated power. (If a speaker has a 4-ohm impedance, which many good ones do, the input power is 2 watts with the same voltage applied.) Most speakers produce a sound level of between 85 and 91 dB SPL under these conditions.

Input-Impedance Curve

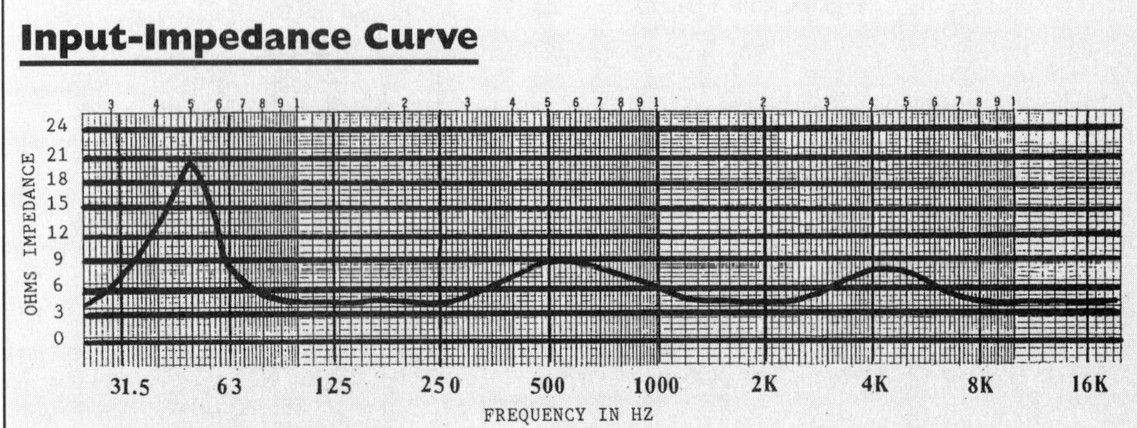

Speaker impedance is a much more complex item than some manufacturers admit. Aside from aspects of inductance and capacitance (caused by the assorted chokes and capacitors in the crossover network, as well as the voice coils of the speaker drivers themselves), the frequency-response measurement is very important. This diagram plots an impedance curve for an easy-to-drive, three-way system. The woofer-resonance peak at 50 Hz tops out at 20 ohms, with 4-ohm valleys in between the midrange hump at 9 ohms and the treble hump at 8 ohms. While some testers might rate this as an 8-ohm system, others would rate it at 4 ohms because of the low impedance valley in the musically busy area between 100 and 300 Hz. A credible manufacturer might describe this system as having a 6-ohm nominal impedance with a 4-ohm minimum. Note that this seemingly erratic behavior is common, and will have no effect on frequency response provided that the amplifier involved has a very low *output* impedance. Fortunately, most do, including many economically priced models.

With this information, it is not difficult to calculate power needs. Surprisingly, really huge loads of electricity are unnecessary to drive ordinary speakers to reasonable levels in ordinary rooms. Every time power input is doubled, the overall output level at the speakers jumps by 3 decibels. If 2.83 AC volts are applied to a speaker that has a sensitivity rating of 87 dB at one meter on-axis, the same speaker will produce 90 dB at 2 watts input, 93 dB at 4 watts, 96 dB at 8 watts, 99 dB at 16 watts, and 102 dB at 32 watts (102 dB SPL is very loud). Typically, you will need *1½ to 2 times* those amounts per channel to get similar volume levels in the reverberant field, when listening to a stereo program on a pair of wide-dispersion speakers in a typical room.

Let's assume that we are installing a set of 87-dB-rated speakers. In most rooms and at normal listening distances, they would have to be fed about 30 watts per channel to produce peak stereophonic output levels of roughly 100 dB. Therefore, these speakers could be adequately powered by a typical low-cost receiver having a modest 45 watts or so per side power rating with a bit of room to spare. Because 100 dB is quite loud—more than most people would care to experience in a typical, reverberant room—this is good news for audio-video enthusiasts of any kind.

cheap amplifers

Unfortunately, *really* cheap amplifiers—with power levels as low as 30 watts per channel—may run into special problems when trying to function near maximum power, when you are listening to some organ music or watching videos with really potent soundtracks, for example. Low-rated amps may not only lack the bandwidth necessary to reproduce the required frequency range but also have severe problems with the fluctuating impedances found in virtually all speakers. Remember, while most speakers are rated at 8 ohms, that figure is not uniform over the entire audible range. Some 8-ohm models may dip as low as 4 ohms at certain frequencies and, under exceptional conditions, they may also present a load that is as low as 2 or 3 ohms—giving amplifiers that cannot sustain high current levels fits, particularly if they are already working near their limit.[1] Consequently, those on a budget should avoid the "$99 special" receiver deals found in some stores.

A good amp manufacturer will note this in their brochures.

Theoretically, an amplifier should have double the power at 4 ohms that it has at 8. However, few if any (for reasons of output transistor or power-supply limitations) approach this ideal. Unfortunately, there are still a substantial number of otherwise good receivers for sale with amplifiers that produce *less* power at 4 than at 8 ohms. A careful shopper (even when on a shoestring budget) should learn the 4-ohm-impedance power rating of any receiver or amplifier being considered for purchase. Even if the speakers to be driven are 8-ohm models, the 4-ohm power rating should be *at least as high* as the 8-ohm rating, and preferably somewhat higher. The more power an amplifier can deliver under demanding conditions, the more certain you can be that it's robust enough to hold up under heavy use. Fortunately, there are a fair number of high- and medium- and even some relatively low-priced receivers on the market (as well as nearly all basic, stand-alone power amplifiers) that satisfy these requirements.

headroom

Buy an amp or receiver with at least *twice* the power you believe you'll need to power your speakers. This guarantees that adequate headroom will be available should a situation arise that demands a bit more "oomph" from the system. As long as the speakers have the capability to handle the extra input, erring a bit on the high side will have no negative impact. Also, if you want to upgrade your system with a more powerful amplifier or receiver, remember that it will be necessary to at least *double* your present amplifier's

[1]A number of speakers have average impedances of 6 and even 4 ohms, and under some conditions a few of those can present dynamic loads as low as 1 or 2 ohms!

Output Impedance

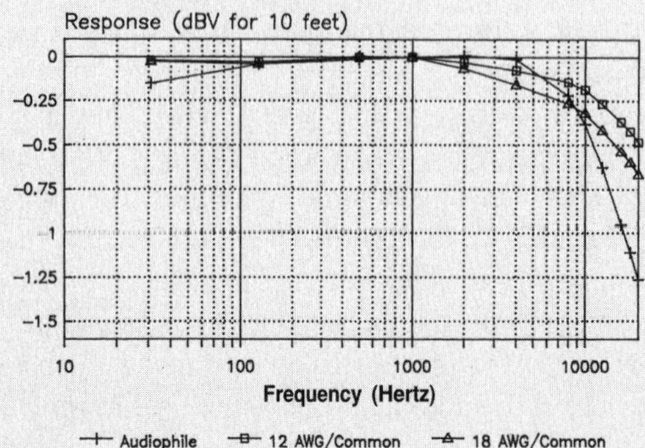

Speaker input-impedance specifications are highly visible in both amplifier and speaker-system sales brochures. They *are* important, because input impedance will determine amplifier stability and output levels. However, an amplifier's *output impedance* is rarely mentioned by amplifier manufacturers, who should be more forthright.

If the amplifier output impedance is not really low (0.2 ohms or less at most frequencies and below 0.5 ohms at low frequencies), typical speaker loads of widely varying impedances will trigger slight system frequency-response irregularities that may be audible under some conditions, particularly if the average speaker-system impedance is quite low and the speaker wire being used has fairly high resistance (say above half an ohm). A high amplifier output impedance at low frequencies can also cause woofer damping problems (this relates to the "damping factor" measured in amplifier test reports) that will affect bass-response accuracy.

Most amplifiers (including the amplifier sections of receivers) have adequately low output impedances, which is one reason that similarly powered, reasonably high-quality units all tend to sound pretty much alike. Even most high-end models are satisfactory in this respect. However, there are a few variants available (most, but not all, are tube units) that measure above 1 ohm over most of their operating range. Certain manufacturers design their products this way intentionally, so that they sound subtly different from more mundane, lower-priced versions. Many so-called "golden eared" audio enthusiasts will hear such deficiencies (that may come across as a more "mellow" or warmer sound) as positive attributes and favor these amplifiers over other designs.

To show the difference between standard amplifiers and these so-called high-end units, a test was run by engineer Fred Davis. The three frequency-response curves show the performance of a standard, economy-grade amplifier working with 18- and 12-gauge wire, and a high-end amp in combination with weighty and expensive "audiophile" cable (a 10-foot pair costs in excess of $2,500!). In each case, the curves indicate the frequency response of the respective amp/wire combinations while driving a standard speaker and *not* a special ballast resistor. You can see that the cheaper combinations perform better at both high and low frequencies than the expensive stuff. However, because the differences over most of the audible range are less than 0.25 dB, only individuals with very astute hearing would be able to detect them—and then only with certain kinds of music or test tones. Note that even lightweight, 18-gauge "lamp-cord" wire performs surprisingly well.

(continued)

Output Impedance *(continued)*

The graph on page 90 is for cable runs of 10 feet only; longer lengths would skew the results accordingly. However, even a 30-foot run would result in differences between the 18- and 12-gauge wire of only about 0.6 dB above 3 kHz, which is still difficult to detect with musical program material. The audiophile amp/wire combination, however, would begin to have significant problems with runs that long (so would your bank account), and "perceptive" buffs would probably consider the effect on the deep bass and extreme highs as being similar to the warm and mellow sound of tube amplifiers. In combination with a more unfriendly speaker load, the differences between the audiophile combination and the cheaper items would be even more apparent. The upshot of this demonstration is that the consumer on a budget need not worry about high-end amplification and speaker-wire combinations putting their modest components to shame.

power to gain any weighty improvement. Even that will net you only 3 dB of additional headroom and will mean nothing at all if your current amplifier always works within its limits during "show-off-my-system" situations. Fortunately, there are scads of reasonably priced receivers (as well as some decently priced basic amplifiers for you aficionados of audio-video excess) that have the kind of power and stability needed to drive common speakers to loud levels.

You can use the chart below in matching amplifier power to some typical speaker configurations; impedance rating would depend on the speakers. The most common setup would be 60 watts at an 87-dB rating.

Preamplifiers

Traditionally, the basic controls in a sound system are contained in the *preamplifier*. However, a true *pre*amplifier is the circuit devoted to raising the voltage of a signal (originally emanating from a phonograph cartridge) high enough

Speaker Rating	Minimum Recommended Wattage per Channel
84 dB	120 watts
87 dB	60 watts
90 dB	30 watts

to be useful to an amplifier. For the sake of convenience, other control functions have traditionally been included within the stand-alone preamplifier unit, and the entire switching and front-panel control section of a receiver or integrated amplifier (a receiver minus a tuner) is called its preamplifier.

Most of the more elaborate preamplifiers on the market can handle just about anything, but some non-A/V models—including separate preamps, integrated amps, or receivers—may not accommodate the quantity of inputs required for good A/V compatibility. Some controls and functions are superfluous or important to only a few individuals; many others are quite important and will be essential for reasonable performance. Fortunately, there are a number of good, reasonably priced components on the market that include more than adequate preamps.

controls

The most basic of preamp controls are those of *volume*, *balance*, and *tone*. Most are controlled either by panel knobs or buttons (two separate buttons or a rocker switch) or by use of a remote control equipped with buttons.

A good volume control must have adequate *tracking* capability—the ability to keep the left and right channels at the same relative level as the gain is adjusted. To evaluate this, set your amplifier to mono and listen to a recording while sitting equidistant from the speakers. Make certain that the central image stays fixed in the center as the level is adjusted upward and downward.

The balance control is designed to adjust the stereo sound spread; it can also be used to correct balance defects in some recordings. I rarely use it, preferring instead to sit in a good location and let the recording dictate the sound stage. However, video surround processing is one area where precise balance is important. If the mono complement of the signal is not reasonably well centered, it may "bleed" into the rear channels and destroy the theater-ambience effect. A good surround receiver should have an input balance control to cancel any center-channel bleedthrough. Some audio-video receivers and outboard processors can adjust themselves automatically, a great feature. Others may have a manual control, but it may be awkward to use when interfaced with some receivers or preamp components.

Tone controls come in a variety of configurations, with the most basic set offering one bass and one treble control that handles both channels simultaneously. If designed correctly, this is about all that is needed. The most common tonal problems encountered during everyday listening are anemic bass or brittle-sounding high frequencies, due to poor program material (on older recordings, particularly), nonoptimum speaker placement, or poor speakers. Single-knob bass and treble controls should handle these anomalies reasonably well.

Midrange controls are becoming increasingly popular on some preamps. However, little is gained by this addition; simply moving simultaneously the

Channel Balance

Just how much of a difference in *level* between two speakers of a stereo pair is needed to shift a centered "image" off to one side?

If you sit some distance away but exactly on the axis between two speakers (in the so-called *sweet spot*), identical sounds coming from each system will form a centered, "phantom" image (provided the speakers are well positioned and working OK). If the sound of one channel is made just one dB louder in level than the other, there will be only a very slight change in the perceived location of the central image toward the louder speaker; one dB isn't very much. However, the following table shows how greater increases can affect your listening if your speakers are about 12 feet apart and you are sitting on the central axis about the same distance back from a line drawn between them.

If you sit off to the side a bit while this is going on, you will see how the balance control will help to keep the central image centered for the best sound under these conditions, by adjusting the relative levels of each channel. Even if your receiver or power amplifier has no level meters to verify balance testing, adjusting the balance control while sitting in a variety of locations will be a good way to learn more about how to "tweak" your system for good performance. Just remember to use a recording with a tightly focused central image, such as a solo singer (test discs are excellent, too).

Just how much of a difference in *time* between the signals of both speakers is required to shift the image to one side?

During stereo playback, with the listener located in the sweet spot, all that is required to move a centralized image strongly toward either speaker is a delay or advance in the sound to one channel of 1 to 1.5 milliseconds. Sound moves at 1,130 feet per second, so a difference of only 1.1 to 1.7 feet will achieve the 1 to 1.5 ms required. Even a difference of a bit less than 0.5 ms (under half a foot) will shift the image 20 degrees off center. This is why shifting the listening position to one side a couple of feet will cause the sound to shift radically toward the nearer speaker. (Moving to one side also causes a small change in the relative sound levels of the two speakers, but the main cause for the image shift is the timing change.)

If a pair of speakers is 12 feet apart and the listener seated at a central position 12 feet from the axis between the speakers, the signal path from each speaker to the listener is about 13.5 feet. If the listener moves 2 feet to the left, the signal path from the left speaker drops to 12.6 feet and the one from the right speaker increases to 14.4 feet, a difference of 1.8 feet. This is sufficient to shift the sounds of transient signals radically to the left and will cause steady-state sounds to have a phasy and diffuse quality, because of comb-filtering effects.

Correcting this can be partially accomplished by using the balance control to increase the gain in the more distant speaker by 6 to 10 dB. (Increasing its level by 6 dB will have the effect of recentering an image that was shifted sideways by a 1.25-millisecond time delay or a difference in distances to each speaker of 1.4 feet.) This will help transient signals stay properly positioned but will *not* eliminate the phasy quality of steady-state signals. When listening critically, there is no substitute for sitting in the sweet spot and keeping the balance control centered.

Difference in Level	Approximate Location of Image
3 dB	Two feet closer to louder unit
6 dB	Midway between center and the louder speaker
10 dB	Within a foot or two of the louder unit
15 dB	Nearly complete collapse of the central image into the louder-playing speaker

Old-Time Balance

Years ago, AR's amplifiers and receivers had an unusual control to aid in setting correct center balance for those times when the listener was in the "sweet spot" and wanted the best left-to-right balance possible.

Typical volume controls are not matched perfectly over their full operating ranges (remember, stereo volume controls are actually *two* controls mechanically connected together). That is, if the left-to-right balance is exact (central image dead center) at one setting, it may be incorrect at other settings. A shift of one or two dB from one side to the other as the control is rotated from "min" to "max" is not uncommon, even with some electronic, push-button volume controls. When you are listening to a recording, it can be quite exasperating to set the balance control to compensate for these deficiencies.

Those old AR units had a unique control to handle the job: In addition to a "stereo" and "mono" setting, the mode control also had a "null" position that treated the left and right channels as difference signals and reversed in polarity. When this happened, central images disappeared completely. Rotating the balance control to either side raised or lowered the volume level of those images. Thus, to get perfect left-to-right symmetry at any given volume control setting and with any recording, all you needed to do was switch to "null" at the desired sound level and then adjust the balance control for minimum sound. Then, when switching back to "stereo," the balance would be perfect for that particular volume/balance setting. It was quick and easy to do and did away with clumsy balance-control twiddling. I know of no unit being made today with such a feature, which is a shame, since it would also be a fast way to get a Dolby-type theater surround system properly balanced.

bass and treble controls in the same direction should mimic the effect of a separate midrange knob.

A number of components substitute multiband equalizers for standard tone controls. While their use can be beneficial under some conditions, their obvious operational complexity may make them less useful for correcting certain program deficiencies.

Some preamplifiers offer separate bass and treble (and midrange) controls for each channel. This feature is useful if speaker placement creates stereophonic tonal imbalances. If one speaker is near a corner and the other is near a large doorway, it may be helpful to selectively boost or cut the bass in each channel separately. However, because bass sounds are mostly nondirectional, a single bass control handling both channels can usually cope with this phenomenon fairly well. Treble irregularities will probably involve asymmetrical wall absorption or reflection, something no tone control can handle properly. These problems should be managed by speaker placement shifts—

Figure 3.2 Basics of Equalizer Controls

This is about as basic a set of graphic-type "equalizer" controls as you will find. As part of a budget-grade receiver, they can do a lot for you if used intelligently. The 100-Hz slider can offer a subtle boost to bass-weak recordings and bass-weak speakers, or in situations (like extreme close-up listening) where treble signals may be subjectively louder than the bass ones. The 330-Hz slider can remove the mid-bass thickness found in certain recordings. The 1-kHz (1,000-cycles-per-second) slider, in conjunction with the 3.3-kHz one, can tame some of the midrange irregularities that occur when a two-way system that is optimized for reverberant-field use is listened to up close; they can also level out certain kinds of recording edginess. The 10-kHz slider can add sparkle to some discs that need it or mellow out the high-frequency stridency found on some remastered CD classical recordings. More complex but similarly designed equalizers can do even more subtle alterations, but require even greater care to operate successfully. For more on equalizers, see Chapter 8.

or room alterations if a proper balance is essential. Separate-channel treble controls may be able to correct minor imbalances, but so can a standard balance control.

loudness control

Many preamplifiers, particularly those included as part of a receiver, have switchable "loudness" controls. These boost the bass (and often the treble) to compensate for hearing sensitivity losses at background-music listening levels.[2] As they are habitually employed, loudness controls have three limitations. First, the compensation may be inexact, because the boost is often fixed, but the ear's sensitivity, particularly to bass frequencies, will *vary* with the volume. Second, there is no way the manufacturer can know the sensitivity of the speakers being employed, so that even loudness controls that take volume set-

[2]As the overall volume decreases, the ear's sensitivity to bass frequencies falls off at a faster rate than its sensitivity to higher frequencies.

tings into consideration cannot work with any guarantee of precision. Third, most listeners use loudness controls to artificially "boost" frequencies at all volume levels, ending up with a system that sounds like a jukebox.

A good set of standard tone controls intelligently used should be all the loudness compensation ability any preamp or receiver needs. If your system is not being played at concert-hall levels and the bass sounds a bit thin and/or the treble seems a mite weak, the problem can be adequately handled with the tone controls. You should not let the absence or inclusion of a loudness switch on a component influence a purchasing decision.

More on surround sound in Chapter 7.

An important feature for some buyers will be the inclusion of surround-sound programming. The ability to deliver surround effects (Dolby Pro Logic, Dolby AC-3 Surround Digital, hall-synthesized, matrixed, etc.) is becoming increasingly important. A control panel that includes a variety of surround-sound functions may be essential to your future listening (and viewing) pleasure, even if you lack the capital right now to take full advantage of it.

Fortunately, a receiver or integrated amplifier without surround-sound facilities will handle future situations with ease if there are *external* connec-

Fletcher/Munson Curves

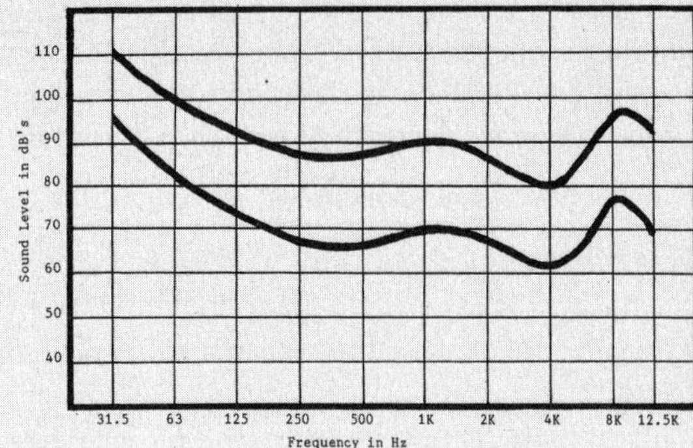

This simple chart, derived from a much more complex one, will give you some idea of what a good tone control should be able to do in terms of making background music more enjoyable. Each curving line indicates the ear's sensitivity at assorted frequencies and sound levels. If you follow the line that intersects the 90-dB level at the 1-kHz midrange frequency over to the 63-Hz bass frequency, you'll see that the signal must be at about 100 dB to sound equally "loud." The ear/brain compensates for this in real life, and many studio engineers equalize their recordings for much higher levels.

Now, if you follow the line that intersects the 70-dB "background music" level at 1 kHz over to the 63-Hz point, it is obvious that the signal must be raised to about 83 dB to sound equally "loud." While the first situation involved a 10-dB difference, the second involves a difference of 13 dB, which is 3 dB more. Therefore, in order for a recording equalized in the studio at the higher level to sound properly balanced at the lower one, it will be necessary to boost the bass range by about 3 dB at the playback amplifier.

tions on the rear panel between the preamplifier and amplifier. These will ordinarily be designated as "preamp out" and "main amp in," will probably be fairly close to each other, and will normally be connected together by simple, U-shaped shorting bars. A surround-sound processor (or any other kind of signal-modifying device, for that matter) can easily be installed by removing the shorting bars and connecting it between the preamp and main-amp sections. These external connections are usually found on more expensive receivers and most integrated amplifiers, but, unfortunately, not run-of-the-mill economy models. If surround sound is something you may eventually want, seriously consider a receiver—or integrated amplifier—with surround processing and four or five (or even more) self-contained amplifiers, even if it squeezes your budget a bit.

tape monitor loop

Most preamplifiers, integrated amplifiers, and receivers have at least one "tape monitor" input/output loop. If you are going do any kind of recording, this will be important to you. If you think you might work with two recorders, check that the receiver control section has the necessary connections. This is not as far-fetched as it seems, because it might not be uncom-

Figure 3.3 Pioneer Receiver (front)

(Photo by the author)

This discontinued Pioneer receiver is a very basic item with minimal controls and flexibility, but with a solid 60 watts per channel into 8 ohms and considerably more into 4 ohms. This last specification is important, because many Japanese receiver manufacturers do not recommend using speakers below 6 ohms with their products; however, most units *can* drive 4-ohm loads with impunity. Japanese precautionary load recommendations are actually designed to prevent owners of two pairs of really low-impedance speakers from hooking them up and running them all *at the same time*. Driving two 4-ohm pairs in parallel, for instance, would result in a combined load of 2 ohms with some, but not all, receivers. A few would not see a 2-ohm load, because some manufacturers wire their "speaker A/B" selectors in series—rather than in parallel—to prevent such mishaps. (Two pairs of 4-ohm speakers in series would total a safe 8 ohms.) Unfortunately, a series hookup can degrade the sound if all the speakers playing at the same time are not *identical* models.

mon for even a modest system to have both an audiocassette recorder and a hi-fi VCR plugged into it. While the VCR could obviously be patched into a basic auxiliary input, hooking it into a second tape input-output loop would allow it (provided it was configured for such use) to be operated as a high-grade audio-only recorder for copying nonvideo programs from a variety of sources.

Videotape loops, outputs, and switching may also be included along with the audio hookups, which is a good thing if you want to dub from one VCR to another or copy a videodisc to tape, or if you own a TV monitor with only one video input. Remember that a second VCR or audiotape-selector function on the unit's front panel may only be referring to an input, not a true input-output loop, which will limit your ability to record a program with both decks at the same time or dub from one to the other in either direction. However, if you never plan to use more than one tape deck or intend to only use your VCR for "off-the-air" recording (in addition, of course, to video playback) then buying a component with only one tape loop will probably save you money.

It is also possible to use a tape loop to integrate a signal processor into

Figure 3.4 Pioneer Receiver (rear)

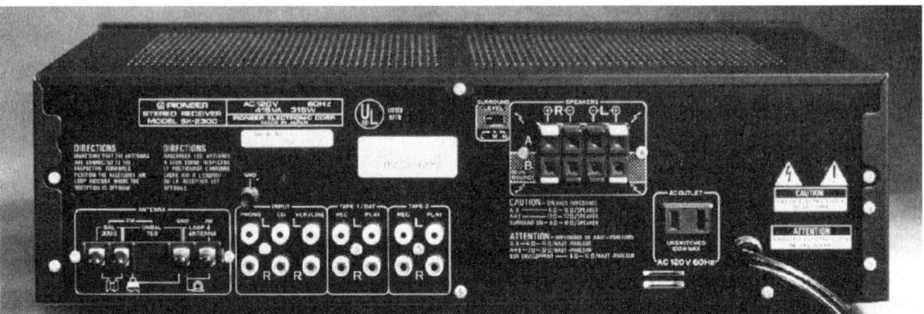

(Photo by the author)

This is the rear panel of the same Pioneer receiver. Obviously, it is very basic and lacks perks such as video picture inputs (and switching, of course), dual tape loops (it does have dual tape inputs but only one loop), and external amp-preamp connections. However, for basic work it has all the stuff that is necessary, including a surround hookup. Unfortunately, this particular surround unit is nothing more than a L–R, Hafler-type, matrix connection (which we will discuss more fully in Chapter 7) and does not include a separate surround amplifier for the rear speakers. Thus, it can simulate some ambience effects but is not a full-tilt surround receiver. For conventional two-channel stereo use, however, this is a fine unit and, within its power limits, will drive even the best dynamic speakers as competently as similarly powered high-end amplifiers.

the system. For example, equalizers can be installed into tape loops effectively and may be less prone to overload than they would be if installed between a preamp and power amp. The very nature of surround-sound processors may limit their flexibility if they are installed in a loop, unless they have a control for overall gain. Note that a few preamplifiers have an *External Processor Loop* (EPL), which, because it is electrically the same as a tape loop, can handle a tape recorder also. While an EPL is supposedly designed to allow you to install an external processor of some kind without labeling confusions or upsetting tape functions, most processors provide redundant tape-loop switching, permitting the installation of a tape recorder via the EPL circuit even with the processor in place.

Radios (Tuners)

The final member of our basic electronic component assembly and the item that actually turns a receiver into a receiver is the radio or *tuner*. Good radio reception is important to a lot of people, particularly if they live in the big city where excellent programming often exists. In addition, the radio part of a typical receiver is essentially a "free" item. Separate preamp and amplifier combinations and most integrated amps (a preamp and a power amp without a tuner on a single chassis) usually cost as much as, or considerably more than, a receiver of similar power output.

Power output aside, it is the performance of the radio that separates good receivers from great ones. Consumers Union, because they realize the basic similarity of amplifiers, heavily weigh their ratings in favor of tuner performance when they rank receivers in their test reports. Simply stated, some tuners work better than others.

Manufacturers list assorted specifications in their brochures that are often helpful. However, if serious FM-tuner performance is your goal and you need more data than what the manufacturers print or what CU briefly lists in their comparison tests, the best thing to do is to look up test reports in magazines like *Stereo Review* and *Audio*. Once you have the test results, or a reasonably honest manufacturer's spec sheet, on hand, check for the parameters presented in the table on page 102.

radio stations

There are other factors that will impact on tuner performance that are unrelated to tuners themselves. Most radio stations do *not* do a good job of producing a clean signal for a variety of reasons. Often this involves simple sloppiness on the part of the engineers at the station, but just as often it is a conscious thing, as when a station compresses its signal to make the quieter musical passages louder, limiting the dynamic impact of the program. Com-

Figure 3.5 NAD Receiver

(Photo courtesy NAD Corporation)

A receiver like this NAD model can serve well in a decent A/V system. It has Dolby Pro Logic, is rated at better than 50 × 3 watts for the three front speakers (8 ohms) and has 20 × 2 watts available for each of the two surround speakers. While this kind of power may seem bush league when compared with many competitive products, most listening rooms will be well served by this unit. In addition, it has substantial front-channel reserve power to deliver output peaks well above steady-state levels. While the wattage available to the surrounds may seem a bit anemic, it must be remembered that because this unit lacks an AC-3 decoder, they will never need to play as loud as the fronts and will not have to produce deep bass or high treble either. Note that this model has video switching, a near-critical feature these days in dealing with multiple video inputs and a TV monitor that does not have enough input connections itself. This unit also has a variety of other features, such as direct-entry FM tuning and simulated hall surround modes. (There's more on Dolby Surround, center-channel steering, and hall-ambience synthesis in Chapter 7.)

pression is done in order to extend the practical operating range of the station. If the dynamic range of the recording were left unchanged, the quiet passages would be obscured by transmission and reception noise in fringe reception areas.

antenna

The antenna will greatly affect the performance of your tuner. Most tuners and receivers come from the factory with a wire loop or folded "dipole" antenna that is not particularly good at receiving distant signals, cannot be aimed properly to get specific stations under most conditions, and is usually not the correct size for good reception. If for reasons of economy you cannot afford anything better, make sure the dipole wire is at least 5 feet above the ground (which can be on a wooden floor in any building from the second story on up) and is positioned straight out at a right angle to a line pointing at the station being listened to. If a local station is too powerful, sup-

Figure 3.6 Old AR Receiver Spec Sheet

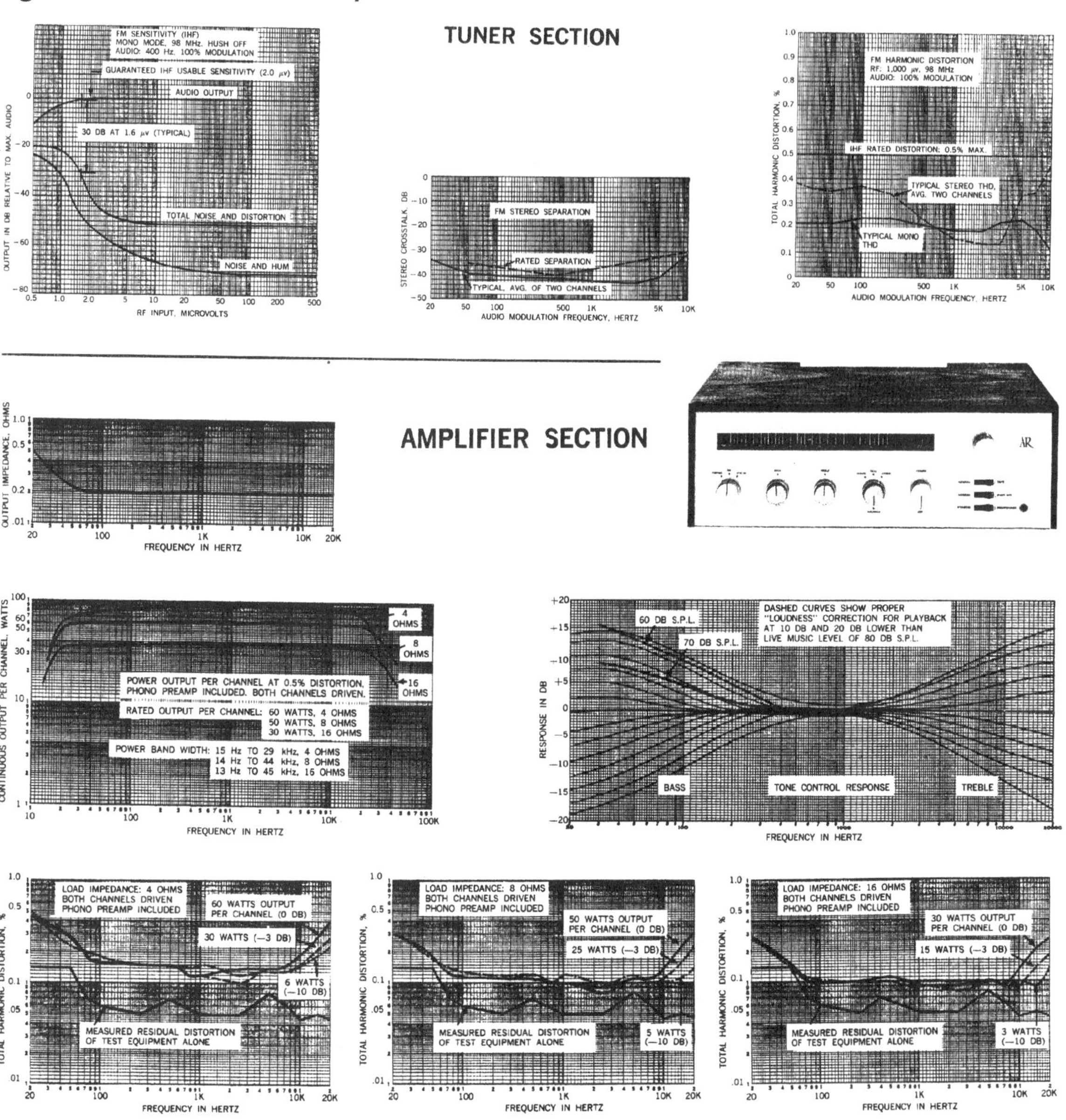

(Courtesy of Acoustic Research)

While not exactly a magazine ad, this excerpt from specification-sheet handout for the "classic" Acoustic Research receiver (produced back in the late 1960s and early '70s) gives you an idea of what a good company makes available to prospective customers who care about the operating parameters of a product. The other side of this sheet had copious data as well, including information on damping factors for specific impedance loads, distortion levels for the *whole* unit (from phono input through to the speaker outputs, rather than just the amplifier section alone), and signal-to-noise specifications measured from input to output (including the phono-preamp stages). One specification not usually seen on current product handouts, let alone advertisements, is the output-impedance curve seen to the left of the "Amplifier Section," which tells you a lot about how an amp behaves while driving a variety of speaker loads. It is almost unheard of to see this kind of candor from current mainstream manufacturers, whether Japanese, European, or American.

Parameters of Tuner Performance

Quality	Definition	Specification
Sensitivity	Ability to pick up weak or distant stations	35 to 40 dBf[1] or better for quieting to 50 dB in stereo; 12 to 15 dBf or better in mono[2]
Selectivity; divided into alternate/local (stations more than 0.4 mHz apart) or adjacent/distant (stations 0.2 mHz apart)	Ability to distinguish among stations that are situated closely together on the FM dial	Alternate-channel: 45 dB or higher; adjacent-channel: 5–15 dB; the higher, the better
Capture ratio	The signal strength needed for a tuner to suppress a weaker station while trying to pick up a stronger one *at the same frequency*; also to control for multipath distortion, which is the result of reflected signals reaching the tuner nearly simultaneously with the main signals; this can happen both in the big city (building reflections) and out in the hilly countryside (hills, of course)	1.5 to 2.5 dB; some really good tuners rate at 1 dB or lower
AM suppression	Similar to multipath distortion	The higher the number, the better, with typical figures being in the neighborhood of 45 to 60 dB
Signal-to-noise ratio (S/N)	The ratio of the volume of the signal to noise	Usually stated in relation to a given dBf input to the antenna terminals and, of course, the higher the number, the better; typical ratings might be made at 65 dBf and would run from about 70–80 dB S/N in mono and 65–75 dB in stereo; at higher dBf inputs, the readings might be several dB greater
Stereo separation	Clarity of the separation between channels	At least 20 dB of separation between channels at most frequencies; typical tuners exceed 35 dB over the range from 100 Hz to 10 kHz, with some hitting 50 dB and more (oddly enough, some stereo TV sets and VCR tuners fall short of even the 20-dB minimum; see Chapter 7)
Harmonic distortion	Additional unwanted signals at multiples of the original	From 0.1 to 0.5% at 65 dBf

[1]Decibels referred to one femtowatt of power at the antenna terminals.

[2]Because tuners with extremely high sensitivity are subject to front-end overload by powerful stations nearby, it is not a bad idea to look for a model with a local/distant switching feature.

pression of its signal can be accomplished by running the dipole parallel to a line pointing in its direction. Most supplied dipole wires are a bit too long and should be properly trimmed to have a 57-inch spread from tip to tip. A quick cut-and-solder job should produce a proper fix.

If you're having serious trouble with FM reception (assuming the tuner involved is up to snuff), buy a good outdoor antenna. Fringe-area reception can be handled well by a high-gain model, and multipath distortion can be controlled (to a large extent) by a good directional antenna that can be aimed. Changing from a dipole wire to a decent outdoor antenna will result in an improvement greater than *any* tuner upgrade could possibly equal (unless the tuner being replaced is a real dog). If you are in an area where lightning strikes are frequent, set up your antenna in the attic. If you can, try to get one that can be aimed by electrical remote control. If you live in an apartment and cannot install a fine outdoor antenna, consider getting a good powered indoor model, like a Terk, Parsec, or Recotron, and follow the manufacturer's instructions. (Beware of devices that claim to turn your house wiring into a gigantic antenna.)

If no antenna fix seems possible, consider tying into your cable TV system or getting a satellite dish. However, check with your local cable service first to see if any modifications are needed to get the best results with FM radio. Often, a cable service will not only get you pretty good sound but also get you some good, distant stations that might broadcast superior programs.

WHY A RECEIVER IS A GOOD THING

There are only a few alternatives to having a receiver as the centerpiece of your audio or audio-video system. One is the full separates route, with a stand-alone amplifier, preamplifier, surround processor, and tuner. Because they are produced in more limited numbers (and even if you are not interested in owning a tuner), combining them will almost certainly be more expensive than all but the most powerful and feature-laden THX-class receiver. In addition, a top-grade receiver will often have more flexibility than the amp-preamp combo, in that it will invariably include surround sound and the surround/center-channel amplifiers to implement the effect. Separate components will undoubtedly take up more cabinet space than even a very large receiver.

integrated amplifier

A second alternative to the receiver is the "integrated" amplifier, which is a receiver minus the tuner. It is almost a given that an integrated amp will be more expensive than a similarly powered (and probably more feature-laden) receiver. At first glance, this just does not appear to make sense. However, the higher the production rate, the lower the per-item costs (quantity more than offsetting the theoretical price differences), and receivers sell like hotcakes compared with integrated amps. It is not at all unusual to see a 100-watt-per-channel stereo receiver (possibly with a pair of 20-watt-per-channel surround outputs and a decent center-channel amp thrown in for good measure) selling for less than a stereo-only, 90-watt-per-side integrated amplifier made by the same company! From any basic-performance angle, not counting all the extra amenities, the receiver will be functionally equal to a similarly priced integrated unit.

The surround concept gives another reason to opt for a receiver over separates. Most receivers are loaded with surround options, particularly video-sound options, including built-in surround and center amps, Dolby processing (with Pro Logic center-channel steering or even Dolby AC-3 Surround decoding), and assorted matrix and hall-synthesizing modes to simulate concert-hall ambience—with a remote control to handle it all from your favorite chair. Even some fairly low-cost models include most of these features, although they may not work as effectively as those found in more costly products, particularly the audio-only surround enhancements. Even an enthusiast with a modest budget can obtain a receiver that will allow for fairly serious upgrades in the video realm down the line. Of course, separates can do this too, but the cost may be substantially higher.

Receivers may also come with numerous other built-in perks, such as equalizers, assorted "memory" tone-control circuits, video as well as the usual audio inputs (including video switching and sometimes video enhancement circuits), a remote control for operation from a distance—and, of course, the tuner. As with surround sound, these options will be found in many lower-cost units as well. Indeed, some of the "econo-special" receivers that are available have more control and input flexibility than some *very* expensive separate, high-end preamplifiers.

power

Power is also no problem with a receiver. There are a large number of low-cost models on the market that produce the 60-watts-per-channel minimum required for good, reasonably loud sound reproduction in most rooms. Indeed, there are a number of fine very low-priced models out there that offer 40 or 50 watts per side, offering nearly identical performance under all but the most demanding conditions, provided they are not called upon to drive low- or oddball-impedance speakers at high levels (some can do that too). Power-hungry individuals can buy receivers with 100+ watts per channel (including the center channel) for only a few bucks more and for only a fraction of what separates with similar capabilities cost.[3]

[3]If you feel that you *must* have 200 or more watts per channel, you will have to go with separates, but you had better know a good banker, have a reasonably well-built and soundproof listening room, be prepared to visit a hearing specialist if you use the amp to its full potential regularly, and, of course, have a loving spouse who is as nuts about your hobby as you are.

receiver disadvantages

In the long run, the only potential disadvantage to owning a receiver instead of separates involves the flexibility that results from being able to intermix and upgrade assorted components. Normally, electronic components are so equal in performance (discounting some of the advance-featured surround processors) that going to the trouble of obtaining expensive separates appears pointless. However, buying separates does have one benefit: if an item breaks, the entire system may not be rendered dysfunctional. If the outboard equalizer or tuner in your rack of separates goes on the fritz, you can still have home entertainment while it is being fixed. If the equalizer or tuner of your receiver malfunctions, you will be without the receiver in its entirety until it is repaired and will, therefore, have no sound system at all. (Of course, if your separate amplifier or preamplifier breaks, you will still be just as much without audio as you would be if the amplifier section of your receiver stopped working.) However, given the past reliability rates for most electronic components, the problem of listening/viewing interruptions due to malfunctions will probably be pretty much the same whether you go with separates or a receiver.

EXPENSIVE VS. CHEAP

There is little difference in subjective performance between most high-quality receivers and "good" (and expensive) separates, unless very high power levels are required or elaborate surround-sound or hall-ambience features are thrown into the equation. Even so, a top-quality receiver, particularly one with Dolby AC-3 Surround, may actually outperform even a fairly elaborate mix of surround-sound-oriented separates in terms of flexibility and ease of use. Separate top-grade surround processors can be bears to set up and use properly.

These days, virtually all A/V-oriented receivers have one big advantage over their older, or at least non-A/V-oriented, brethren. They will usually have Dolby Surround Sound *and* Dolby Pro Logic center-channel steering, which, provided you have a place to locate a center speaker, can enhance the movie-watching experience. Pro Logic helps to keep the center dialogue sound properly positioned, even if the viewer is sitting well off to the side.

However, for basic music listening, steering logic of any kind, although surprisingly effective with some conventional recordings, may create more problems than it solves. With most nondedicated software, it will compress the lateral width and depth of the sound stage. Unless an audio-only recording has Dolby Surround encoding (which a fair number have, I should note) or a very wide stage spread to begin with, the center-channel-steering feature

See pages 249–77 for basics of Dolby Surround

should be shut off. Fortunately, standard Dolby Surround[4] can enhance some audio-only recordings because of the effect of the surround speakers. Those who buy more up-to-date or expensive hardware should not fret, however, because every A/V receiver with Pro Logic will offer the ability to switch to "normal" stereo functioning. Receivers with Dolby Surround or Dolby Pro Logic can dramatically enhance any good hi-fi videotape or videodisc soundtrack. In addition, many models have a "matrix" mode, which can enhance the recording's large-hall ambience, even when a tape was made in conventional stereo.

However, if you are not interested in movie reproduction or surround effects of any kind (and a lot of "purists" are in this category), almost any decent receiver with power in the 70- to 100-watts-per-channel range can deliver the goods in even the largest typical living room (3,000 to 4,000 cubic feet), and models with 40 to 50 watts per side will work in more modestly sized rooms (2,000 to 2,500 cubic feet). Those lower power levels are also more than adequate for large-room listening, provided you don't require demonstration-level sound power. In terms of standard, two-channel stereo sound quality, I would not hesitate to match any medium-grade, discount-house receiver (even a Radio Shack model) against the very best separates that money can buy, provided that the receiver is not called upon to deliver sound levels in excess of its design limits and the speakers do not present an oddball load.

Usually, more expensive receivers offer greater flexibility than cheaper models from a control and input-output standpoint. However, most people, including many really diehard A/V enthusiasts, do not need these capabilities. Complex tone controls, memory circuits, extra flashing lights, multiple tape inputs and outputs, vast numbers of auxiliary audio and video inputs, and even a remote control are not required to get good *sound* from a receiver. I have seen very complex and expensive receivers installed to power modest systems, where only a fraction of their potential was tapped; their owners could easily have gotten along with much smaller, cheaper models.

tuner performance

Amplifier performance aside, expensive receivers may outperform cheaper models in terms of tuner performance. If you need good radio reception, then a more pricey receiver may be in order, although there *are* cheaper models out there that deliver fine tuner performance. Budget-minded radio buffs should closely read the reports published by Consumers Union and A/V magazines. Fortunately, in most geographic locations, a typical tuner is more than able to pick up any (reasonably powerful) station.

Patriotic-minded enthusiasts will often be disappointed if they want to "buy

[4]Available on the early Dolby processors and receivers, Dolby Surround does not have a true center channel (it uses a "phantom" center, just like regular stereo).

American." I know of no inexpensive or even medium-priced amplifiers built today in the United States; the entire market has been ceded to offshore manufacturers. It is hard to believe that the U.S. audio industry, which still produces excellent loudspeakers—and invented AC-3, DSS, and MPEG video—cannot figure out how to be competitive in the modern, electronic-hardware mass market.

SHOPPING: NEW VS. USED

Shopping for a receiver, an integrated amp, or a separate amp and preamp is easy. Although used gear is available, the fact that good, cheap, new stuff is readily available makes going out of your way to get upscale used equipment unnecessary. These days, receivers in the 60-watt-per-channel category can be had for under $200. Surround-sound versions, with 100 watts on each of the two main channels, nearly as much for the center, and 50 watts for the surround channel, are readily available for under $350. This is what receivers of similar potential (minus the surround sound, of course) cost back in the 1970s. There's been no inflation in the audio receiver market! The smart shopper will opt for new equipment over similarly priced but more feature-laden (or more, but excessively, powerful) used gear.

One reason that prices are so low is that they reflect what you will pay at a typical discount house or from a mail-order catalog. While list prices are often somewhat inflated, the amounts charged by the mass-marketing people are always substantially below list. It is not unusual to see electronic components from major Japanese manufacturers discounted 30% or more, and discontinued models, which often are in no way inferior to their newer replacements, are sometimes found on sale for half list price.

When purchasing new gear it is important to check the store's or mail-order house's return agreement; most offer at least a week's grace period. During that time, I suggest you run your new receiver (or amp and preamp if you have gone on a binge) constantly to find any bugs. Most electronic equipment fails either during the first few hours of use or after many years of operation. Some dealers will offer up to a month of return time, a good deal for new owners who don't have the time to run the component(s) for hours at a time every day for a week.[5] It is a good idea to get the return policy stated in writing on the dated bill of sale.

Mail-order houses are excellent sources for fine new equipment. While I support the local dealerships in my area, there are times when mail-order prices are just too hard to resist. If you live far from big-city retail outlets,

[5]You can, of course, leave the stuff on all night, with the volume all the way down, but make sure that no electrical storms are moving in before retiring for the night.

mail order is quite simply the *only* way to buy good electronic items. The main problems with mail order are that specific products may not always be in stock and items may be damaged in shipping or arrive with a mechanical problem. Fortunately, hardware is usually quite dependable and is generally packed well enough to limit shipping damage. Perhaps the best advice to those who contemplate a mail-order purchase is to go easy at first to see how the sale is handled. The better companies take good care of their customers because they realize the value of repeat business.

used gear

Are used amplifiers, preamplifiers, or receivers a good buy? Under most conditions reasonably new ones are, because they hold up well and defects can be fairly easily pinpointed prior to a purchase. However, I see no point in buying used "economy class" gear unless it is priced *very* low. After all, brand new, full-factory-warrantied equipment is available very inexpensively. For example, I might pay $100 for a decent used 60-watt-per-channel receiver if it appeared to be in pristine condition and if I was fairly certain that the owner did not operate it in a bad environment (wet basement, hot and humid back porch, etc.), did not own a music collection consisting primarily of heavy metal, and/or did not keep saying "what" when I spoke softly. However, a new version of the same thing can be had often for about $180, so the used one would have to be in fine shape indeed. If it were being sold by a dealer, it would have to include an in-house warranty of some kind, and the dealer would have to have the facilities to make the warranted repairs.

high-end gear

Some excellent buys can be found if you're looking for high-end gear, particularly at some of the specialty stores where used equipment is commonly traded for still more expensive items. Audio buffs usually take good care of their equipment (and they frequently do not keep their hardware very long), so it often will be as good as new. (Possibly, it will be better than new, because it will have passed safely through the break-in stage that highlights initial defects.) However, you will probably not get a factory warranty (the store may offer an in-house guarantee, which may be OK if it has a first-rate repair shop), and the equipment will probably still be more expensive than equally well performing new but more mainstream discount-store products. A good, used high-end amp-preamp combination may cost one-third of its original list price but still be considerably more expensive than a new discount-house-marketed receiver with only slightly less power (but still enough to suit your needs) and a tuner.

In spite of the risks involved, I shop for used gear myself and often marvel at the deals available at some stores. Nevertheless, it would probably not be a good idea for novice shoppers to seriously consider used gear. However, once a certain amount of expertise is developed (or the assistance of a competent fellow enthusiast is obtained), there is nothing wrong in taking advantage of the deals available in used hardware.

USE, CARE, AND LONGEVITY

Amps, preamps, integrated amplifiers, and receivers are pretty tough, just like speakers. As with CD players and computers in general, equipment failures usually occur either in the first few hours of use or after many years of use. The nature of the "electronic" parts involved is such that if they survive the first few hours of operation, they will usually last for a long time. The nature of the more "mechanical" parts (such as switches, knobs, etc.) is such that they will develop problems at a rate that is proportional to the amount of use they get: they rarely fail early in their lives and will usually last many years, even if operated quite a bit. Interestingly, these audio components do best if they are left on continuously, which will prevent any moisture-absorption-related deterioration. Unfortunately, in many parts of the country, lightning storms can make this operational procedure risky.

common problems

The two most common problems in older amplifiers are electrolytic capacitor deterioration and analog meter failures (if the unit has meters). The usual symptom of failing capacitors is increased noise. These anomalies can be subtle and may not be easy to pinpoint at first, but look for deterioration to begin happening at about 10 years of age. Mechanical meters will usually start to stick when they get really old. Nonmechanical, "LED"-type meters have no such problems, because they have no moving parts. However, by nature, they are less accurate, even when brand new. The fix for either problem is to visit your local repair facility or return the equipment to a regional manufacturer's repair center.

Another problem that may show up with power amplifiers of any age is power transistor burnout, which will usually be the result of shorting the speaker leads together while the unit is running with the volume turned up. This is a catastrophic occurrence that will result in no sound at all coming from the offending channel. Many units have built-in protection circuits to limit this kind of mishap, but I certainly would not want to put these devices to the test on purpose.

volume controls

The most common problem with older preamplifiers involves noisy volume controls and switch failures. After about seven or eight years of normal use, many knob-type controls will begin to emit "scrubbing" sounds through the speakers as they are rotated, the result of a wearing away of the contact surface of the wiper arms within the control and/or of accumulated residue. This can sometimes be handled by using well-aimed compressed air on the control or liquid electrical cleaners, although these fixes are rarely permanent. My experience has been that when a control of this kind is getting noisy, it is also wearing out and deserves to be replaced. The same goes for switches, which may begin to deliver intermittent behavior after about the same amount of time, assuming they are getting regular use. Note, however, that switches that are not used much at all may display similar effects due to internal oxidation of contact surfaces. This problem *can* often be fixed with liquid cleaners, or even by repeated back-and-forth switching to wear away the built-up oxides. (It is a good idea to regularly exercise all unused switches on the front panel to rub off internal contact oxidation.) Any Radio Shack or other electronic parts supply store should have suitable cleaners.

cleaners

Typical FM radio receivers usually hold up for a long time, but the tuning controls may begin to drift in time and need realignment. This happens mostly with older analog-dial models with a mechanical tuning knob; those with solid-state tuner circuits are very durable and often have fewer internal adjustments.

repair shops

When hunting for a good repair service, try to find a dealer that is an "authorized repair station" for your brand of equipment. While local shops may have the ability to do a good-as-new repair, particularly when it comes to fixing amplifier problems or preamp knob or switch malfunctions, a factory-authorized facility is certainly going to be more able to get parts quickly and will have had a lot more experience with your particular brand than a shop that does general repairs on a wide range of equipment. If no authorized repair shop is nearby, it would be a good idea to contact the factory. If the unit is produced offshore, contact the national repair facility listed on your warranty card. Follow their advice to the letter, because they will usually know the closest and most able repair business for your component and will see to it that any warranty work is done free of charge.

heat

Amps, preamps, and receivers may run fairly warm to the touch, but rarely get hot—tube units being the exception. If you run your hardware for several hours, the chassis should get no more than good and warm, even if fairly high volume levels are involved. Some separate power amplifiers and big integrated units can get fairly hot, but I would be wary if it was getting any hotter than a hot-water pipe while playing music at any level. That should *not* be happening. Contact the manufacturer as soon as possible if chronic high heat levels are showing up in your gear, even if it is otherwise working fine. An exception would be a normally hot-running, expensive, high-end "class-A" operating-power amp that you may have been lucky enough to purchase for a reasonable price used.

How long will a properly cared-for component last? Well, I once had an AR receiver that was twelve years old when I sold it to a friend. He ran it for several more years until it developed electrolytic capacitor problems and a noisy volume control. As a corrective, he bought replacement parts from the factory and fixed it, and it continues to operate well after more than twenty

Tube Amplifiers

Thirty-five years ago, every hi-fi amplifier was a tube model; transistors had yet to be perfected or mass-produced. Other than having higher-than-desirable output impedances—which would tend to skew the frequency response slightly with more demanding speaker loads—the better, more expensive ones might be a sonic match for any number of good mid- and low-priced A/V receivers available today.

Consequently, it is hard to imagine why anyone would be interested in using a tube amplifier, what with the heat, inherent tube (and sound) deterioration over a period of time, slow warm-ups, and obvious placement problems. However, there are a few revisionists out there who swear by them, and they justify their beliefs with references to test reports in numerous alternative-press audio journals and their own propensities to believe that things esoteric, exotic, and expensive have to be better than mass-market products. Consequently, of late there has been a "renaissance" of sorts in tube amplifiers, and even though most of them are outrageously expensive (while having performance that is measurably inferior to decent, less-expensive transistor models), they have been selling fairly well—at least to types who are "serious" about high-end audio.

However, there is absolutely no rational justification for spending money on a tube amplifier. The so-called "tube sound" that some rave about is mostly the result of the high output impedance that is a prime characteristic of the design. The frequency response tends to parallel the input impedance of whatever speaker systems are being used, and the result is usually a slightly fuller bass than what would be provided by even a perfect transistor amplifier and a slight muting of the highs. This is the smooth, liquid, and mellow tube sound that advocates rave about, and it has nothing to do with accuracy or high fidelity.

years of use. It still meets factory specs! While this is admittedly pushing the reliability envelope, there is no reason why more modern components of the same type should do any worse. Therefore, the main reasons for trading in older electronic components should be the need for additional features or cosmetic considerations, not performance deterioration.

Of course, you shouldn't feel like an ecological wastrel if you resist using your components until they literally fall apart. There is nothing wrong with trading up to new hardware because the new models look good; just be adult enough to admit why you bought new stuff. Poor sound quality, at least with electronic components, is rarely a real consideration.

4 All Ears

MANY PEOPLE PURCHASE high-quality speakers and amplifiers in the hopes of improving their home listening experience well in advance of converting their system to a total audio-video home theater. For them—as well as for died-in-the-wool audiophiles—the audio playback and recording equipment is more important than video. And even those who are setting up a true home theater will want to be able to use it to listen to their favorite recordings—whether they be LPs, CDs, or cassettes. This chapter will focus on the audio-only aspects of home theater, beginning with playback (LP, CD) and then focusing on recording (cassette, DAT, DCC, and MiniDisk) equipment.

AUDIO RECORDINGS: FROM LP TO CD

Prior to 1983, there was no such thing as truly "high" hi-fi for anyone who did not have a *lot* of money to spend. The sonic limitations of the long-play (LP) record, along with the shortcomings of the analog recording process, resulted in less-than-satisfactory sound quality. Even those who could afford super-quality playback equipment were hamstrung by the limited quantity of high-quality recordings available.

Then, in 1983 Sony and Philips jointly introduced the *Compact Disc* (CD), and the audio world was turned upside down. Unlike the LP record or the analog cassette, which use electromechanical or electromagnetic processes to duplicate real-world sounds, the CD is a digital *software program* that encodes sound as nothing more than bits and pieces, or, rather, as a series of "binary" on-and-off signals. If this computer programming is done rapidly enough and with great care during both the recording process (the

binary signals

The Compact Disc

A CD is a sandwich of print-covered lacquer, extremely thin aluminum, and hard, polycarbonate plastic. The polycarbonate layer (the same stuff some motorcycle helmets and scratch-resistant spectacles are made of) covers the playing side and is the thickest material on the disc. Contrary to popular belief, the fragile part is not this surface at all but the thin, lacquered one (the "label" side). Indeed, the coating is so fragile that some early labeling jobs actually damaged it and the aluminum underneath, because of chemical reactions between the colored ink and the coating. Manufacturers have since learned to be more careful. On most players, the disc is loaded label side up, and the laser scans from the inner circumference outward—backwards from the grooved vinyl LP.

When the beam hits the playing surface of the disc, it is about 800 micrometers wide, but the polycarbonate is designed to focus it down to about 1.7 micrometers by the time it passes through to the encoded aluminum layer underneath. It is this feature that makes it possible for a pick-up beam to "focus through" some surface blemishes, because the blemish "shadows" are reduced in size to the same extent. A typical player's error-correction circuitry can correct a maximum of 220 errors per second and can easily compensate for blemishes up to .02 inch in diameter. Blemishes larger than this may be handled by error-concealment circuitry, which patches in the missing data by duplicating material adjacent to the scratched or obstructed sections. One reason some manufacturers prefer not to produce discs with long (70+-minute) playing times is that the closer to the edge the laser must track, the more "warp" it encounters—increasing the possibility that marginal-quality players would mistrack.

The disc rotates at anywhere from 200 rpm (when tracking the outer circumference) to 600 rpm (when tracking the inside); this variable rotational speed produces a uniform linear speed of from 1.2 to 1.4 meters (3.9 to 4.9 feet) per second. A playing disc delivers over 1.4 million bits of information per second to the playback system, and a 70-minute disc will contain over 5.5 billion bits. The spiral pit track of a 70-minute disc is over 3½ miles long, and if we expanded its surface so that each bump measured 1 inch long, the resulting area would be about 2¼ miles across.

analog-to-digital conversion) and the playback process (digital-to-analog reconversion), it will be impossible to hear defects other than those existing in the analog part of the chain. At the production end, these analog factors include the studio or concert hall's acoustics, the microphones used to make the recording (their types, quantity, and placement), and the audio mixing console (although digital consoles are becoming the industry standard). At the playback end, the analog factors include the quality of the playback system, particularly the room/speaker combination.

Binary Numbers

Humans normally count by means of the "decimal" system. That is, we start at 1 and continue 2, 3, 4, 5, 6, 7, 8, 9, and then jump to 10, where "10" is really a notation for 1 plus 0. We then continue counting, 11, 12, 13, etc., but those additional numbers are actually just 10 plus 1, plus 2, plus 3, etc.

Musical computers (CD players, as well as DAT, DCC, and MiniDisc player-recorders, are really just computers) cannot make efficient use of decimal counting. They can only operate in an electrical "yes/no," or "on/off," manner. Therefore, they make use of the even more efficient "binary" system, whereby the only numerals used are 1 and 0, with the 1 designated as an *on* condition and the 0 for an *off* condition. The first-place 1 is still 1, but the number 2 in the decimal system is read as 10 (one, zero) in the binary system, the number 3 reappears as 11, 4 is 100, and 5 is 101. Nine would be read as 1001, and the number 40 would be 101000. While counting in a binary manner would be a nightmare for us 10-fingered humans (and while it also seems like a huge amount of work for even a computer to handle), it is actually easy for a rapidly calculating machine that depends upon on/off electrical activity.

The etched aluminum surface of a compact disc is read by the laser as a series of bumps and flat areas. When the light hits a bump, its reflection is scattered and weak and is read as a binary 0; when it is strongly reflected back to the receiver in the player, it is read as a 1. All those 1s and 0s add up to the program that is eventually reproduced as music.

Old-Timers' Corner: The LP

When I heard my first CD recording in 1983, I knew instantly that the vinyl record's days were numbered. However, many people still have a large collection of LPs and are still enjoying listening to the music on them, despite the audio imperfections. There is also a diehard crowd of LP enthusiasts and collectors who relish their "vinyl" as a superior recording medium. For those of you who want to maintain a connection to the LP era, here's a few tips for buying and maintaining equipment and recordings.

TURNTABLES, TONE ARMS, AND CARTRIDGES

changers

There are essentially two kinds of LP record players: changers and single-play models. No serious audio enthusiast should own a changer. When records are stacked, grit is pressure-sandwiched between them and acts as a microscopic surface-etching compound. In addition, while it is a good practice to clean the stylus prior to playing each record, it is nearly impossible to do this effectively when playing a stack of records. As a result, the stylus is often caked with record-damaging grime by the time it gets to the third or fourth disc.

LPs that have been repeatedly stacked and played on a changer (particularly if the stylus has not been properly maintained) are nearly unlistenable when heard on a good hi-fi, as countless record-owning individuals who have finally purchased decent audio gear have discovered to their horror. A changer may make playing records convenient, but the price of that luxury is an eventually ruined record collection.

tone arm

All changers and most single-play machines have a pivoting *tone arm* that traces a short arc across the disc as it tracks the record. While this arc does not parallel the radial-cutting action of the lathe that made the "mastering" disc, the small amount of "tracking distortion" that results is inconsequential with a properly mounted arm—at least compared with other LP defects. Radial-tracking models, which have the entire arm crab sideways as it tracks the disc—keeping the cartridge/stylus assembly nearly perpendicular to the groove—are still available from a few companies. However, they perform no better than well-made pivoting-arm models and usually have serious problems of their own—the most obvious being that it can be awkward to clean the stylus before playing a record. In addition, the complex nature of the tracking mechanism makes it more breakdown prone.

wow and flutter

A good turntable should maintain accurate speed, adhering to the standard 33-, 45-, and 78-rpm average rotation rates. This means it should have minimal "wow" (slow speed variations) and "flutter" (faster, more warble-like variations). The majority of the models on the market are "direct-drive" units, and those—even the cheaper ones—nearly always have adequate speed accuracy. The direct-drive turntable, unlike belt-drive versions, includes a low-rpm motor assembly that is an integral part of the turntable platter itself.

warp wow

Most of the speed-related distortion heard on recordings (particularly of piano) involves the hill and valley variations caused by warp, called "warp wow," or the more annoying irregularities caused by a misaligned center hole. A good tone arm can control warp wow to some extent because its vertical pivot is located at the plane of the record.[1] A quality arm will also be low in mass, reducing the tracking-pressure variations that occur when a warped disc is played. As it reaches the top of a warp rise, the arm will tend to continue moving upward because of its momentum, and tracking pressure will decrease; as it reaches the bottom of a warp valley, it will tend to bottom out and increase tracking pressure. If those pressure variations exceed the design parameters of the cartridge or the arm, distortion and increased record wear will result.[2]

[1]If it is located above that point, as is the case with most changers, the effect is made somewhat worse.

[2]One nice thing about most of the radial-tracking models I have seen is that their tone arms are lower in mass than most pivoted arms. Unfortunately, the low mass is usually accomplished by making the arms short, which aggravates warp wow. There is no such thing as a free lunch!

The business end of any player is the cartridge/stylus combination. Early phonograph styli were conical, but these could not reproduce both high and low frequencies well. All modern, quality cartridges have an "elliptical" stylus, which has a somewhat oval—or even chisel-shaped—contact point at the groove surface. The wide-but-thin shape allows it to handle both the bass and high frequencies with equal aplomb, particularly at the inner grooves, where cutting angles can become extreme.

cartridge

A stylus/cartridge combination is nothing more than an electrical generator. It has coils and a magnet assembly that move in some kind of combination to produce electrical impulses. Those impulses are routed to the preamplifier, which equalizes and boosts them to workable quantities for the amplifier to further boost to feed your speakers. Although moving-coil cartridges are available, the moving-magnet design is more common today, because it is cheaper and easier to maintain while offering equal performance.

SHOPPING: NEW VS. USED

I do not recommend purchasing a turntable/arm/cartridge/stylus combination if you do not own any LP recordings or if your LPs are so worn out that no amount of TLC or cleaning will restore them. A start from scratch requires concentrating on CDs. If, however, you already have a large LP record collection and have taken good care of it, then it will certainly pay to get a decent player/cartridge combination to make those recordings sound their best, even if you plan only to play each of them one time as you copy them to tape.

I advise against purchasing a used turntable/arm combination for two reasons. First, used equipment can have problems due to wear-and-tear, particularly with center-bearing and/or motor-drive-shaft wear (resulting in noise and "rumble"). The tone arm can also be misaligned or have binding problems, due to lubricant losses or mishaps and carelessness. Second, new turntables are so cheap that, unless the used one is in excellent shape and is being sold to you for next to nothing, there will be no significant savings. While a used "high-end" unit might be an interesting acquisition and a decent investment if it has been well maintained, it may still suffer from the same wear-related problems as a cheaper model.

While I can safely recommend purchasing a good medium- or economy-grade turntable, it is worthwhile to spend a few extra bucks on a top-grade cartridge/stylus combination, particularly if you want to preserve a cherished record collection. Even if you plan to transfer your collection to tape for general use and to store the records as master sources, the use of a top-notch cartridge will determine the quality of the recording job. You don't have to buy a super-exotic, super-expensive cartridge, but a $25 special is not recommended, either.

I most emphatically advise against purchasing a used cartridge/stylus combination. While a cartridge can be very rugged (particularly if it is a moving-magnet design) and can last for a long time, the stylus that plugs into it may have been damaged over its lifetime by even the most conscientious previous owner. If the stylus has been scrupulously maintained, it is still a good idea to replace it if it has more than 200 hours of use on it, even if the rest of the ensemble looks pristine. If in doubt, get a new cartridge. They are not *that* expensive.

mail order

New turntables are readily available by mail order or from local discount stores for under $200. Virtually all of them are Japanese in origin or at least in design. Given the state of typical LP recordings, they will be more than adequate to handle even the most beloved of record collections. A good cartridge, like the classic Shure V-15, Type 5, will cost about the same as the turntable but will be worth it. Thus, a typical player assembly will run about $300, or more than twice what a budget CD player costs.

MAINTAINING YOUR COLLECTION

No LP recording is going to last forever, even if it is stored properly and rarely played. It will also probably not last as long as any CD, even one that sees hard use. The LP record must be physically touched by the playback stylus, limiting not only its fidelity but also its longevity. Nevertheless, LPs can last a long time if properly handled. Here are some pointers—not necessarily in order of importance—for those of you wishing to make your collection semi-immortal.

1. If you play discs daily, change your stylus once a year. If you play discs only occasionally, change it at least once every two years, because the suspension material may stiffen up even if the stylus is rarely used.

2. Never purchase a bargain-basement replacement stylus.

cleaning

3. Clean the stylus with a good brush before each record side is played. Cut an artist's-grade small brush so that its bristles are about three-eighths of an inch long (make sure that none of the bristles are sticking together), and stroke the stylus tip from back to front (in the direction of record-surface movement) a few times. Every few plays, soak the brush *lightly* with a solution of pure alcohol (available from hardware or drug stores). Clean just the stylus tip, as you do not want to get the pliable suspension elements of the assembly wet. The alcohol will quickly evaporate.

4. Clean your records before each play with a fine bristle-pad cleaner like a Discwasher, not one of the "silicone" cloths sold in record shops. Do not give them a bath in the sink, with dishwashing detergent, unless they are

really filthy. Make sure that you do not get the center label wet. If you must give them that kind of cleaning, plan on taping them immediately after they dry, since the soaking may cause deterioration of the plastic material over a period of time.

5. If possible, clean your records *as they play* with a device like the old Watts "Dust Bug," which resembles a small tone arm but has a brush and pad on the end that cleans the disc as it rotates. There may still be copycat devices of this kind available in some stores.

tracking pressure

6. Set your arm's tracking pressure to a point in between the extreme ranges deemed acceptable by the manufacturer. (Premium-grade cartridges usually have a range running from about one gram to a gram and a half.) Tracking toward the light side of acceptable may cause problems with warped discs, as the pressure drops too low at the tops of the warps. Oddly enough, tracking a bit on the heavy side should not cause problems if the cartridge is a premium-grade model. If a disc is tracking too lightly, there will be a chattering, buzzing, or harsh sound during loud passages.

7. Handle your discs by the center-label area and the edges; never touch the grooved parts. Clean your hands before handling a disc (this includes CDs, too, by the way).

8. Store discs in their sleeves inside their jackets. If the sleeve gets worn out or lost, get a replacement. Discard the cellophane wrapper on the outside, since it may act as a moisture trap and promote mildew.

9. Avoid wire-type record-storing racks. Store your LPs as you would good books, on a large, sturdy bookshelf, steadied by book ends.

10. Keep records away from heat, sunlight, and humidity.

11. The best way to preserve any LP recording is to store it properly after carefully copying it to audio tape (preferably digital), MiniDisc, or even hi-fi video tape—each of which is capable of capturing all the nuances of any long-play recording ever made—keeping the original as a "master" for possible future recopying should the copy be lost or damaged. Remember, someday you may not be able to purchase a good replacement stylus for your top-grade cartridge, which means that your days of LP ecstasy will be over if you have not made copies.

12. If you smoke, stop: the fumes from cigarette combustion settle on discs and gum them up. They do the same thing to your lungs.

13. Never lend your LPs to anyone!

CD Technology: How Does It Work?

Now that you've thrown your LPs into the dust-bin of history, you're ready to enjoy the world of CD audio. While it's not necessary to understand the nuts-

Figure 4.1 Sampling

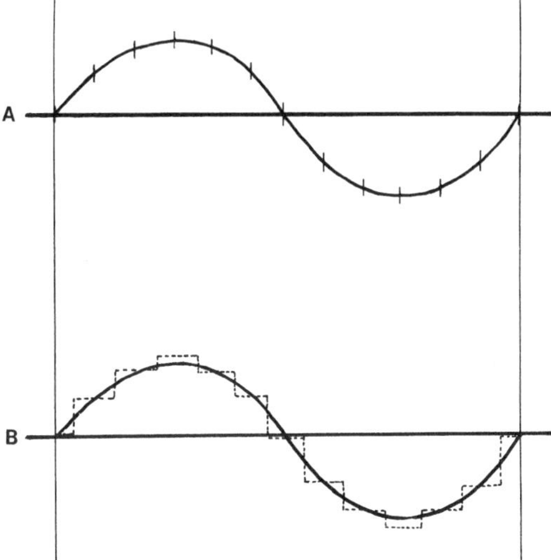

This basic, extremely simplistic illustration of sampling will give you an idea of why digital works. The top sine wave represents the kind of assorted analog sounds that exist in the audible world. The little hash marks on the wave are hypothetical "sampling" points that are encoded into the digital program on a CD after an analog-to-digital conversion. On the disc surface, the marks exist as binary information manifested as concentric bumps and flat areas and bear no resemblance to any kind of "wave" or analog signal.

The bottom stepped wave form (dotted line) is the signal that emerges from a playback unit after the digital-to-analog converters have changed it back to a signal that we can hear. The sine-wave overlay, identical to the initial wave, is what results when that stairstep wave is smoothed by the final low-pass filters. Advanced players, with digital filtering, will have a finer stairstep pattern and require less filtration to achieve a smooth wave.

and-bolts of CD technology, here's a quick overview of the technical information behind this audio revolution.

DEEP WATER

The digital encoding system used by the CD, as well as MiniDiscs (MD) and prerecorded versions of both the Digital Audio Tape (DAT) and Digital Compact Cassette (DCC) formats, samples the analog electrical signal from any input source at 44.1 kHz (that is, 44,100 times per second).[3] This may appear to be inordinately high, considering that most people cannot hear much above 15 to 20 kHz and little music goes beyond 12 to 15 kHz. However, the nature of the digital operation requires the sampling frequency to be

[3] See "Ditigal Technology" on page 135 for additional data on the sampling rates and data-reduction systems used by DAT, DCC, and the MiniDisc formats.

at least double the highest frequency that must be reproduced; in this case, that's 22.05 kHz.

PCM

Now, that is still higher than needed, but the process must use anti-aliasing devices (low-pass filters) to prevent ultra-high-frequency artifacts inherent in the digital *Pulse Code Modulation* (PCM) recording process from feeding back into the system and affecting sounds in the audible range. The elaborate and expensive analog filters used in most of the early players (the second- and third-generation units that followed hot on the heels of the initial models used somewhat better and cheaper digital filters) could not cut the high frequencies with absolute finality, so the rather high upper limit allowed the system to begin rolling off the treble frequencies at a slightly lower 20 kHz with sufficient attenuation at 22.05 kHz and above to suppress ultrasonic feedback. The digital filtering used in later-generation players allows simpler analog filters to be used in the final stages, which saves money and also cleans up the measured specs—to further satisfy equipment testers and other audio journalists.

filtering

Although (if it is properly built and adjusted) the digital recording mechanism is theoretically capable of reproducing the audible range of signals with subjective perfection, there are a number of minor glitches possible in the record-and-playback chain. CD players work by converting digital information (contained on the CD) into analog information that the rest of your stereo system—amplifiers, speakers, and so on—can understand. This is accomplished through a *Digital-to-Analog Converter* (DAC) circuit. This digital information is contained in "bits" (binary digits); the maximum number of bits that the DAC can handle is 16, and thus you will hear of 16-bit converters.

16 bits

If 16 bits is the upper limit, why do some manufacturers build players with 18-, 20-, or even 24-bit capabilities? When a CD player is not perfectly aligned, it will not take full advantage of the DAC's 16-bit potential. By using an 18- or 20-bit converter, a manufacturer can get away with less stringent tolerances in the adjustment of a player's final output-trimming circuits. The circuits "select" the *Most Significant Bit* (MSB) from the DAC to output to the system. This is particularly important at very low output levels (below 60 dB). If the player is misaligned, the sound can abruptly cut off or become distorted when the music fades into the below-pianissimo range; by adding extra bits of information, the mass-produced player has a better chance of accurately reproducing these soft sounds.

First-generation processors did not exceed 16-bit capacity, because early high-bit converters were very expensive or even unavailable. Indeed, some early Philips designs used 14-bit processors, because the company felt that 16-bit converters were unreliable. To compensate for the 14-bit DAC's limitations, the initial Philips players were designed to oversample the signal. The

Channel Separation

Typical CD players have an isolation level between channels (separation) of anywhere from 80 to 110 dB (the hypothetical limit is about 100 dB, but some advanced circuit designs can exceed this). Power amplifiers do nearly as well, sometimes also having in excess of 100 dB.

When CD players are tested by some enthusiast magazines, the reviewers often discuss extremes of channel separation as if it makes a great deal of difference. Some attribute tight imaging or an expansive sound stage to the very high separation measurements found on some super-grade (and super-expensive) models. However, for all intents and purposes, a channel separation of 30 dB is "perfect" (at least with musical program material), and it is nearly impossible to detect any benefit from separation levels in excess of even 20 dB. Look at it this way: a 10-dB difference in sound level is a power difference of 10 to 1; 20 dB adds up to 100 to 1. That is, if one channel is putting out 1 watt and the other is 10 dB louder, the latter is generating 10 watts; 20 dB louder equals 100 watts.

Even a neophyte will realize that if one channel is producing 1 watt and the other is cranking out 100, it will be impossible to hear what the weaker channel is playing, even if the listener is sitting nearly on top of the softer-playing speaker. A mere 20 dB of separation generates level differences so large that going beyond that point is academic. This is why typical phonograph cartridges, which rarely have much better than 20 to 25 dB of separation, are adequate to reproduce stereo.

While players that cannot achieve separation levels of at least 80 dB may have analog-circuit design defects or a malfunction that precipitates other audible problems, people who, for psychoacoustic reasons, place much importance on CD-player channel separation that is notably better than that are chasing shadows.

very high sampling frequencies involved (two or three times the standard 44.1-kHz rate) allowed the use of more economical digital noise-shaping filters that got near 16-bit noise performance from a 14-bit converter. Later second- and third-generation, multibit players made use of both 16- and more-than-16-bit converters *and* oversampling, to reduce low-output-level distortion even further while still reducing production costs.[4]

oversampling

An even more elegant design is the now commonly used *single-bit*, or *bit-stream*, converter. Imagine, in this simplified, admittedly flawed analogy, that

[4]A number of high-end players on the market have dual DACs. On most players, a single DAC handles both channels and operates so quickly (within a few microseconds) that the inherent timing lag on one channel is analogous to moving its speaker backward a fraction of an inch further away from the listener. Properly functioning dual converters clean up the phase shift specs a bit but have no practical audible effect whatsoever. Indeed, if the dual units are not well matched, they may create more problems than they solve.

a standard 16-bit DAC is akin to 16 lightbulbs flashing on and off in a series of sequences to simulate a digital output. Each bulb has to have the same brightness to produce a "distortion-free" series of pulses. Unfortunately, this rarely happens in a mass-produced product, which is why more expensive multibit systems have hand-adjustable trimming circuits. The single-bit system simulates a single lightbulb that can flash off and on very rapidly, equaling the potential output of all 16 bulbs in the "conventional" system. Provided the system can flash fast enough—which is possible because of recent advances in computer technology—the fact that the single bulb always has the same brightness means that any series will be very low in certain kinds of nonlinear behavior. Single-bit systems do not all use exactly the same coding techniques, but the assorted "Pulse Density," "Pulse Width," and "Pulse Edge Modulation" configurations now in use are superior to most early, PCM versions in terms of low-level distortions. This type of player now dominates the digital-audio scene, a fact that budget-minded enthusiasts should celebrate.

single-bit systems

Dither is another term commonly used in reference to CD player performance. A dithered signal has a very low-level amount of random "noise" added to the input signal during the recording process. In spite of the seeming contradiction, this process actually improves fidelity. Some digital recording systems have inherent problems when dealing with conditions of near-silence,[5] and a very low input may cause an abrupt loss of *all* very low-level signals as the system searches for barely detectable sounds. Adding a very small amount of dither noise fools the system into behaving properly at very low input levels. In some cases, the small amount of background hiss generated by the recording microphones themselves will be sufficient to simulate dither and eliminate the need for a specialized input. However, dither cannot be successfully added by a playback unit to the sound of a disc that has already been produced.

Interestingly, most of the serious problems with CD sound, at least during the early years, had mainly to do with mistakes in the recording process, rather than in the manufacture or design of the units themselves. Many inexperienced recording engineers failed to understand some of the limitations of early digital recorders, particularly how they would respond to overload.

At all levels below their overload points (an indicated "0 VU" on calibrated recording-level meters), digital recorders are far superior to their analog counterparts in terms of distortion. However, while a typical analog unit will have a smooth and linear increase in distortion as recording levels reach and then exceed saturation levels, a digital recorder will operate cleanly until it abruptly slams into its upper limit. When this happens, the recorder will generate massive amounts of distortion. Think of a race car of top quality as

[5]This problem is called "quantizing distortion."

an analog system. It will go 150 mph on a certain racetrack but becomes progressively more unstable as it approaches 150. Certain drivers can perhaps get 155 mph, or maybe even 160, on a good day. The digital system is a new design that can go 168.2 mph (maybe it runs on magnetized rails). Any and every driver, even novices, can attain 168.2. Just set the throttle to the known top speed and go. However, no one, no matter how good they are, can do 168.3. Anyone who tries crashes.[6]

Another problem with early CDs came from the source material itself. Many early CDs were reissues of LP recordings that were remastered from the second- or third-generation "cutting tapes" used to drive master cutting lathes. The problem with using these tapes (which, unfortunately, are often the only ones available) is that they were equalized to compensate for the inherent limitations of the LP medium, meaning that they often sound nothing like the "original" master tapes. In some cases, after complaints from consumers and the artists themselves, these reissues were withdrawn and replaced by superior products based either on the (by then located) original master tapes or by reengineered cutting tapes to improve sound reproduction.

digital harshness

Many of these technical teething problems have inspired the myths of digital harshness, jitter, coldness, lack of ambience, and so on, that have been promulgated by the antidigital establishment. These defects are said to be inherent in many—and maybe even most—mainstream-produced CDs. Unfortunately, all too many individuals think that these problems can only be corrected by supposedly superior, exotic (and, needless to say, expensive) playback units.

The digitalization of recorded sound has resulted in an "either/or" situation: either the player works to subjective near-perfection, or else it distorts stupendously (or does not operate at all). Defective players or discs do not misbehave as graciously as typical analog playback devices but react quite audibly to electromechanical malfunctions or pressing defects. Digital imperfections will usually manifest themselves as skipping, popping sounds, rapid repeats, no sound at all, or even a total refusal on the part of the player to operate. The flip side of this dilemma, however, is that once a machine, even a low-cost model, is functioning correctly—or a defective disc is replaced by a good one—you will have near-perfect sound from the digital part of your system that previously was unobtainable.

Finally, digital record-and-playback systems have one characteristic that virtually none of their analog counterparts have: inherently accurate speed. Whether we are talking about timing accuracy or speed variations ("flutter"), digital systems, because they temporarily hold all processed signals and then release them under the control of an internal clock mechanism, will not have

[6]This analogy is courtesy of audio writer Tom Nousaine.

any audible speed problems at all. No analog system—whether turntable, reel-to-reel tape machine, or cassette player—can match this performance.

FEATURES AND OPTIONS

While a really bare-bones player, assuming it does not crash after only a few weeks' use, can produce subjectively perfect sound and is more than good enough for most users, there are a number of features available on slightly more expensive models that might justify shopping for something a bit more upscale. The following table outlines some of the better ones.

SHOPPING: NEW VS. USED

One of the nice things about the digital-audio revolution is that anyone—even a complete "technophobe" or an enthusiast on a restricted budget—can be confident that even a low-cost CD player, if assembled with reasonable care, will perform nearly perfectly. Shopping for digital playback gear is so straightforward that even a bumbling audio or discount-store clerk will not be able to steer you into buying something that does not produce good sound, although they certainly could talk you into buying something too

Figure 4.2 **Sony CD Player**

(Photo by the author)

This discontinued, second-generation Sony player has few fancy programming features and even lacks a remote control, although it does offer shuffle play. This machine has standard, 16-bit processing and, like many budget players, has slightly audible distortion at *very* low output levels. However, these anomalies are rarely noticed with even the most demanding musical program material, even when played back through top-grade speakers. Players like this one are often available used, but be careful to check if the one you are looking over can play slightly blemished discs or those over 70 minutes in length, where even slight warping at the edge can tax a worn playback mechanism.

elaborate or expensive. The digital revolution is the best thing ever to hit the economically minded audio enthusiast and music lover.

New CD players are so cheap that it is hard to justify buying a used one at all, unless it is an "up-to-spec" budget model selling for twenty or thirty bucks or a "not-too-old" high-end version that is selling for no more than one-fourth of its original price. The new single-bit models are both cheap

Feature	Cost	Pros and Cons
Remote control	\$30–\$50 above normal price	Allows you to quickly shift from song to song or skip an unwanted selection without leaving your seat
Shuffle play	Usually nothing	Changes the order of the tracks on a CD randomly to enliven the listening experience; best used with pop-music recordings
Programmability	Minimal	Allows you to automatically bypass tracks or change the order of play; can be complicated to use (easier to make changes on the fly with the remote)
Digital output connection (standard cable or fiber optic)		Allows you to connect to an external DAC, supposedly for better sound; has no real use except perhaps for capabilities to work with future systems
Carousel changers	\$75 and up	Holds multiple CDs in a revolving table, allowing for hours of uninterrupted listening; complex mechanism more prone to break down
Cartridge changers	\$75 and up	Holds multiple CDs in a cartridge; convenient for those who have car CD players and suitable home machines and want to transport discs back and forth; awkward to load and cumbersome to use when listening to a single disc; not all systems are compatible

Figure 4.3 CD Changer

(Photo courtesy Matsushita Consumer Electronics)

CD changers now outsell single-disc players, and no wonder, since they often cost just about the same amount. This carousel changer by Technics is one of their single-bit MASH models, all of which use a technology somewhat more advanced than the multibit processing of first-, second-, and even third-generation designs. In addition to those who favor background music and want uninterrupted party sound, indolent classical-music lovers who listen to multidisc operas and long Mahleresque symphonies will appreciate a changer.

High-End Players

Many of the ultra-elite players on the market are nothing more than standard models in cosmetically different boxes, with maybe a few analog circuit changes to "improve" the sound. These modifications usually have no audible effect—unless they are negative! Some high-end companies modify the outputs of their players so that they have built-in minor frequency-response aberrations (usually in the form of a slight high-frequency rolloff) that some enthusiasts may judge to be an improvement. Unfortunately, in most cases, these so-called enhancements cannot be bypassed.[1]

To be fair, a number or *really* high-end players are custom made and feature military-grade mechanical and electrical innards and very expensive, proprietary, DAC circuits. A few high-end players are built with the transport mechanism on one chassis and the DAC circuitry on another, with both connected by costly cables. The physical separation is said to reduce interference effects between the transport and the electronics; the sonic benefit is nil.

[1] An exception involves the signal-manipulating circuits within some early Carver Corporation players, because those could at least be shut off. On some older, poorly mastered CD reissues of LP originals, the Carver circuitry, by subtly altering interchannel phase relationships and the frequency response, could make the program more listenable. Similar results could often be obtained by simply cutting back on the high frequencies by using the preamplifier's treble control.

and of high quality and surpass the first- and second-generation multibit models in terms of measured performance and, most importantly, durability. However, a pristine older-design model should be equal to any of the newer players in terms of everyday listening satisfaction.

An older model, particularly if it is more than two years of age, may be well on its way toward some kind of future malfunction that will defy any attempts to get it repaired—outside of an air-freight shipment to Tokyo. Because even the most high-tech and robustly built machine contains fragile mechanical parts that are susceptible to wear and tear, I do not suggest purchasing a used CD player unless it is the deal of a lifetime and/or is coming from an absolutely reliable source. Spend a hundred and fifty bucks and get a new one.

used CD players

If you absolutely must purchase a used model (because it is a top-grade specimen or a "classic"), make sure that it functions in every way it is supposed to. Play a variety of recordings on it to make sure that it operates well, and try out a slightly defective (scratched, fingerprinted) disc to see if the player can handle the blemishes. Play the last parts of a couple of very long

Cheap Music

While most CDs are more expensive than all but the most arcane tweak-oriented LPs or the most specialized cassettes, they are still better buys than any other playback medium—including digital tape—for a number of reasons.

First, a number of non-chain-owned stores throughout the country will have a selection of used CDs for sale. Due to wear and tear, a used LP is usually not a good buy; but a used good-as-new-sounding CD should be available for as little as half the new-disc price. Second, a few companies—Naxos and LaserLight come to mind—have released extensive catalogs of very low-cost but decently high-quality classical-music recordings by lesser-known but still competent European orchestras and individual performers. Listeners interested in building a collection of classical titles can do so at a very reasonable cost.

Third, if you live near a library that lends recordings, you can find a large selection of good music. The library that I use has over 6,000 discs, and they lend them to patrons for up to a week at a time. Again, unlike LPs that could be quickly worn out by butterfingered patrons using less-than-state-of-the-art playback equipment, CDs are not wear prone, and loaners usually perform as well as brand-new samples. Individuals living near a lending library need only purchase a player to be digital ready. Add a cheap amp and a small pair of good speakers (or even more economically, a pair of decent headphones), and big-time audio is waiting to happen.

New Labels

The CD revolution has spawned a plethora of new, small, but often fast-growing and interesting record labels.

Prior to the introduction of the digital disc in 1983, the audio software industry (and audio in general) was in an indolent slump, with a few large and stagnating recording operations dominating the industry. However, since the CD hit its stride, the number of pint-sized record companies has multiplied like bacteria. The total now tops 1,200. While many of these outfits are still owned by the industry giants, others are fully independent and produce novel, artistic, and often great recordings. Several are becoming powerhouses in their own right. Top-notch labels like Delos, Telarc, DMP, Geffen, GM, Gothic, Harmonia Mundi, Chesky, Reference Recordings, and GRP are growing and have an enthusiastic constituency. What's more, ethnic specialists like Buda Musique, ARN, Auvidis, and Chant du Monde are making it possible for us to appreciate musical styles from around the world, and European outfits like Argo, Chandos, Erato, Eurodisk, Hungaroton, Nimbus, and Teldec are producing classical pieces of very high quality. There are many small labels that focus on special musical styles like Arhoolie, Biddulph, Centaur, Discuba, Famoso, Fone, Hyperion, Private Music, Rounder, Rooster Blues, Titanic, and True Venture. In addition, companies like Blue Note and Pablo have made significant contributions to the modern and traditional jazz catalogue.

(70+ minutes) discs to make certain that the player can track clear out to the more warp-prone edges. If you have a test disc, use it to test critical aspects of the player's performance. Make sure that *every* feature of the player works, because if one function is in trouble, the unit may be well along the road to the junkyard. These same techniques are also useful for checking out that brand-new player you just took home.

Get it in writing.

If you purchase a used unit from a dealer, see what kind of warranty is offered. The dealer should offer at least some guarantee that the equipment will be working for a couple of months, which will be important should bugs not discovered at the store materialize later on at home. If you purchase it from an individual, you will probably be on your own if the device malfunctions later. Caveat emptor, in this case.

SERVICE AND STORAGE

When the CD first appeared, it was touted as a "near-indestructible" medium. Some claimed that CDs could be manhandled, scratched up, stepped on, roasted, covered with dust, and even glopped with assorted food products—and still work. However, today almost everyone, even the ever-optimistic disc manu-

CD Rot

Despite numerous scares, there are only a few documented cases of rapid CD deterioration, and they are the result of a failure to properly implement the industry standards. For instance, some time back, when demand was outstripping supply, a Philips facility in England used a wet-process-applied silver coating (similar to what is used in photographic films) instead of vacuum-evaporated aluminum to form the layer that holds the digital data. While aluminum is fairly stable and protects itself with a self-generated coating when attacked by most kinds of surface pollutants, silver does not self-protect and can tarnish. This should not happen if the plastic and lacquer coatings are properly applied, but if they were not, particularly if the disc was stored in proximity to paper or cardboard materials containing sulfur byproducts, the silver would gradually turn black, starting at the disc's outer edge and working inward. (Some of the cardboard containers that replaced the plastic jewel box were notorious sulfur-byproduct generators.) Fortunately, this wet process has been abandoned, but there may still be a few stray discs on the market. If your disc was produced by Hyperion, ASV, Unicorn, or a few other British labels and you see "Made in the UK by PDO" printed around the center hole, you may be listening to ready-to-blacken silver instead of aluminum. I suggest you contact the disc manufacturer about getting a free replacement.

facturers, have come to the conclusion that the CD is anything but bulletproof and should get the same kind of TLC required by tapes and LPs—although CDs at least do not require the same preplay cleaning that vinyl recordings demand. Of course, people with any sense have known this all along.

The good news is that while the CD is anything but immortal, there is no doubt that it will outlast any of its competition, particularly if heavy use is contemplated. The LP record, even when touched by the most pristine of diamond styli, will physically wear as it is played. A cassette tape, be it analog or digital, must contact a record or playback head and, consequently, will have its oxides slowly eroded as it is repeatedly used. The CD, because it is touched by nothing more than a weak light beam, cannot "wear" out.

rough handling

CDs can, however, be damaged by rough handling or by storage in an unfriendly environment. Also, there is just no way to determine the long-term durability of the plastics, metals, and lacquers used in their manufacture. The chances are very good that the plastics used in even the best CDs will develop enough microscopic surface imperfections over a period of, say, 15 to 25 years—even if not played at all—to become a problem for even the best playback error-correction systems. Of course, compared with LPs or cassette tapes, this is a fairly impressive life span.

With this in mind, it behooves us to exercise reasonable care in the use of our recordings and to be prepared to *not* will them to our grandchildren. Here are some rules to follow for those interested in keeping their discs in decent shape during their lifetimes.

1. Store CDs in their containers, and do not leave them lying around loose. The original plastic "jewel" boxes that are so vilified by some writers for being hard to open and damage-prone are superior to many of the new versions (especially the cardboard boxes), because they support the disc in the center and only air touches the encoded areas.

2. Avoid touching disc surfaces, top or bottom. Finger sweat even on the top of the disc may slowly react with the thin lacquer coating or printing inks and migrate through to the aluminum. Hold the disc by the center hole and/or the outer edge. Make sure that your hands are clean before handling a disc.

3. Never clean a mistracking disc unless you can actually *see* dirt on it. Never clean a whole disc when you can get away with cleaning part of it. To clean a disc, use a Q-Tip or lens-cleaning tissue to do a local-area job. Sometimes breathing on the surface will put down enough condensation to supply the needed moisture. If more is required, try distilled water. Cleaning solvents can be dangers. They may damage the disc surface if applied with too much vigor. *Never use cleaning solvents on the label side of the disc.* Photographic lens cleaner—the kind made to clean plastic as well as glass lenses—works well.[7] Whatever solvent you use, always clean a disc with "radial" strokes, center-to-edge or edge-to-center. Circular cleaning, recommended for LP recordings, is a no-no, because it may put small, radial scratches above the encoded material, overwhelming the error-correction ability of your player.

solvents

4. Discs that get scratched on the *playing* side can possibly be repaired by use of one of the scratch-removal kits on the market. However, be prepared to write off the disc should things go wrong. The better kits make use of varying grades of extremely fine sandpaper working together with polishing compounds to first sand down a bad scratch and then buff out the left-over microscratches created by the sandpaper. If a scratch is not too deep, the polishing compound may be able to buff it away by itself. In any case, under some conditions it might be easier, cheaper, and certainly less time-consuming to simply buy a replacement disc. *Do not* attempt to remove a scratch from the label side, because you will probably damage the aluminum underneath.

scratches

5. Avoid bending a disc—which may be difficult to do when removing it

[7]At one time, Kodak made a photographic film cleaner (the still-film variety, not the motion-picture mixture, which contains lubricants) which, after doing an excellent cleaning job, would evaporate almost without a trace. Unfortunately, environmental concerns have resulted in its discontinuation.

A Few Other CD Formats

As any computer enthusiast can tell you, the CD can be used for more than just musical playback. Few of these formats deal with music or home theater, but it may be important to know just what some of them do.

CD-ROM (Read-Only Memory) has clearly been a runaway success; almost every PC made today features a CD-ROM drive. This format lets you retrieve information quickly from a very large database, along with photographs, sound, and even video and film. There is little doubt that uncompressed CD-ROM, as well as newer, data-reduced DVD versions based upon MPEG digital-video technology, are riding the wave of the personal-information future.

CD-I (CD-Interactive) and CD-G (CD-Graphics) have, in addition to music material, stationary and sometimes moving graphics, as well as textual notes that you can look over as the program moves along. Special players that hook up to your TV set are needed to play these discs. I don't care to read program notes or composer biographies while listening to music. However, individuals with money to spare or children (or both) may find the concept appealing, and it does have educational uses.

The Sega CD system holds software for the Sega game system. It's for really serious game enthusiasts only.

The Photo CD system stores about 100 still images from slides or prints that can be displayed on most CD-ROM hookups. To me, the image quality of this format is vastly inferior to what even middling 35-mm slide images can reveal.

The CD-V system had approximately 5 minutes of analog video and digital audio, followed by about 20 more minutes of straight digital audio. It was a dismal failure, mostly because it required the purchase of a LaserVideo combi player, which was much more expensive than a standard CD player.

from certain types of tight-fitting storage spindles—because the stress may put microcracks in the lacquer coating that will eventually admit enough corrosive moisture to attack the aluminum substratum.

coatings

6. Never layer your discs with "protective" coatings like Liquid Wrench, Armor All, or green (or any other color) ink. Coating the *edges* with ink has been a fad for a time (some nitwits believe that it improves the sound) but may damage the edge plastic. Other lunatic-fringe fads that come and go involve such things as cryogenic freezing of discs for improving their sound. In a perverse reversal of all things reasonable, there is now even a special CD available with test tones to "demagnetize" your entire system.

7. Never use snap-on edge "stabilizers" on your discs, because they offer no sound improvement whatsoever and could conceivably overwork the player's mechanical parts and overload the error-correction circuits. (They

Dog-Bone Blues

Anyone who has purchased even a few CDs has certainly encountered the dog-bone shaped, metallic-looking, plastic-tape seals that, once the cellophane shrink wrap is removed, further secure the unopened jewel-box package from unauthorized tampering. These little (1¾-inch long) silvery pieces of often hologram-decorated tape wrap around the edge of the jewel box and make it impossible to open it without either cutting the strip or peeling it off.

Many people try to peel back the little seal from one edge with their fingernails, rapidly tearing it loose once a grip is established. However, this often leaves a sticky, silvery-colored residue behind. I have seen a number of suggestions about removing this tacky stuff, and the most common one is to use some kind of special solvent, like "Goo Gone," to remove it. This makes a mess, because some solvents will fog up the plastic of the jewel box. Here are some more practical suggestions for dog-bone tape removal.

Cut the bone in the center with a sharp knife, removing each half by pealing the cut ends back *slowly*. Most of the newer-designed seals will come off cleanly if you use a bit of finesse. (The older versions would separate and leave ample residue behind no matter how careful you were.) If a small amount of residual "stickum" remains behind, you may remove it by simply using the tape pieces themselves. Just "blot" up the residue with their still-gummy surfaces. However, if all or most of the glue remains on the box, you may need a more potent piece of sticky tape to do the residue-blotting job, such as a heavy wide plastic packing tape.

More and more companies have been using much more formidable clear or labeled plastic tape strips (roughly 1 × 5 inches) that fold over and encapsulate one of the four edges of a jewel box under the shrink wrap. These strips usually have "pull" tabs on them, but some do not. Needless to say, they usually come off in several extremely sticky pieces. My only suggestion for removing such formidable "seals of quality" is to work slowly and stay calm. Most of these strips that I have removed at least come off cleanly once you finish the often fingernail-bending lifting process. If any residue remains, you'll have to go through the goo-removal techniques to clean it up.

will really cause a sonic Donnybrook if they come loose while the disc is playing.) The same thing goes for weight-type stabilizers that fit on top of a disc in a player.

8. Store the discs vertically, because there is a possibility that gravity will gradually curve them if they are stored horizontally over a long period of time.[8] Vertical storage should also improve access. Avoid storing CDs near excessive heat or humidity, as this may damage the discs.

[8]This is unverified, I must admit, and probably the result of a rumor begun by some paranoid CD-ophile; but, better safe than sorry!

9. Finally, if a disc is playing and you want to pull it out of the player, hit "stop" first, and then wait a moment before hitting eject. If you hit "eject" without first hitting stop, the disc may still be rotating slightly when it is lifted to the eject position in the drawer. The result of such repeated "Frisbee-like" landings could be a buildup of radial scratches on the playing surface. This might eventually confuse the error-correcting circuits of some players, particularly cheaper ones. To be safe, always hit stop first; this will insure that the disc is fully immobile before it is kicked out.

Audio Recording Systems

There are basically two types of recording machines available: analog, which records music as electrical waves stored on a magnetic medium such as recording tape; and digital, which converts an analog signal into computer binary code. Although analog records (LPs) have been quickly replaced by digital CDs, the analog recording medium—particularly the ever-popular cassette—remains popular well into today's digital age, thanks to its low price, convenience, and relatively good sound.

The chart on page 134 gives a quick overview of the recording mediums we will be discussing, with their pluses and minuses.

ANALOG TECHNOLOGY

For years, the *analog cassette*, with Dolby B, C, and (later on) S noise reduction, has been the most popular format for both home recording and prerecorded playback, particularly in automobiles and portable players (Walkmans™ and boom boxes). The standard home-cassette recorder-player is reliable, compact, easy to run, and cheap to own—although there are complex and expensive models out there for "serious" users. It is also the hands-down sales champ for automotive use, which has boosted prerecorded cassette sales and promoted the sale of low-cost home decks for making tapes. Finally, the analog cassette offers decent sound quality, good enough to satisfy many listeners.

open reel

The analog *open-reel recorder* (with the large, dual tape reels sitting side by side, Mickey Mouse–ear style), predates the cassette recorder by two decades. Although it is still with us and is certainly an impressive and viable medium if properly used (particularly when combined with a Dolby or dbx noise reduction device), it has only residual use in the current market. It will continue, like the LP record, to be utilized by a few diehard enthusiasts, mostly those already owning large collections of open-reel tapes.

Format	Advantages	Disadvantages
Analog		
Cassette	Cheap, convenient, well-established, reliable, common in automobiles, decent sound (especially with Dolby C and S)	Noisy (if not equipped with Dolby C or S noise reduction); limited deep-bass reproduction; tapes can be damaged; wow and flutter can be audible
Reel-to-reel recorders	Better fidelity than cassettes, particularly with Dolby or dbx noise reduction; easily edited; flexible recording formats (2-track, 4-track, etc.)	Cumbersome tape system; speed-control (wow and flutter) problems; hard to find parts and tape
VCRs (for audio only use)	Long recording time; ideal for transferring LPs of operas or boxed sets	Often no ability to adjust input levels, limiting dynamic range; not designed for high-quality music recording
Digital		
DAT (Digital Audio Tape)	Studio-level sound quality; ideal for live recording	Expensive; no prerecorded tapes available; can only handle 2-track stereo recording
DCC (Digital Compact Cassette)	Can play back "normal" analog cassettes; cheaper than DAT	Hardware and software hard to find; cannot record on analog cassettes
MiniDisk (MD)	Portability; more durable than DAT or DCC tapes; easier search functions; very easy to edit	High data reduction results in slight loss of fidelity; not compatible with regular CD players; more expensive than DCC, although cheaper than DAT

VCR

The high-fidelity *videocassette recorder* (VCR), when used for "audio-only" work, has been a workable high-quality sound recording format for several years but has never caught on among serious users, probably because this capability has not been adequately explained by VCR makers. It can do a remarkable job on most kinds of home recording—and is certainly acceptable for copying old LPs, particularly really long boxed sets, such as operas, and is more than competent to handle radio broadcasts and most pop CD recordings. This FM-based system can produce reasonably high-quality

Figure 4.4 Rotel Analog Cassette Recorder

(Courtesy Rotel)

Units like this straightforward analog cassette recorder are universally available, inexpensive, and more than adequate for most home users. Although this particular model has only Dolby B and C noise reduction and Dolby HX for recording purposes, most users will find it sufficient for recording and/or playing back almost all types of music. Decks with Dolby S can approach digital quality in terms of dynamic-range. An analog deck's biggest limitation involves speed accuracy, because even the best Dolby-enhanced models do not have any kind of digital-electronic buffer to control running speed. Thus, with piano music in particular, speed variations such as wow and flutter may sometimes be apparent.

Lots more on the VCR in chapter 6.

results that are difficult to distinguish from the original source. However, most current, midpriced hi-fi models lack adjustable input-level controls, which compromises their signal-to-noise ratios, and even the best contemporary units may be plagued with modulation-noise artifacts when faced with demanding source material. What's more, some units will not record sound unless a video signal is also present.

PCM adaptors

As a footnote, digital *PCM adapters*, designed to interface with hi-fi video recorders (either Beta or VHS), can achieve outstanding results and *have* been used by professionals for some time. However, they have never appealed widely to the amateur market, because they are complex, cumbersome, and expensive.

DIGITAL TECHNOLOGY

Unfortunately for those craving stability, the world of audio recording continues to be in a state of flux. For some time, three additional digital formats have been battling for user loyalties. While one is somewhat expensive and professionally oriented—and will almost certainly remain so—the other two are fairly reasonable in cost and even more convenient to use than the analog cassette. In addition, they have experienced a drop in hardware costs similar

Figure 4.5 Panasonic Hi-Fi VCR

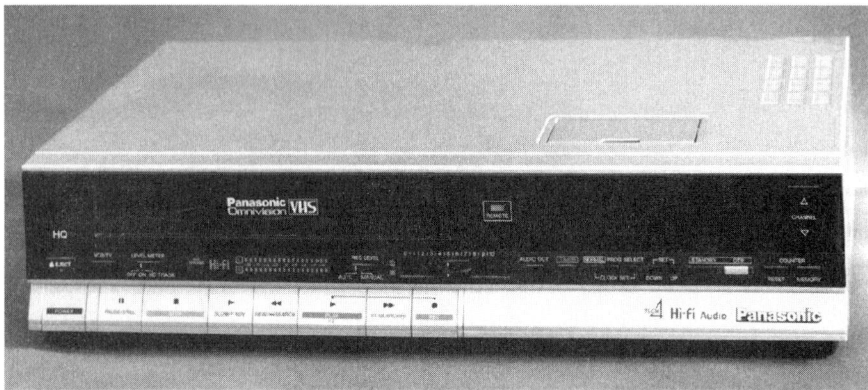

(Photo by the author)

I must admit to being partial to the hi-fi VCR as an audio-only recording device. While it cannot match any digital-audio recorder in terms of ultimate performance, this format can put up to 6 hours of sound on a tape (8 hours with a T-160), making it ideal for holding a very long piece of music, lengthy background-music compilations, or PBS money-drive telethons. It's a near-ideal device for copying long operas from either CDs or LPs. And it's a bargain when you consider that anyone buying a VCR to play and record videos (its most common use, obviously) will also be obtaining a fine audio recorder for free.

This Panasonic model is long out of production, but it has a now-uncommon feature that allows hi-fi video to achieve ideal record and playback performance: record-level controls. Most current video recorders have automatic circuits for controlling audio-input levels, compromising the signal-to-noise ratio and sometimes even dynamic range. With manual controls, the record levels can be carefully set for optimum performance, leaving the incoming audio signals uncompressed. Better sound results, particularly if the source material has a wide dynamic range. The deck also has an audio-only record feature, which allows it to tape signals from nonvideo sources, without the problem of potentially excessive wow and flutter that would result from recording without the video-synch signal that VCRs use to stabilize their recording speed. (Some VCRs cannot record an audio signal at all unless a video signal is also present.)

These days, you may have to purchase a substantially more expensive model to get these features.

to what occurred with CD players between 1982 and 1989. All three of these digital formats deliver performance well above what's possible with even the best analog cassette deck.

DAT

DAT (Digital Audio Tape) was the first of these upscale formats to appear. Actually, it has been around for some time, at least in Japan and Europe, but suffered a delayed introduction in the USA, because the American commercial-recording industry was opposed to consumer-oriented digital recorders from the beginning. They feared that "perfect" audio recorders would allow individuals, particularly tape pirates, to expertly steal material, which would deny performers (and of course record companies) their income. The Philips-designed SCMS (Serial Copy Management System)—which allows a user to make a single copy, which cannot itself be recopied—diminished those fears, and DAT was released for sale.

SCMS

Even though DAT is available, it is an expensive format and appeals mainly to professionals and well-heeled "advanced" amateurs who do live recording. The DAT system uses a rotating-drum system similar to that found in videocassette recorders to play back and record onto special cassette tapes; this system is expensive to make and requires careful upkeep. In addition, prerecorded tapes for this format are now unavailable, because there was never enough hardware sold to justify tooling up to make them. The limited number of titles that appeared initially were more expensive and less convenient to use than the sonically equal CD.

Nevertheless, if you want to go first class and particularly if you intend to do live recording, nothing tops a DAT recorder, unless more than two tracks are needed.[9] The measured performance of DAT can exceed that of the CD, since a 48-kHz sampling rate is available,[10] giving it the kind of effectiveness that performance-oriented enthusiasts dream about. A DAT tape can hold 1,300 megabytes of information, and data are retrieved at a drum-scan speed of 10 feet per second, compared with the 3.9 to 4.9 feet per second linear speed of the compact disc.

DAT works extremely well and can copy a top-grade CD with no degradation, although there is no advantage to making such a copy. It can also equal the best performance available from a VCR/PCM adapter combination.[11] However, DAT's complexity has made it difficult for manufacturers to lower the price enough to make it attractive to those on a limited budget, although seasoned professionals and serious hobby types love it.

DCC

Digital Compact Cassette (DCC), although not exactly a marketing triumph, may still be worthy of consideration by bargain-oriented digiphiles. While this format has not become a replacement for the analog-cassette tape, it still has a lot going for it, especially as a good clearance-sale buy or if the

[9]Elaborate digital recorders and mixing consoles can easily handle more than two dozen tracks, while the DAT unit is locked into the standard stereo pair—eliminating it from discrete-channel surround-sound work.

[10]The CD operates at 44.1 kHz, and a DAT deck can use this also, as well as 32 kHz, for programs with limited-bandwidth needs.

[11]Theoretically, DAT tape should be more dropout-prone than VCR-based systems, because of the very small tape size, but that has not proved to be a real-world problem.

hardware is purchased used, even though a large selection of hardware and software may not be readily available. Sales flop or not, it rates a look-over because it is so technically interesting.

DCC is backward-compatible with the analog-cassette system, meaning that while conventional analog-cassette tape recorders cannot play back a DCC-made tape, DCC recorders *can* play back analog-cassette tapes. What's more, the speed error of an analog tape played back on a DCC deck may be lower than what is typically found in analog-only cassette decks, because the digital requirements of the format mandate an advanced speed-stabilizing servo control. It would be nice if DCC decks could also *record* analog tapes for, say, automotive use. However, this option was found to be technically very difficult to achieve.[12] Analog-playback compatibility was supposed to have made DCC attractive to individuals with large analog-cassette tape collections who felt uncomfortable with another round of software obsolescence. The idea was that potential customers would be grateful that their money had not been previously blown on a collection of suddenly obsolete, analog-cassette recordings.

data reduction

DCC was also somewhat cheaper than DAT from the beginning, because it does not make use of a complicated rotating-head assembly. This system works because digital *data reduction* fits a large amount of musical information on to a tape traveling at a conventional 1⅞ inches per second past a stationary, "thin-film" head assembly (similar to those used in computer hard drives).[13]

masking

The Philips system is called *PASC* (Precision Adaptive Sub-Band Coding) and makes use of the effects of "masking" in real-world listening. Anyone who has tried to listen to a conversation at a loud party or raucous sporting event has experienced this effect: individual words or parts of words often get buried in all of the surrounding noise. The same thing happens during musical performances. Many of the sounds made by the performers are not actually heard by the audience. PASC takes advantage of masking, ignoring signals that would ordinarily not be heard (passing over as much as 75 percent of them, as a matter of fact), because those sounds are being masked by louder ones. This allows the system to handle only the audible sounds, result-

[12]The problem was where to put the analog erase head. No erase head is needed for DCC, because digital recording does not require a signal to be erased before a new one is laid down on a previously used tape. The playback heads in the analog section of a DCC deck are included within the same molded assembly as the digital recording and playback companions, with their left and right channel gaps off to the side.

[13]Note that the data-reduction process we are discussing here has occasionally been called data "compression." However, a digitally compressed signal can be completely recovered during retrieval, while a digitally reduced signal *cannot* be completely retrieved.

ing in an information reduction of up to 4 to 1.[14] This level of reduction allows a small quantity of tape, moving slowly past the fixed-head assembly, to carry enough workable, digital data to subjectively reproduce even a complex musical signal.

Whether the PASC system produces distortion that is *readily* audible is not a question that anyone other than high-strung equipment junkies ask. Philips points out that subjective experiments during the development of the format guided engineers in producing a system that stays within the masking-effect "envelope" that exists in any live or studio recording. Most listeners agreed, and it would be an astute individual indeed who could reliably hear PASC distortion. In fact, double-blind ABX tests conducted by Dave Clark and others pretty much proved that DCC was subjectively indistinguishable from full-data digital.

However, many musicians—who as a group are notoriously uninterested in audiophile preoccupations—are still suspicious of data reduction. This is because—hidden effects aside—it manipulates the ineffable. (When I told one serious musician about PASC, he said, with a suspicious look in his eye, "What do you mean by, 'It eliminates some of the music?'") For better or for worse, data reduction seems to run counter to the whole premise of high fidelity, even if it has no negative "audible" effect.

The DCC system also has advanced error-correction ability, made possible by the existence of eight simultaneously playing "data tracks" along the tape path, with a ninth track for sub-codes, bringing the total to 18 tracks for bidirectional operation. The signal is divided into 32 separate bands spread over the eight tracks and, because there are no specific signals assigned to any one band, free bits are assigned to the more active bands to assist them. Because of the way information is distributed among the tracks, large amounts of information can be obscured on individual or multiple tracks without affecting the sound quality. Conventional error-correction circuitry is also employed if the multitrack system becomes overwhelmed. Whatever you think about its marketing success, you cannot help but be impressed with the DCC concept.

Philips had hoped that DCC would replace the analog cassette and become the de facto standard for home recording and low-cost prerecorded music, as well as automotive sound reproduction. Poor sales have pretty much scotched this expectation. In part, this is because of the (admittedly limited) success of the Sony-designed *MiniDisc* system. The MiniDisc is

MD

[14]The amount of data reduction results in 384 kilobits of information per second (Kbps) being handled by the recorder at its 44.1 kHz prerecorded-playback sampling rate or 768 Kbps at its highest (48 kHz) recording rate—compared with the approximately 1.4 megabits per second (Mbps) handled by a CD player and roughly 1.5 Mbps handled by a DAT recorder operating at its 48 kHz sampling rate.

slightly larger than half the diameter of a CD and is encased in a flat, protective box, similar to a 3½-inch computer disc. A sliding shield protects the disc access-point surface, securing the disc from careless-use misfortunes that have plagued the CD from the beginning, especially in automotive situations.[15] MiniDisc portable players are much smaller than any comparable CD device, and automotive versions take up less space in a dashboard; players designed for home use are also small.

The MiniDisc (MD) has two characteristics that are of interest to both music lovers and techno-freaks. Like DCC, it makes use of data reduction to put a lot of music on a small area. The MD goes even beyond the 4-to-1 reduction of DCC and operates at 5 to 1, reducing the amount of data to be recorded by 80 percent. This is pushing the envelope pretty hard, but Sony claims it is "almost" as good as the CD.[16] For most people, that is good enough, and it is certainly adequate for most automotive and portable situations—and the MD's performance definitely surpasses that of the analog cassette.

recordable medium

The MD, unlike the CD, is a recordable medium. *Prerecorded* versions use optical technology similar to the CD, but they will not work in standard CD players because of the encapsulating case and data-reduction encoding. However, like the CD, the prerecorded MD is a no-wear item during playback and storage and will last longer than any magnetically recorded, head-contact medium (DCC, DAT, or analog tape) or mechanical one (LP). *Recordable* versions are magneto-optical and can be erased and rerecorded. They use a system that magnetically structures the disc surface (utilizing a 200°-C laser) so that it can be read by a laser pickup. It too is potentially more durable than analog- or digital-recording formats, because there is no mechanical contact between the pickup and the disc surface and no gradual erosion of the magnetically receptive oxide.

access time

The MiniDisc has other endearing qualities as well. Like the CD and other disc formats, it has faster "access time" than any tape medium.[17] This means little to those of us who simply want to listen to a program from beginning to end. Nevertheless, there are times when even the most conservative listener will want to zip forward to a selected program, and tape can be maddeningly slow under such conditions—especially when you have already experienced the speed of the laser-read formats. In addition, repeated fast

[15]The DCC cartridge also has a movable protective shield to keep wayward fingers from touching the tape.

[16]The amount of reduction results in a 44.1 kHz, 300 Kbps data rate for the MD, compared with the 384 Kbps playback rate for DCC and 1,400 Kbps for the compact disc.

[17]DAT is pretty fast but cannot match even a cheap CD player, let alone an MD unit. DCC, although quicker at program searching than the analog cassette, is somewhat slower than DAT and really slow compared with all laser-read disc formats.

searches can shorten any tape cassette's life, while they have no negative effects on a laser-read disc at all.

error correction

Another interesting aspect of the MiniDisc is its error-correction network. Like the better CD players, MD players have a well-designed stable transport mechanism with excellent error-correcting capabilities. However, MD handles the really *big* problems (such as jolts and bumps) with a powerful "buffer" circuit (initial designs utilized 4 megabytes of RAM) that delays the pickup-to-playback interval by 10 seconds. This gives the error-correcting computer plenty of time to straighten things out when vibration-induced mistracking occurs. (Some of the better portable CD players also have this feature.) Prior to the format's formal introduction, some of Sony's demonstrations actually involved removing and quickly reinserting a disc *while it was playing* with no loss of musical playback data. How's that for error correcting? Even now, the MiniDisc is a technological tour de force.

Finally, this format has the ability to record without accidentally overwriting important other material already on the disc. If you have previously recorded several selections and you wish to replace a deleted short selection with a longer one, the player-recorder will fill the newly empty space and then, if additional space is available anywhere else on the disc, put the remaining material there. During both recording and playback, the time interval caused by the laser having to skip to the new section to record or play back the material is handled by the buffer. The song moves along smoothly and continuously, even though it may be located at two or more areas of the disc. Obviously, editing is a snap with the MiniDisc.

While Sony has not experienced overwhelming success with the MiniDisc, it has done better than Philips has with DCC, notwithstanding the latter's modest sonic edge. In spite of its less aggressive use of data reduction and its appeal to serious tape users, there is little chance that DCC will survive as a workable format, even though hardware has dropped so much in price. Although the MD may still catch fire and be a viable portable, automotive, or computer-oriented medium, this is unlikely, even with the clout of Sony behind it. That company has failed in the marketplace before: witness the demise of Beta, the failure of DAT as a home-consumer and automotive format, and the El-Cassette fiasco some years back.

The analog cassette recorder is still king, in spite of the current spate of high-quality digital hardware. Most music lovers do not require ultra-high-quality performance from their home recording equipment. Indeed, except for those who copy material for automotive use or do live recording, very few enthusiasts—on a budget or otherwise—make recordings at all. Sound systems in a moving automobile will not provide concert hall (or even good home stereo) ambience, so for anyone but dedicated auto-sound buffs

Format Face-Off

All ratings typical for the type of equipment at optimal settings

	Analog cassette	*Hi-Fi VCR*	*DCC*	*MiniDisc*	*DAT*
Typical frequency response	30Hz–15kHz ± 3 dB	20Hz–20kHz ± 2 dB	20Hz–20kHz ± 1 dB	20Hz–20kHz ± 1 dB	20Hz–20kHz ± 1 dB
Typical wow and flutter	0.2%–0.05%	0.01%–0.005%	None	None	None
Signal-to-noise ratio	60–80 dBA	80–90 dBA	90+ dB	90+ dB	90+ dB
Channel separation	40–50 dB	70–80 dB	80 dB or better	80 dB or better	80 dB or better
Maximum playing time	60–120 minutes	6–8 hours	90–120 minutes	About 75 minutes	2–4 hours
Operating method	Linear analog	Linear frequency modulated	Data-reduced PCM	Data-reduced PCM	Linear PCM
Access time	Very slow	Very slow	Quick	Nearly instant	Quick
Hardware costs	Low to high	Moderate to high	Low to moderate	Moderate to high	Fairly high
Hardware durability	Good	Good	Good	Fair to good	Fair to good
Prerecorded availability	Ubiquitous	Video only	Poor	Fair	Nil
Blank software costs	Very low	Very low	Medium	Medium to high	Medium
Prerecorded software costs	Low to very low	Audio NA; video low to high	Medium	Medium to high	NA
Prerecorded software durability	Fair	Good	Fair	High	NA
User-made software durability	Fair	Good	Fair	Fair	Fair
Sound compared with CD	Poor to good	Good to very good	Near equal	Good to equal	Equal

involved with DSP ambience-synthesizing technology there's little use in replacing an inexpensive cassette machine with a top-of-the-line digital setup.

There is no reason for using an "ultra-grade" digital-recording device to duplicate your own CDs. While "borrowed" discs can be copied, the savings are slight when you consider the price of the recorder and blank tapes or MiniDiscs. (What's more, copying someone else's material for use on even your own system may, in some cases, be illegal.) Granted, it is possible to put together a dubbed collection of good material from a group of different discs, but the job is expensive, tedious, and time-consuming. This "customized" listening experience can easily be handled by a decent disc changer, either preprogrammed or monitored from the listening position with a remote control. There is also little sense in purchasing prerecorded DAT, DCC, or MiniDisc material, assuming you can find it at all, for high-quality home playback, because the CD already has that function covered for a lot less money. (Just compare CD player costs with that of any digital recording-and-playback device.)

The analog cassette *can* do a fine job of recording high-quality material for home use, depending on the caliber of the software and hardware, which means that getting really good sound can be expensive. However, a mid-priced or better analog cassette deck, if good tape is used and care is exercised, can produce sound that is subjectively "close to CD" in quality, particularly if the recorder has dbx, Dolby C or—for the very best in analog-cassette sound—Dolby S noise reduction.

SHOPPING: NEW VS. USED

New analog-type cassette tape recorders are a bargain in today's market. You can open any mail-order catalog or visit any discount-house sales outlet and find decent units for under $200. Budget-grade players will not offer state-of-the-art, "close-to-CD" performance, but for the way most of us will be using them such units are satisfactory. All current models offer Dolby-B noise reduction, and most include Dolby C. In addition, while early versions of those decks with the very high-quality Dolby-S function were nearly as expensive as some digital recorders, reasonable-cost models are now available.[18]

Virtually all automotive analog-cassette playback units use Dolby-B noise reduction only, and for two good reasons: Dolby B requires less alignment

[18]A few "new" but out-of-production analog decks may also have the discontinued dbx noise-reduction feature, which, although offering better background-noise quieting than any of the Dolby versions, never made it in the marketplace for a variety of reasons, the most important of which were a tendency to produce background noise "pumping" and frequency-response errors when certain kinds of music were recorded.

The Hiss Test

A random-noise or hiss test is an efficient way to evaluate a recording machine's performance and to discover if, at least with an analog-cassette or open-reel recorder, the most compatible tape is being used. While it is no substitute for a good, professional bench test, the random-noise test will allow a surprisingly exact comparison between the "source" and the "copy." Indeed, it is a good way to see if a professional bench test is necessary at all.

Good-quality hiss—which is "white" noise (equal energy per frequency) or, better yet, "pink" noise (equal energy per octave)—is to audio measurements what a good test pattern is to video measurements. Trying to pinpoint recording-device deficiencies while listening to regular program material can be tricky, as well as misleading. Hiss has a steady-state uniformity that enables you to do a precise comparison between the source and the copy. It will tell you whether the recorder is operating at its best within its own, self-contained, record-playback loop. The hiss test, unfortunately, will *not* tell you if your analog deck has compatibility problems with material recorded on other machines. For that, the professional bench check, using precisely calibrated alignment tapes, is required.

The noise test requires good "hiss" source material. This can be the between-station noise found on the FM dial when the sensitivity switch is set to high and the muting is off, or the more exacting white or pink noise found on some test discs. (Unfortunately, the latest, digital tuners will not tune between stations—so you may have to use test discs by default.) For a video recorder, the audio noise found at unoccupied-channel dial settings can be used. (Note that some VCRs have a mute feature that suppresses the sound when a signal is not present, again requiring you to use a test disc.)

Simply record the noise you have found for a minute or so at a typical input-setting level. If your deck has level meters, they will probably be reading quite low (below –10 or –20 VU). This is fine, because the test is mainly going to pinpoint high-frequency saturation limitations or tape-dropout flaws that do not depend on input levels approaching 0 VU over the entire audio bandwidth. After recording the noise, rewind the tape (or re-index if you're testing a MiniDisc recorder) and play it back while comparing its sound to the original source—first with one channel and then with the other. If your recorder has a "monitor" feature, you can make the comparison while the recording is taking place. If everything is operating properly, the recorded hiss should sound identical or nearly identical to the source hiss. (Remember that you are looking for "quality" differences, not "level" differences, so make every attempt to match the playback loudness between the source and the recording.)

The results of this test will vary with the format. A digital recorder should give a perfect copy of the noise, even when the record-level meters are run right up to 0 VU. An analog recorder will not be perfect and may exhibit obvious high-frequency attenuation and in some cases display subtle midrange colorations or a lack of sound uniformity, the latter due to uneven tape emulsions. An analog deck will fail miserably if the meters are set anywhere above –10 VU during the test, with

(continued)

The Hiss Test *(continued)*

severe muffling of the higher frequencies. (With analog, run the test both with and without the noise-reduction circuitry engaged, in order to help pinpoint what part of the device might be causing problems.) A hi-fi VCR might have a slightly "ripply" sound with this test, depending on the quality of the tape (particularly emulsion uniformity) and whether the unit is able to record in an "audio-only" mode. (If the VCR does not have an audio-only or "audio 2-channel" switch, it may not be adaptable for audio-only use.)

It is important to remember that the random-noise test is very demanding. Analog decks, in particular, may have a tendency to disappoint their owners. Consequently, to get an idea of what normal performance should be, it will be expedient to perform the test on more than one recorder. Check out those owned by friends (who should also be interested in the outcome) or even at dealer showrooms to gain a point of reference. After you learn to judge what normal performance is, the test will be an even better tool. It is important to remember that less than "perfect" behavior may still be adequate for most kinds of recording—particularly "dubbing" jobs for automotive use or casual listening.

This test can be used as a "quick check" procedure to tell you whether your recorder is in good enough shape to do an important bit of recording. The test will tell you whether the tape heads are dirty (which you can deal with yourself), whether the record-head alignment is out of whack (which, depending on the deck model and your technical acumen, you also may be able to correct yourself), whether the tape being used is compatible with your machine (which you can ascertain by purchasing several samples to compare), and whether the recorder itself has some kind of subtle electrical or mechanical problems (which will almost certainly require professional servicing to fix). In each of these cases, the main symptom will be a loss in the high-frequency part of the noise signal in one or both channels. If going to a different brand of tape or carefully cleaning the heads does not correct the problem, a bench check at a repair facility will be in order. Finally, the test will also tell you if any bench check you just paid for was worth what it cost.

precision than Dolby C, and even a Rolls Royce is noisy enough to mask the subtle background-level differences between Dolby B and the higher-quality Dolby C. This is why prerecorded-tape manufacturers have stuck with Dolby B, and why you still find B-type circuitry on even the latest analog-cassette recorders.

Frequently, dealers will offer brand-new, medium- or premium-grade units at stupendous discounts. A discontinued $400 (list) deck selling for $250 is a better deal than a $250 (list) unit being discounted at $180. The extra $70 is money well spent. While the more costly deck may not produce tapes that are significantly better, it will probably last longer under hard use and should require less in the way of long-term maintenance. However, given the current costs of some cut-rate-priced DCC decks as well as MD and even DAT recorders, I would be disinclined to purchase a really expensive ($500+) super-grade analog unit, no matter what its stated attributes.

heads

The number of heads is something to consider when purchasing a moderately upscale analog-cassette deck, either new or used. Budget decks will have only two: one for erasing and one for record and playback. Classier, more expensive models will have separate heads for erasing, recording, and playback. Three-head designs have a pair of advantages over two-head versions. With two heads, you can listen to the original signal coming from the

Sound on Wheels

There are many general-product A/V emporiums and specialty electronics shops that specialize in automotive-audio aftermarket equipment. Auto-sound systems are hot right now, and some companies produce big-ticket gear to satisfy the enthusiast who wants music on wheels.

Yet, I feel that the phrase "automotive hi-fi" is somewhat misleading. The interior-noise level within a moving car or van precludes any chance of securing the near-realistic sound reproduction promised in magazine ads. Automotive interiors, behaving as inside-out woofer enclosures, do allow even moderate-sized bass drivers to reproduce impressively deep tones, but there is more to hi-fi than smashing bass response. An automotive system may be impressively undistorted and loud and have a very wide, flat frequency response—but the sound will still not compare with a decent home hi-fi system in a typical, reasonably quiet living room.

The acoustic environment for auto audio—especially if the car is moving—is all wrong. The small space and the close-proximity effects of multiple wide-spaced automotive speakers make realistic and balanced imaging difficult for more than one listener at a time in anything but the very best (and most expensive) automotive systems. Automotive speakers—positioned as they typically are and located in an environment that is designed for the ergonomic comfort of the driver and passenger rather than for audio use—do a very poor job of direct-field sound reproduction.

Yet, I have heard (and, for that matter, have owned) automotive sound systems that were satisfying to listen to and worth every penny of their modest cost. In each case, they were factory installations. Those systems lacked the monumental bass and peak-volume potential of some of the aftermarket systems (I have seen 15-inch automotive-system woofers that could handle 200-watt peaks with ease, and I have heard of auto amplifiers that had that output), but they were easier to use, took up less space, looked better, and produced sound that was nearly as "realistic" sounding, given the limitations already noted, as some top aftermarket systems.

Generally, auto-audio systems are impressive because people are used to unclear, buzzy,

(continued)

source and monitor the input signal with the record-level meters—but you cannot *audibly* check to see if the recorder is properly functioning until the program is over and you rewind and listen. Three-head configurations allow you to monitor the recorded program *by ear* as it is being made, which can prevent some serious recording-level errors. More importantly, a three-head deck can produce better-sounding tapes. Optimally, the record-head "gap" should be wide, while the playback-head gap should be narrow. Dual-function heads use a compromise gap size, which serves neither role optimally. If you are serious about home tape-recorder sound and have some extra cash for a new deck or a better-quality used model, get one with three heads. (Fig-

Sound on Wheels *(continued)*

and anemic sound from typical car audio sets. You cannot help but be overwhelmed by the impact of a 10-speaker car or van stereo with 100 watts (or more) of amplifier power producing 25-Hz bass. While it is certainly possible that automotive-audio DSP processing, in conjunction with an intelligently installed array of quality speakers, could come close to recreating the ambience of a live music environment, such a system would be prohibitively expensive. A "super-grade" automotive system would only work at its full potential in a vehicle that was stopped and had its engine turned off or was (like a Rolls Royce, Mercedes 500 SEL, or top-of-the-line Lexus, Lincoln Town Car, or Cadillac) designed for very quiet operation. Then too, while moving, even luxurious chariots could not do justice to a good recording of works which, like, for example, a symphony by Shostakovich, are noted for their often spectacular dynamic contrasts.

Many luxury (and even midpriced) cars now come with optional CD players. This is certainly convenient for those people who want to listen to their music on the road and don't feel like buying cassettes (or making tapes of their collection). However, when it comes to practical sound quality, automotive CD players are not much better than any decent (and much cheaper) analog-cassette player, at least if your automotive sound system is a typical manufacturer-installed design with medium-fi speakers.

If you do get a car CD player, remember that automotive single-play units are probably more reliable than changer models for the same reasons as home versions are. Also note that the safest ones (from a theft-security perspective) are those that can be easily removed from the car by the owner or have some kind of antitheft feature like a removable front panel.

Another point on automotive CD systems: Never leave discs in a hot car. While CDs are surprisingly heat resistant, the life of the plastic coating and the lacquer will be shortened by repeated heating and cooling. (This is one reason that I don't like automotive-trunk-mounted disc changers; can you imagine any place on earth hotter than the trunk of a car—especially one painted a dark color—sitting in the sun on a summer afternoon?)

ure on paying two to three times as much for a three-head deck as for a two-head model at typical new-deck discount-price rates.)

dual cassette decks

Another feature you will have to consider is dual-cassette capability. While recorders with this feature allow you to produce two tapes at a time from the same source or copy one tape from another tape without having to use a second deck, many models are cheaply made. The gain in convenience is often more than offset by poor sound quality. First-generation tapes played on cheaper dual-well decks often sound unsatisfactory, and dubbed copies sound even worse. In the real world, people rarely copy cassettes to other cassettes (most tape dubbing involves transferring information to another format) and the

Noise Reduction

Digital recording devices do not need noise-reduction electronics. However, analog-cassette recorders can't offer hi-fi performance without it. (While fine analog open-reel decks can do a good job without noise reduction, adding an outboard processor helps them also.) Even hi-fi VCRs need noise reduction, and virtually all models (including the digital 8-bit 8-millimeter format) employ dedicated compression/expansion circuitry.

The most popular noise-reduction system is Dolby, and it comes in a variety of types. A good recording unit will offer about 50 dB of noise reduction on its own; the Dolby systems add additional "weighted" (depending on the noise's audibility) reduction. The old standby is

Dolby Noise-Reduction Systems

Type	Specification	Use: Pluses & Minuses
Dolby B	About 10 dB of subjectively "weighted" noise reduction	Oldest system; used in portable machines and in car stereos
Dolby C	About 20 dB of weighted noise reduction; does so over a broader frequency range than Dolby B	Difficult to align properly but is impressive if set up correctly
Dolby S	About 24 dB of noise reduction	Easier to align than Dolby C tapes; somewhat compatible with common Dolby B players; very impressive performance
Dolby HX	A variable-bias *record-only* process for reducing high-frequency distortion	Very effective and reliable

(continued)

dual-well feature is mostly not used. Unless the deck being eyed for purchase is an upscale model (one rich-looking but reasonably priced job I saw had Dolby S, in addition to B and C, an indication that the manufacturer was taking good performance seriously), the dual-well concept is an invitation to subpar sound.

speed accuracy

The single greatest problem with analog-cassette decks is speed accuracy. While some used units may be superior to similarly priced brand-new budget models, you must remember that they are still "used" and may very well be on their last legs when it comes to such all-important qualities as speed accuracy, head wear, belt wear, capstan wear, alignment, and tape handling.

Before spending money on a used deck, take the time to run a few quick

Noise Reduction *(continued)*

Dolby B, which is not totally effective for recording material with a lot of dynamic range but is still used for nearly all prerecorded tapes. Dolby C, similar to professional Dolby A, is a later version, and Dolby S the most-current and best of the three. *Dolby HX* is used only for recording.

A few years back, the *dbx system* was the main competition to Dolby. Although some felt that the process introduced audible pumping or "breathing" sounds during dynamic musical moments, there is no doubt that its 27 to 30 dB of noise reduction allowed the quietest home-produced analog tapes possible.

All of these systems except Dolby HX depend on complementary signal compression and expansion to reduce tape background noise. During the recording process, parts of the audible spectrum are increased in level, or "compressed" upward, during quiet passages. This raises output enough to allow low-level signals to "mask" the quiet but still-audible background noise inherent with analog tape. During the playback of a such a tape, mirror-image expansion is applied to the signal, and the quiet passages, along with the background noise, are "expanded" downward to their original levels. Thus, the musical dynamics are restored, and the subjective noise level is reduced at the same time.

While Dolby noise-reduction circuits can be quite complex and may require fairly precise adjustment to work their best, the dbx system is broad band and is much easier to align. Indeed, the *MTS broadcast video system* employs a variant of dbx noise reduction. Because the dbx design was more basic than that of Dolby, it lent itself better to installation in low-cost decks, and a number of those may still be found for sale used. However, dbx tapes are not compatible with any other format, so individuals who have a large collection of homemade ones should be prepared to replace them once their decks wear out and replacement versions or outboard-mounted processors disappear. While the dbx name is still attached to some budget-level products, the company that produced the good consumer-grade stuff has been out of business for some time.

tests. In addition to looking for reasonably good reproduction accuracy in the frequency domain, do a flutter test by copying a steady-state tone of some kind and playing it back while comparing it with the original source. Slow piano music is a good choice for an evaluation of this kind, but a compact disc with proper test tones may be even better. Head wear can be critical; you can check for it by *carefully* running your fingernail over the edge of the head, or heads, to discover any tape-wear "notches" etched into them. Any irregularities in the head surface make them unacceptable. (This procedure can also be used to check for wear on the capstan and pinch roller, although wear effects may not be so apparent.) A well-cared-for deck of recent vintage that passes the hiss, fingernail, and speed-accuracy tests, and which is being sold by a reputable dealer offering a reasonably good warranty, may be a wise purchase.

slob factor

As with other mechanically oriented audio products, if you are purchasing from an individual, take into consideration the "slob factor." If the unit looks

Analog Tape Types

Type	Required EQ Setting	Advantages	Disadvantages
Type I, "normal-bias"[1] (ferric oxide)	120 microseconds (ms)	Cheapest; default setting for most players; good for recording background music, low-dynamic-range rock, and voice	Not the best fidelity; may be inconsistent in quality
Type II, or "high-bias" (originally chromium dioxide; now better-quality ferric oxide than Type I has)	70 ms	Better high-frequency performance than standard, ferric-oxide tapes; have lower background noise	None
Type III, the original "ferrichrome" design		NOTE: Discontinued	
Type IV, or "metal" tapes	70 ms[2]	Best overall performance	Most expensive

[1] Bias is a user-selectable ultrasonic signal available from the deck, which is mixed in with the audio signal to reduce distortion and keep the response linear.

[2]Because both Type II and Type IV tapes use the same playback equalization, they can be interchanged on a deck having a "metal" tape *playback* selector switch. However, Type IV tape requires a higher "bias" setting than either of the other types when recording, which means Type IV tape cannot be *recorded* to good effect on a deck with no provision for it. Many decks have a three-position bias switch, to allow for roughly proper accommodation to the three tape types. These switches, however, rarely offer precise equalization for the brand you may be using.

(continued)

Analog Tape Types *(continued)*

There are various different types of recording tape available for analog reel-to-reel and cassette machines. These tapes are usually classified by number "types." These are arbitrary assignments and attempt to tell the consumer what level of record-and-playback quality to expect.

Not all tapes within any category are the same, because each manufacturer has formulated what it considers the best combination of materials to achieve good performance at each price level. In addition, different brands may perform differently on various decks. For general-purpose recording, particularly if automotive playback is your main goal, you can pretty much set the record-playback equalization switch on your recorder for the type of tape you have purchased (Type I or Type II), and enjoy. I suggest Type I tape for the vast bulk of your work, particularly if your car player has no manual or automatic tape-type selection feature. Remember, shopping for the best and most expensive tape will not automatically guarantee the best sound. A deck not properly set up for a given Type IV tape, for instance, might perform worse with that item than it would with a Type II tape that closely matched its alignment characteristics.

For critical work, I recommend that you purchase samples of several tape brands and types that are in your price range and then evaluate them by using the "hiss test." Once you have discovered the specific tape/equalization/bias combination that most closely duplicates the source "hiss," purchase a decently substantial quantity of that tape and stick with it for most of your work. If you are philosophically committed to a specific brand and type of tape that does not work well with your deck, you will have to have the unit fine-tuned at a service facility to match that tape (self-adjustments invite disaster).

Analog cassette tapes come, by the way, in a variety of lengths. For a long time, the three most common sizes were C-60, C-90, and C-120 (giving 1-, 1½-, and 2-hour playing times, respectively). Newer formulations include the C-100 (100 minutes). I have long considered the C-120 risky to use because it is so thin that it tends to break and jam recorders, particularly those that are a bit out of mechanical whack or cheap to begin with. The C-60 and 90 are better bets, although the more recent C-100 apparently is as durable as the C-90, because of advances in tape material.

like it has been roughly handled or if other items in the owner's abode look run-down, don't buy it. Personally, I would never purchase a used tape recorder from a private party unless the person were a friend of mine and had a reputation for being meticulous. Common sense says that "the buyer must be aware."

Shopping for digital recorders is less complicated, because few are available used (although some DCC owners may jump ship and unload their units) and prices are high enough to make cost the sole consideration for enthusiasts on a tight budget. However, I can offer two suggestions. First, stick to major name brands, because they will be more reliable, should be easier to get repaired, and will be lower in initial cost. Second, remember that even though hardware

Owner's Manuals

Reading the manuals that come with home-entertainment products, particularly those made in Japan, is about as much fun as listening to an insurance salesman—or getting a tooth crowned. Owner guidebooks are often examples of the very worst kind of obfuscation and corporate thoughtlessness.

Manual reading can be exasperating for the owner of any new recording device, because one needs to use the machine correctly to justify its purchase in the first place. This is especially true with analog recorders, because it is necessary to operate them within a rather tight performance envelope to insure decent sound quality. Many of the more highly touted machines on the market have instruction sheets that shortchange the owner in terms of maximum operational potential.

When shopping for a recorder (or any other piece of home-entertainment equipment, for that matter) look at a copy of the owner's guidebook *before* the purchase. This will allow you to see if the people who built the device were thorough enough to formulate decent operational instructions. While the recorder may be well-made and even operationally friendly in the hands of an expert who may not even need a manual, the owner's guide will often help a novice decide if the unit is worth buying in the first place.

costs may take a dive due to marketing pressures, software availability might end up being more important than recorder performance. DAT tape availability should continue to hold, because of semipro use; however, DCC could be at a real disadvantage here, and MD may end up being in a similar fix. Ultimately, the most prudent thing may be to take a wait-and-see attitude.

At one time, my choice for an all-around, *super*-quality, home-oriented format (assuming that an enthusiast was ready to spend a few bucks more than what it would have cost for a low-cost analog-cassette recorder) was DCC. This was because high-end analog decks have wow and flutter, DAT was (and is) just too software limited and expensive, and the MD has slightly inferior "measured" electrical performance. Plus, DCC was a good choice for the individual who wanted to ease into digital and yet already had a large collection of analog cassette tapes. Obviously, not everybody agreed with my views, since the format has turned out to be a marketing washout.

TAPE, DISC, AND RECORDER CARE

You can treat a MiniDisc player much as you would a CD player or any other reasonably delicate electronic component. Don't get it wet, for example, and avoid temperature extremes. This may be easier said than done, because a prime location for these items will be inside automobiles (where it may be hot) and out in the weather (portables). Prudent users will exercise common sense when using these rather costly devices under adverse conditions. A potential problem with MD recorders, one shared with every CD player, is that the laser assembly must be correctly aligned; if not, the unit will mistrack. Unfortunately, extended use, particularly in a vibration-prone environment like a car, will probably make them more prone to misalignment.

The discs themselves are more user friendly than CDs, because they are covered with a neat protective case. Prerecorded versions are the toughest, because their programs are permanently molded into them at the factory. The recordable ones are more susceptible to strong magnetic fields than analog recording tape, so avoid storing them near loudspeaker systems or electric motors. However, there is no need to be wildly paranoid about stray magnetic fields, because it will take prolonged exposure to a strong field for the disc (or any magnetic tape) to suffer any major degradation.

See page 199 for a diagram of rotating head assembly.

The DAT recorder is complex, mainly because of its use of a small, VCR-like, cylindrical, rotating record-and-playback head assembly, making it the most fragile of all digital recording devices. However, since DAT is mainly a professional or advanced-amateur device, I assume that anyone who purchases a DAT recorder will know how to care for it and will have the resources to do so.

Open reel, analog cassette, and DCC are similar in one respect: they all move tape over stationary heads. While these three types of tape recorders are as "mechanical" as a CD player, MD recorder, or even a DAT deck, I see them as more reliable, because they have less-complex construction. For instance, none has a high-speed rotating head or requires a precisely controlled laser-tracking assembly.

While caring for any tape recorder can be tedious, anyone with a reasonable amount of intelligence can keep one running quite well. Maintenance usually involves nothing more than periodically cleaning the tape path. Open-reel units are the easiest to clean, because of their larger size; however, DCC and standard analog decks are not much more difficult.

cleaning

Clean the tape path by carefully using a Q-Tip and some pure alcohol (not the oil-laced rubbing variety). Lightly dampen the cotton wadding, because too much alcohol may actually impair its dirt-absorbing qualities. Go over the heads and tape guides carefully, checking the Q-Tip as you go to see if you are picking up brown oxides. When the used "tip" looks clean, so are the deck parts; the alcohol residue on the heads will evaporate quickly. The rotating pinch rollers and capstans are more difficult to clean; it is best to actually run the deck while holding the Q-Tip against the moving part.[19] If there are oxides on those parts, the Q-Tip will take on a brownish stain. Keep

[19]This may be tricky, because your machine may have a built-in tape-path pressure relay, so be sure to carefully read the instructions in your owner's manual before cleaning these parts.

Analog Adjustments

There are three adjustments on most analog decks that can radically affect the quality of their behavior.

One, involving head azimuth, may have to be done by a service facility, but is quick and easy for any competent individual to perform who has the requisite shop manual. When a playback head is not aligned at exactly a 90° angle to the tape path, any recorded signal *not* made on that machine may sound muffled. This rarely happens with tapes made on the deck itself, because the factory alignment between the record and playback heads should be OK and will automatically be perfect with the combination record/playback head found on two-head cassette decks or with three-head decks having the record and playback gaps in the same magnetic head assembly. By using a quality test tape with lots of clean high-frequency energy (random noise can also work), you can mechanically align the playback head to produce the cleanest sound. Once the playback head is properly aligned, the record head can then be aligned by the same method, using material made on the machine *after* the playback alignment is completed. This will require a continual readjusting of the record head until material recorded by it sounds proper when played back with the now-aligned playback head. Unfortunately, if the deck had prior azimuth problems, any tapes made on it before the alignment is made will now sound muffled.

The second adjustment, involving zenith, may not be possible on some cheaper decks. It involves aligning the head or heads in such a way that their contact surfaces are exactly parallel to the moving tape surface. If zenith is off, the tape will contact the head assembly unevenly and may wear out prematurely. The channel balance may also be off, and the quality of the signal on one of the channels may be substandard.

The third adjustment, involving bias calibration, can only be done on decks with a user-operated bias control (be it the continuous kind or a multiposition switch). This is easily done by the owner of the deck, provided the right kind of signal, such as random noise, is chosen. The procedure simply involves using the blank tape of choice and recording white or pink noise while adjusting the control for the most accurate reproduction. On a three-head deck this is easy, since the control can be operated while the monitor switch toggles back and forth between the input and output. On a two-head deck the job is more tedious and requires rewinding the tape to do the A/B comparison. However, even a two-head deck can be accurately calibrated if you use a bit of common sense.

cleaning with additional clean tips until they pick up very little color. Do not use too much cleaner, because it can dribble off and get into the bearings.

There are a number of aftermarket tape-path cleaning cartridges out there, and all of the ones I have seen appear to work pretty well for light-duty cleaning. However, serious work requires the Q-Tip process. Never overuse any kind of cassette-type cleaning cartridge. Once it becomes dirty itself from repeated cleanings, it may do more harm than good. A typical open-reel or cassette deck (DCC or analog) should be cleaned after every 10 to 15 hours or so of use, provided that you use premium-grade tapes.[20]

demagnetizing

Open-reel decks need their heads "demagnetized" at 8- to 10-hour intervals. This procedure is not complex but it requires a demagnetizer, which should be available at any electronic parts store, like Radio Shack. Cassette decks should be demagnetized at 15- to 20-hour intervals; DCC decks may have their own schedule, so check your owner's manual. Cassette decks have less working room in the tape path than open-reel models, so make certain that your demagnetizer will fit.

Generally, any tape path can be demagnetized (or degaussed) by first *shutting off the recorder*, plugging in or turning on the demagnetizer while it is a foot or two from the deck, bringing it slowly to the metal parts in the tape path, and then moving it at about 1 inch per second over, around, and very close to those components. When finished, move the demagnetizer at the same slow speed to a couple of feet away from the deck and unplug it. Failing to regularly demagnetize an analog deck can increase recorded background noise, impair high-frequency performance, and increase hum.[21]

tape care

Magnetic tapes and discs require little care. Keep them (particularly digital versions) well away from strong magnetic fields, like loudspeakers or other motors, excessive cold, and heat. (While moderate cold will not hurt, going rapidly from a cold environment to a warm one may result in condensation within the tape shell, an important consideration for those of you who play a lot of automotive tapes.) Leaving a tape in a played, "tail out" condition keeps it packed more uniformly than is the case after it has been rapidly rewound. The latter puts uneven stress on a stored tape, which could cause problems over the long haul. So, after a tape is played, do not rewind it unless you plan to play it again within a few weeks. A good, premium-grade "analog" tape, by the way, should be able to handle over 100 "passes" through a deck before it begins to show serious deterioration. Please note that magnetically recorded digital tapes and discs will have less storage and playback durability than analog, because of their lower initial levels of magnetism and catastrophic-failure potential once error-correcting circuitry has been saturated. They can be rerecorded many times, however.

[20]Please do not buy supermarket-sold, budget-grade tapes to play on these machines; they may deposit oxides and generally dirty things up in the tape path at a much faster than normal rate.

[21]While video-deck care will be discussed in Chapter 6, it is a good idea to note here that VCRs should *never* be degaussed by the procedures above.

5 Television Sets and Monitors

THE LAST MAJOR COMPONENT of a home-theater system is the television set or monitor. Obviously, without some way of viewing video material—whether it be broadcast TV, cable or satellite transmissions, or videotapes or discs—home theater would not exist. This chapter on basic television sets and video monitor/receivers will be our introduction to video. We will lead off with the basics and then dig into more complex (and sometimes more expensive) areas as we move through this chapter and into the two subsequent ones. Hopefully, this information will encourage you not to be afraid of more deeply involving yourself in video.

One problem with serious video is that really good products can be pretty expensive, making it quite a challenge to obtain a really fine audio *and* video system for a bargain-basement price. However, it is possible to do well for a reasonable amount of cash, provided you shop carefully and don't expect to obtain a world-class system at near-giveaway prices. Shoestring-budget video can be more than adequate for the ordinary viewer. Even a medium-sized television set can deliver a program with substantial impact if it is supported by a good, but still surprisingly low-cost, sound system.

Unlike the owner of an audio-only system—who must spend gobs of money on software—CDs, MiniDiscs, tapes, etc.—the owner of a good A/V system already has a huge collection of material waiting at the local video rental store. While it costs money to rent videotapes, the cash outlay is considerably less than if all of your program material had to be purchased. What's more, if you spring for a decent satellite dish system, the only thing to worry about after the initial investment is the monthly costs for the assorted software packages.

The money that you save from not building up a large collection of video

Table-Model Home Theater

Unlikely as it may seem, a vintage 19- to 20-inch table-model television can function as the centerpiece of a decent audio-video home theater, even if it lacks direct A/V connections and has no stereo hi-fi capabilities. However, you will need a good audio system and a hi-fi VCR. Also, a workable home theater needs no high-grade, antenna-produced source material at all, because rental tapes—and discs when available—can provide the movie-house experience better than nearly any broadcast source short of a digital-satellite hookup, particularly when audio quality is important. A 20-inch set "views" well at distances up to about 5 to 7 feet, which makes it workable with any good, small audio system. In time, as funds become available, the start-up set can be replaced by a bigger, monitor-type model without disrupting the existing configuration. The smaller set can then be moved to a back bedroom or den to continue its standard video duties.

software can be applied to the purchase of better video—and, if necessary, audio—hardware. With the exception of surround sound, most people—who already have modest A/V systems with reasonably potent speakers—will get a greater return on their entertainment dollars if they spend their money first on top-grade video gear—particularly a larger television monitor or hi-fi VCR—rather than on more potent audio equipment.

TYPES, FEATURES, AND SIZES

In the video "stone age," there were only two technical questions to consider when shopping for a TV set: (1) "color or black-and-white?" and (2) "large or small screen?" (Large in those days—the 1960s and 1970s—was 21 to 25 inches.) Audio considerations were pretty much ignored, because the TV networks did not broadcast in stereo, and their signals were not "hi" enough in "fi" to warrant building sets with good audio performance. Black-and-white sets are pretty much gone now (except as mini-monitors inside camcorder viewfinders), and the operational principles and size categories are more varied than ever. Audio, of course, has become quite important. Current sets come in a multitude of styles, screen sizes, and capabilities.

console sets

Most video consumers recognize just two basic types of television sets: consoles and table models. Console sets are easy to identify: they look like furniture. Some are really nothing more than cheap TV sets in fancy cabinets, but a lot of them are top-grade video performers. Many offer nothing in the way of decent video sound, lacking even stereo, but an ever-growing number

Figure 5.1 Large-Screen, Direct-View Monitor

(Photo by the author)

"Almost" big-screen, direct-view television monitors like this 4:3-ratio 27-inch Sony console are quite common today and deliver performance that was only dreamed of two decades ago. Sets like this one have horizontal-resolution numbers in excess of 400 lines (some get above 600), making them ready to deal with nearly any video format a normal person can afford. Models in this category will have most of the bells and whistles, including remote control, MTS stereo decoding, multiple video inputs, dual antenna inputs, and reasonably good sound. Some have PIP (picture in picture), built-in surround-sound amplification, and even Dolby Pro Logic steering.

Sets this size view best from about 7 to 10 feet, depending on the quality of the source material. If you sit too close, scan lines and other video artifacts will be obnoxiously evident. If you wish to sit back farther and still have a theater-like experience, a 30- to 35-inch direct-view set or an even-larger projection model may be an expensive but enjoyable alternative.

Be forewarned that really large direct-view sets may have somewhat inferior picture quality than smaller sets. The larger set can have less sharpness at the edges and more geometric distortion. The main disadvantage will involve color tracking. Large sets may present good color balance with one scene and then not-so-good balance in the next one, due to the change in scene brightness. While these problems are not monumental, it is nice to know that those of us who must purchase smaller and cheaper models at least have something to feel good about.

table-model sets

are full-bore stereo models with decent sound quality including surround-sound processing. Table-model sets are more basic looking—and sometimes very cheaply made—but are frequently more "high tech" than consoles and can be just as—if not more—expensive. Table sets are usually pretty small, but some are massive and must be placed on muscular stands rather than roll-away TV carts. In most cases, this makes them just as difficult to move around as consoles. Cheaper sets (be they console or table-model) can be integrated into a good audio system and will perform admirably, provided that at least a high-fidelity VCR is included in your setup.

Figure 5.2 Simple TV-VCR Hookup

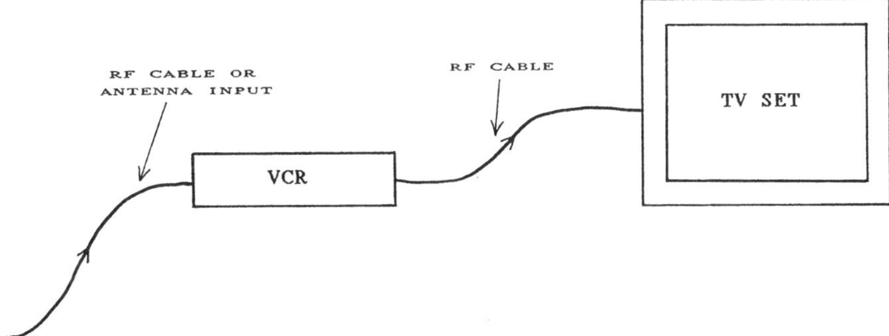

Most people hook up a TV set and a video recorder this way. The antenna or cable input attaches to the VCR, and the latter is then linked, via a (usually short) additional length of antenna-type RF (radio-frequency) cable fitted with "F" connectors, to the antenna input of the television set. By pushing a button on the front of the recorder (usually labeled "TV/VCR"), the viewer can select to watch the antenna or cable input, or a prerecorded video. In TV mode, the television operates normally with its tuner picking up whatever channel is selected. To watch a videotape, the button is set to "VCR" mode, and the TV tuner is adjusted to channel 3 or 4 to receive the taped program by means of a "retransmitted" RF signal from the VCR. A signal splitter within the VCR allows the viewer to tape one show while watching either it or another channel.

In either TV or VCR mode, the sound comes through the TV speaker(s) and, in the case of a taped program (even a superior, prerecorded one), is neither hi-fi nor stereo. Even if the VCR is stereo hi-fi model and the set is equipped with stereo speakers and a stereo decoder, it would be impossible for a hi-fi stereo *taped* program to be enjoyed via the channel 3 or 4 television inputs. No VCR has stereo "retransmitting" circuitry; the signal sent to the RF output of the VCR is derived from the lo-fi, analog-audio heads and not the hi-fi heads, if the unit is so equipped.

While this particular set-up is fine for watching talk shows, the news, old reruns, and some sitcoms, it is not appropriate for watching anything with impact, particularly if the program includes theater-like sound effects or well-recorded music.

TV Sets vs. Monitors

basic TV sets

monitor-type sets

While neophyte video fans categorize sets as either consoles or table models, ardent and knowledgeable video enthusiasts see them either as basic "TV sets" or as "monitors." Basic TV sets receive programs over an antenna or cable via a self-contained video radio-frequency (RF) "receiver." Any additional video input signals (like those from a VCR, camcorder, or videodisc player) must also be routed through those same antenna or cable inputs. While this is the most straightforward way to hook up these components, it does not result in the best video (or audio) performance. Monitor-type sets have (in addition to the antenna and cable RF inputs found on basic sets) separate user-selectable "video" and "audio" inputs that increase cost and complexity but improve potential picture and sound quality. The other internal components of typical monitors, particularly the picture tube, are also normally of better-than-average quality, and the sets may also employ digital (three-line) comb filters, better high-voltage regulation, and special video noise reduction circuitry. These extra features allow monitors to take advantage of the cleaner, direct A/V signals available.[1]

A monitor's switchable, auxiliary audio-video inputs (numbering anywhere from one to even four, with one on the front panel for camcorder use) allow the user to route VCR, camcorder, and disc player signals directly to the monitor's signal-processing innards, bypassing the RF "*de*coding" circuitry of the set and also eliminating the signal-degrading antenna RF "*en*coding" electronics circuitry of the VCR, camcorder, and/or disc player. Indeed, a hi-fi VCR or videodisc player cannot deliver a stereophonic signal to a monitor without the use of direct-audio inputs.[2] Most monitors also have video and audio outputs, allowing them to route signals to other components, such as audio receivers. Stereo-audio outputs will be particularly important if you connect the TV monitor to a hi-fi system.

Direct View vs. Rear Projection

For the greatest audio-video impact, it is necessary to have a fairly large TV picture. There are two types of large sets available. *Direct-view sets*, which have

[1]A true television monitor, like those found at TV stations, is designed to "monitor" direct-feed signals as they are broadcast and has no RF inputs at all. So the consumer product we are describing is actually a *monitor/receiver*, although most people—including your author—often use the terms monitor and monitor/receiver interchangeably.

[2]An MTS-equipped set can decode a stereo signal from the antenna or cable source, and a VCR can pass that signal through to the set unmodified in its "TV" or "bypass" mode. However, VCRs and disc players do not have internal MTS "encoders," which means that all VCR tape and videodisc sound sent through the RF cable to the set will be in "lo-fi" mono, even if the source material was recorded in high-quality stereo.

Figure 5.3 Budget TV/Hi-Fi VCR/Hi-Fi Audio Installation

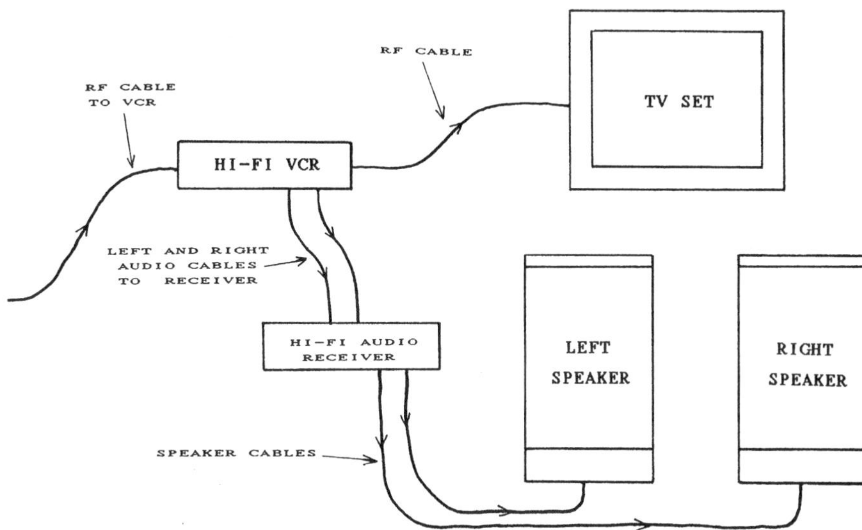

Here's how to dovetail a standard TV set into a good audio system for low-cost but decent-quality home theater. Building upon the standard videotape setup (Figure 5.2), we substitute a hi-fi VCR for the standard one. (See Chapter 6 for a discussion of hi-fi VCR performance compared with regular mono or linear-stereo versions.) While an RF cable still carries signals to the TV to supply a picture, the special stereo audio outputs of the hi-fi VCR are routed (via standard left- and right-channel audio interconnect cables) to a hi-fi audio receiver's "aux," "video," or "tape" inputs. When watching an antenna- or cable-received show with high-impact sound, the VCR's tuner is used, the TV tuned to channel 3 or 4, and the audio played through the hi-fi. Ditto when watching smash-bang adventure hi-fi videotapes. If the hi-fi system's speakers are too widely spaced for the small screen, the TV set's speaker(s) can be turned up a tad to provide a "centering effect" for the dialogue. If a program does not demand hi-fi sound, the receiver can be left off and the TV-VCR combination run as in Figure 5.2.

a single picture tube (whose large, reasonably flat glass face is the screen surface we look at), are most common in a 25- to 27-inch diagonal, 4:3-ratio size. However, it is not uncommon to find direct-view sets with measurements as large as 30, 35, and even 40 inches, with some wide-screen versions available at 34 inches. Large-screen, direct-view models can cost plenty, but are worth it if you want a more home-theater-like experience. A number of 30- and 31-inch models are available for less than $900 (discounted), but the 35-inch models are pretty expensive, with many selling for upwards of $2,000.[3]

[3]The tubes in the 34-inch wide-screen or 35-inch or larger 4:3-ratio models are expensive to build, and the positioning of a direct-view set that massive is so critical that a rear-cabinet "earth's-magnetic-field-compensation" adjustment control may be included.

rear projection

An alternative to the large direct-view set is the rear-projection monitor. Standard, 4:3-ratio versions for home use come in diagonal sizes of 40, 45, and 50 or more inches, continuing on up past 70 inches in varying increments, with wide-screen versions running from about 40 to 60 inches. The big advantage of a rear-projection set is that the cost of a modestly sized one is lower than that of most of the large direct-view models. Projection sets are popular in bars, but they also work well in a home installation when the viewing angle is not too extreme and they are combined with a large audio system. Their pictures are more than bright enough, and they can subjectively hold their own against all but the very best direct-view monitors.

There are some disadvantages to owning a rear-projection set. Even the more diminutive models are still pretty big and may not fit in with the decor of a living room. You either learn to live with the set's intrusive size, build it into a wall recess (needless to say, this requires a major room modification), or opt for a front-projection set that works with a pull-down or portable screen along the lines of the home-movie set-ups of old. Unfortunately, front projectors cost a lot. Also, even though the initial outlay for a rear-projection set may be lower than for a direct-view monitor, the long-term costs of owning a projection-type set can sneak up on you. The projection-tube "guns" do not hold up as well as standard direct-view-set picture tubes, because of their high operating temperatures. (Most are liquid-cooled, but even those run hot.) The tubes are expensive to replace, and every set has three of them. Usually, when one gives out because of long-term wear and tear, the other two may have to be replaced to properly balance the color.

front projection

My suggestion for the prospective projection-set owner is to also have at least one decent direct-view set on hand with which to watch regular "TV." The projection set would then be reserved for special occasions, like major films on tape or disc and maybe broadcast extravaganzas like the Superbowl, the Indy 500, or the Olympics. It is obvious that owning a set of this kind can be financially daunting. Still, it is worth it if you are a real movie nut and can afford the outlay without giving up food or sacrificing your medical budget.

Audio Capabilities

Most standard TV sets offer substandard audio performance because of their poor internal speaker systems and small amplifiers. However, monitors usually have fairly decent audio-electronic componentry, and most now have dbx Corporation–certified *MTS decoders* (for Multichannel Television Sound) that allow them to receive stereo signals over the air or via cable. Quite a large number of stations, even those in many so-called rural areas (thanks to cable and satellite TV), broadcast in stereo. All the major networks now offer stereo

MTS

Figure 5.4 Toshiba Rear-Projection Monitor

(Photo courtesy of Toshiba)

A big rear-projection set like this 16:9-ratio, wide-screen model by Toshiba makes use of three picture tubes or focused electron "guns" within the cabinet simultaneously converging their overlaid images to the back of the big, translucent screen. Each gun handles a different color (red, green, or blue), and the composite image has all the colors needed to reproduce a full-color program. (A front-panel control allows the owner to easily adjust the alignment of the triple overlay.)

While early rear-projection sets offered poor picture quality, current models subjectively rival the best direct-view sets. A 45-inch 4:3-ratio monitor views best from about 10 to 11 feet, and with conventional program material the pictured 56-inch wide-screen model will perform similarly. If viewed from any closer than this, the "scan lines" that make up the image may be visible, as will videotape artifacts. However, some individuals feel quite comfortable viewing from even closer distances.

All wide-screen sets offer image-enlargement functions to fill out the screen when wide-screen programs are presented, and a variant of this process—lateral expansion—is very effective with anamorphically squeezed Digital Video Discs or DSS satellite presentations. However, with radically "letterboxed" films on conventional LV discs or prerecorded tapes that are strongly expanded to fill out a set's screen, the scan lines may be evident even at reasonable distances. In such cases, either the viewing distance will have to be adjusted, or you will have to live with the imperfections.

This model is one of the largest currently available, with a picture that is slightly taller than that of a 45-inch 4:3-ratio set—but over a foot wider. While physically quite large, models like this one will often have a surprisingly shallow depth; some enthusiasts capitalize on this by custom-integrating them into existing rooms via a false wall built flush with the cabinet front. Although equipment like this is not cheap, slightly smaller versions, particularly in the standard 4:3 ratio, are not unreasonably expensive. This type of set is mandatory for serious video enthusiasts who want maximum horizontal and vertical resolution, particularly if they enjoy viewing wide-screen digital videodisc programs (see Chapter 7 for more on DVD).

programming, as do the pay movie channels, like HBO, Cinemax, and Showtime.[4] The owner of an older or budget-grade television set can still have quality stereo audio by obtaining an MTS-equipped VCR, because the latter can operate as the TV-tuner section for the system. (If a VCR is operated as a tuner only, with the tape transport not running, its wear-and-tear level is negligible.)

TECHNICAL CONSIDERATIONS

Analyzing performance claims made by the various TV manufacturers can require some savvy on the part of the consumer. However, you don't need an advanced degree in engineering; some basic understanding of the workings of video componentry along with common sense should help guide you to making an informed—and economically sound—decision.

Horizontal Resolution

Consumers can be fooled by advertisements for upscale TV monitors (be they for direct-view or projection designs, wide-screen or 4:3-ratio) that list awesome *horizontal resolution* specifications in the neighborhood of 500, 600, or even up to 900 lines. However, it's important to remember that the quality of your picture will be limited by the input source: video player, cable or satellite system, etc. The limits of these formats are set by their video frequency-response characteristics. Once your set surpasses the capabilities of your input source, its additional horizontal resolution potential is gilding the lily. The following table outlines these limitations. Note that prerecorded material often delivers 10 to 20 percent *less* horizontal resolution than the theoretical maximum.

contrast

Horizontal-resolution measurements are not a hard-and-fast specification in any TV set. A set's resolution potential is a mixture of its actual resolving power and its ability to differentiate contrast. Testers generally use a black-and-white test pattern, which features a black line pattern printed against a white background, to measure this capability. However, individual testers might "read" this test differently. For example, one monitor may be able to resolve 500 lines at very high contrast; the lines would appear dark black against a stark white background. Another set might resolve the same number of lines, but the image would lack contrast, showing alternating fuzzy

[4]Small-dish satellite video broadcasts are in high-quality digital stereo, and the sound can be routed directly to a set's audio inputs or to an audio receiver just like that of a hi-fi VCR or videodisc player.

Stereo TV

For a while, two different stereo TV systems were available in home TV receivers. In addition to a stereophonic decoder, the *MTS system* employs a dbx Corporation–licensed expansion circuit that complements the compression encoding of any MTS-equipped television-station transmitter. The MTS system delivers the 15 to 20 dB of separation between channels needed for good stereo reproduction.[1]

A now-defunct design produced by Thomson was not really stereo at all and was used in some budget-level sets manufactured by (or for) GE, Sharp, Philips, and RCA that may still be available used. This system, called "XS Stereo Sound," used out-of-phase audio information to "fake" stereo performance. The effect was similar to the "synthesized" stereo found on some reprocessed old-technology monophonic audio recordings. XS Stereo Sound created an extra-wide audio sound-field illusion from the rather narrowly placed speakers on the front or sides of the set. The XS system could also alter the sound of a taped signal sent to the set via a VCR's RF-antenna feed, imparting a stereo-like effect to the monophonic signal. While the effect was pleasant under some circumstances, at least if the set was operated by itself and not hooked into a larger audio system, it was not true stereo.

If you want the terrestrial antenna or cable-transmitted programs you watch to be true stereo, make certain any "stereo" set you purchase (new or used) is an MTS model. (A true MTS-stereo set, by the way, will have an SAP—second audio program—feature, which cannot be managed by the narrow-channel-separation XS system.)

[1]Note that not all receivers in either stereo TV sets or stereo VCRs are properly aligned to achieve this level of performance. What's more, for good Dolby Surround Sound performance, channel separation should be in excess of 25 dB, which only a well-aligned MTS decoder can achieve.

Input Source	Maximum Resolution (Horizontal Lines)
Videotape players	240
Super-Beta	260
Broadcast TV (standard cable or antenna)	330
Analog laser video (LV) players	400 plus with premium players
Super VHS (S-VHS) or digital VHS (D-VHS)	400 plus with top models
DVD	450 plus

Video Splitters

Need to attach additional video components to your cable or antenna? Get a splitter (available at Radio Shack and other outlets) and follow the instructions that come with it. Most splitters are three-way versions, although four- and five-way designs are also available for more ambitious networking jobs. Devices of this kind must be properly hooked up, with one leg specifically designated as the "in" hookup and the others configured as "out" connections.

However, before going hog-wild with a splitter, remember that each additional active video component (VCR or TV) attenuates a cable feed by 3 dB. Therefore, using multiple three-way connectors anywhere within your video system (or even a single four- or five-way model) to link together myriad pieces of video hardware may result in too much signal attenuation, particularly if the original signal was weak to begin with. The consequence will be poor reception by some or even all of the involved hardware. Remember, a VCR siphons off as much signal strength as a TV set.

When hooking up a splitter, make sure all the cable "F" connectors are tightened snugly (but not excessively). A loose connection with just one component may affect picture quality in all of the other video hardware within the network, even items that are attached properly. In addition, if a splitter has an unused connection, you should secure it with a terminator cap to insure a balanced load. This will eliminate multiple images (ghosts) and other possible artifacts. Terminator caps should be available at places like Radio Shack.

lines of dark and light gray. How different manufacturers and equipment testers rate the amount of contrast between those test-chart lines as they get narrower and closer together will be a determining factor in how much resolution they claim for a given TV set. Fortunately, most modern middle- and upper-level sets achieve all the high-contrast resolution needed to reproduce today's video source material, and even economically priced sets should be able to do justice to conventional rental-grade videotapes.

Vertical Resolution

Rarely is the *vertical resolution* of a TV set (or a VCR or LV player either, for that matter) mentioned in product ads, because—for all conventional sets (that is, those that are not HDTV models), wide-screen or otherwise—it was set at 330 lines by the American National Television System Committee (NTSC) many years ago. The NTSC vertical resolution figure is identical to the horizontal resolution limit for broadcast signals, which are set by law.

In sets using picture tubes, the amount of vertical resolution is deter-

Figure 5.5 Using a Splitter

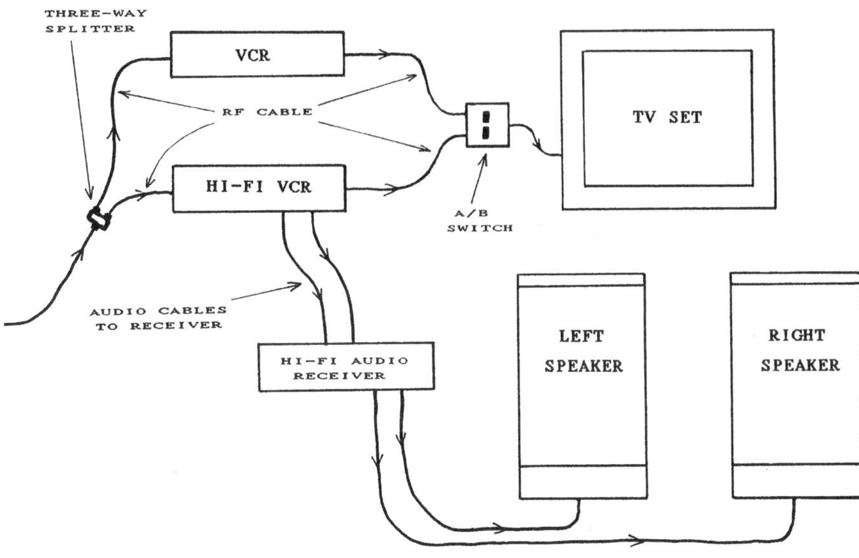

If you upgrade by adding a new hi-fi VCR to your existing TV installation, what do you do with your old VCR? Well, with cable and standard antenna (but not satellite-dish) systems, you can use a three-way splitter in combination with a selector switch, which is a splitter with selectable routing to prevent interference between different sources, to route signals to both the old VCR and the new one. The switch allows you to choose either unit for playback, and also allows either VCR to tape a program while you watch the other VCR or any antenna or cable input. As a matter of fact, both VCRs can simultaneously record different programs while you watch a third one (either VCR can further divide any incoming signals and send one to the TV and record a second). A four-way splitter and three-position switch will allow still another VCR or descrambler box to tie in. The more stuff you get, the more you can do. Welcome to the audio-video smorgasbord!

Kell factor

mined by the number of image-producing "scan lines" zigzagging across the face of the picture tube from the upper left to the lower right corner. In any NTSC-spec set, even wide-screen versions, a total of 525 lines are used, but only 483 of them form the image.[5] Thanks to a little-known psychovisual phenomenon called the *Kell factor* (which may be the result of the way the lines are laid down), only about 330 lines are subjectively visible. This is the maximum vertical resolution any standard set can deliver.

[5]The 42 others exist in the vertical blanking interval between picture frames—the black bar that is visible if the vertical hold is adjusted so the picture shifts upward or downward—and are used for timing purposes, closed-captioning codes, program-guide information, copy protection codes, and additional information, with a few simply blanked out for the beam's return swing to the upper left corner.

Cables, Antennas, and Ghosts

There are two types of RF (antenna/cable) connections available for a typical home TV set or monitor. One is the wide, flat, 300-ohm twin-lead wire that has been in use for years to bring signals from an antenna (either a small indoor unit or a roof model) to a TV set. Most sets are configured to receive either VHF (very high frequency) or UHF (ultra high frequency) signals from a 300-ohm antenna hookup.

The second type of connection is the 75-ohm "shielded" cable found on all home cable systems and better antenna setups. Most new TV sets these days will have a 75-ohm connection for cable use, but many older models do not. (Three-hundred-ohm to 75-ohm antenna input adapters are available for those older-model sets.) The main advantage of the shielded cable is, of course, its shielding, which limits any kind of extraneous electrical interference.

If you receive all your broadcast video via local-station transmissions over the air, you will find that a far cleaner-looking picture is received if you have a roof antenna. The latter will do an even better job if it is a "directional" model, making use of a motor-operated rotation device. The motor will allow you to aim the antenna at any station you wish to receive.

One common feature of signals received over the airwaves, particularly in hilly country or in skyscraper-landscaped cities, is *ghosting*. This annoyance exists because signals reflected off nearby surfaces are received by the set a very short time after the main signals and are processed separately by the RF-decoding circuitry (the tuner may not be able to fully reject the "image" of the ghost). Because the reflected signal is weaker than the main signal, it appears as a secondary group of images adjacent to the main ones.

In some cases, the reflected signal—because the main one is blocked by another hill or building and the ghost is not—can be quite strong. The best weapon against both audio and video ghosting is a highly directional antenna with a rotational aiming feature. Ghosting—sometimes called "ringing"—may also occur with direct-video inputs if the set has incorrectly adjusted internal video levels. Note that the term for ghosting in audio equipment is "multipath distortion."

One advantage claimed for cable TV is that it eliminates the causes of ghosting. Cable signals cannot be reflected off buildings or hills. While in theory this is true, ghosting sometimes occurs in cable systems. Usually it may be the result of problems with the hardware at the transmitter's central signal source (called the "headend") or because of internal glitches within the cable line itself. A good TV set or monitor should be able to limit ghosting, but even the best of them will show reflections if the cable or antenna feed is substandard. A few high-tech monitors have digital circuitry to limit ghosting, but the best and cheapest solution to the problem is a good directional and rotatable antenna or a conscientious cable company.

The scan lines that appear on an NTSC-configured TV set are reproduced in two complete screen sweeps of 1/30-second duration. Each sweep lays down only 262.5 lines, about 165 of them image-resolving, because of the Kell factor. Each successive sweep puts its set of lines in the intervals *between* those laid

interlacing

down previously. This procedure is called *interlacing* and supposedly gives the subjective effect of lines being laid down at 1/60-second intervals. (Sixty fields must be generated to create 30 frames: a 1-second image.) However, the actual 1/30-second timing interval often results in the interlaced lines appearing to quiver a bit or to simulate dot-like artifacts, often called "dot crawl," moving across the screen along the edges of objects, particularly if the interlacing is not perfect, and finely spaced, horizontal images, such as venetian blinds or herringbone suits, are being viewed.[6] Some television sets, particularly those with "digital" comb filters, are much more adept at limiting dot crawl than others.[7]

Line-Doubling Circuitry

There are an increasing number of NTSC sets on the market (with both wide-screen and standard width-to-height picture ratios) that employ what is sometimes called line-doubling or line-accumulating circuitry; another term for this is improved-definition television (IDTV).[8] These often very expensive pieces of equipment use digital circuitry to hold the scan-line fields in memory and then feed them to the picture tube together at 1/60-second intervals. This gives the effect of a full 483 visible scan lines appearing nearly simultaneously. Line doubling, which simulates computer monitor and HDTV progressive scan, can reduce interlacing errors such as dot crawl and can give the impression of a smoother picture. Because of aspects involving the Kell factor, it can also improve the subjective vertical resolution from 330 to about 450 lines—or even more. The digital circuitry usually limits horizontal resolution to about the same amount as the vertical, or 450 lines, although at least a few line doublers seem able to surpasses this figure. However, 450 lines should still be enough horizontal resolution to handle nearly any NTSC video source material.

Sets with line-doubling capability may have pictures that are dimmer than interlaced sets, because of the need to keep the beam-generating circuitry from overheating. In addition, when reproducing certain kinds of action sequences, some of the earlier IDTV sets or outboard processors could actually subjectively decrease resolution, add artifacts, and increase video noise, at least with a large-screen monitor. However, the best contemporary line dou-

[6]Video "comb filters," which help to separate the color and picture signals in modern sets, can also trigger dot crawl, which in this case may be referred to as "cross-luminance."

[7]Some program material, particularly that found on analog-video laser discs, is processed so that edge sharpness is keener than what would be encountered with a straight print. When this happens, some sets, even those with good interlace, may display more interline flicker, or "aliasing," than is usual when scenes feature a lot of horizontal lines. A notorious example of this was an earlier, non-THX, CBS/Fox release of *My Fair Lady*.

[8]There are also a few outboard-mounted line doublers (and even quadruplers) available for use with top-quality projection sets that have the necessary inputs and can accommodate the higher scan rate.

Satellite TV: Advantages and Disadvantages

There are several different types of satellite TV available. Traditional, large-dish, C-band satellite systems (with antennas ranging from 5 to 12 feet across and especially designed to pick up microwave transmissions) have over 2 million subscribers, and offer very diversified program material. More recently, medium- and high-power small-dish Ku-band direct-broadcast satellite (DBS) systems—including one that employs a dish a mere 3 feet across (available as a rental item from PrimeStar Partners), two that employ dishes about 2 feet across (AlphaStar and EchoStar), and one that requires a dish with a diameter of only 18 inches—have entered the field. This last system, called digital satellite system (DSS), sends signals to dishes and receiver-decoders produced by Thomson Consumer Electronics (sold under the RCA name), Sony, Toshiba, HNS, Uniden, and others. DSS delivers programming from Direc-TV and USSB (United States Satellite Broadcasting). Both have standard news and entertainment channels, plus first-run, pay-for-view movies appearing one after another all day long—with electronic programming guides like StarSight (see pages 212-16) to help you locate and/or record them.

Disadvantages notwithstanding, there is little doubt that satellite television, particularly systems that employ a small receiving dish and digital transmissions, is here to stay. Because of the nature of satellite operation and the ease by which such a system can be integrated into an existing national home A/V network (no new cables, fiber-optic conversions, distributor infighting or politicking, etc.), it is likely that the bulk of digital HDTV and wide-screen transmissions will be available in their best and least-expensive forms via satellite systems.

Advantages	Disadvantages
Allow you to have television reception even if there are no cable or local TV facilities nearby	Initial setup costs and monthly fees can be daunting
Often a huge number of channels, including network feeds, specialty channels, and pay-per-view movies; no waiting at retail outlets	Local stations may be unavailable; movie selections cannot begin to match multitude of options at typical rental outlet
Dozens of audio-only programs, running the gamut from classical, to rock, to country	Additional fee for radio service; subscribers living in large cities may already have ample free music available via FM radio
Superior sound and picture quality compared with standard cable and antenna sources	Dish is fragile and can be damaged by bad weather; aiming and clear line of sight is crucial; service may be affected by weather or electromagnetic interference from the sun
Videotape copies made from satellite feeds can be of high quality (better than those from antenna or standard cable feeds or what is available commercially prerecorded)	May be required to pay to descramble tape-jamming code, particularly with pay-per-view programs, in addition to viewing fee
Multi-aspect ratio compatibility: pan-and-scan 4:3 ratio, letterboxing, and anamorphic operation available from some programs	Can't watch one program and tape another (or tape two at the same time) without extra, often-expensive additional gear; cannot use a dish and receiver combination for one DBS system to receive signals from another system

blers can produce an instant picture upgrade for every source, including tapes, videodiscs, cable—and even digital material. If you can buy a new Lincoln Continental with pocket change, this technology may be for you.[9]

Brightness and Color Controls

Adjust your set.

After horizontal resolution, the brightness level of a set's picture is often the most touted specification mentioned by advertisers. Manufacturers frequently factory-adjust their sets so that the normal setting of the brightness (black level) and/or contrast or "picture" (luminance or white level) controls allows the set to catch a potential buyer's attention in a well-illuminated showroom. However, as Joe Kane of the Imaging Science Foundation points out, these adjusted sets may not deliver the best picture in a typical living-room environment, especially if the room lights have been lowered for best dramatic impact. The picture may "bloom" somewhat, with the more vivid colors bleeding into lighter ones, and the sharpness of the image may fuzz up a bit. On some sets, the picture may actually change size as scene brightness varies, a sign of poor high-voltage regulation. When viewing a program, if you back off the set's brightness level a tad and then dim the room lights to compensate (particularly when viewing a rear-projection monitor) the result will usually be improved picture sharpness, size consistency, and background smoothness. In addition, cutting the set's light output a bit will extend picture-tube life, an important consideration with both projection and direct-view sets.

color noise

Visible color "noise" is a common fault with many sets. This manifests itself as a grainy quality in smooth, solid backgrounds—particularly red—and may become gratingly apparent when a set's color control is adjusted for too much saturation or the sharpness control is turned anywhere but all the way down or close to it. Unfortunately, as with the brightness control, many manufacturers configure their sets so that the indicated "normal" positions of these controls oversaturate colors and increase graininess. While this factory setting may improve program sources with weak color and make the set appear splendid on a brightly lit showroom floor, in a home environment it is sometimes better to back off these settings a bit to subdue oversaturated colors and reduce background-area graininess. Backing the sharpness control nearly all the way down should be standard operating procedure during the preliminary adjustments of any new set. This control adds noise and edge ringing, not detail, and may aggravate interline flicker.

sharpness control

[9]Why use interlacing in the first place? Well, at any given video transmission bandwidth, interlacing will actually produce a sharper picture than progressive scan. However, it can only do this *if* the playback television monitor has perfect interlacing abilities—not usually the case.

The color and brightness controls are important tools for users who want to tailor their set to specific program material. Anyone who has scanned through the channels of a typical cable TV system while watching a decent set will be aware of the need to sometimes fine-tune the color saturation and brightness levels. I normally leave the color strength and brightness controls of my three system monitors, and even my kitchen set, turned down a tad unless I encounter a program that is having trouble with its color transmission. All four sets have their sharpness controls permanently turned to the full-down position.

hue

The color-to-color balance or "hue" of most TV pictures is controlled by an automatic circuit that may or may not complement a specific program. The circuit attempts to render skin tones accurately, but often distorts the other colors in the process and may also obscure detail, because of the loss of small gradations in color. This feature is often defeatable, and I suggest you turn the control off at least once, particularly if you are watching source material of high quality, and try adjusting the color purity yourself to see if the mechanism is working the way you want it to. Most sets work fairly well, and if the picture is in the ballpark in terms of color accuracy and is not too grainy, the brain tends to "normalize" the color balance after a few minutes of viewing. Hue will also be affected by the overall color "temperature" of the

Descrambler/Converter Complications

A cable hookup becomes more complex when a pay-TV "converter" (or descrambler) is involved. If it is plugged in ahead of the television set and/or a VCR, it is impossible to record off one station while watching a second. A creative bit of cable and connector manipulation can partially solve this problem, as can owning a set with a descrambler "loopthrough" feature, but the fix may be more complex than many people care to fool with.

A big problem with converter boxes is that they must "decode" pay video signals and then retransmit them on channel 3 or 4 to the cable input of the TV set, which cannot help but degrade the picture. Worse yet, all the non-pay signals are also visually degraded because they too must be retransmitted. One solution for this problem is to use a three-way splitter and an A/B switch that will allow you to use the descrambler strictly for watching pay shows. When non-pay programs are watched, the signal can be set to bypass the converter and feed directly into the cable-ready tuner of the set. Many TV sets have a similar-functioning, decoder-bypass feature built in.

Another headache is that the multichannel "cable-ready" features of any advanced-design TV are mostly nullified by the use of a converter. Couch potatoes will be irked to note that the TV's remote control is rendered useless as a station selector, as is any "picture-in-picture" function.

picture. Most sets produce a picture that is slightly cool or bluish, although a handful of others tend toward the warm or reddish end of the spectrum, a phenomenon that is readily apparent in set-to-set comparisons. Some rear-projection monitors display slightly different color temperatures at different horizontal viewing positions. As with the subjective impression of hue, most viewers tend to let their brains normalize color temperature after a few minutes of viewing.

RF Limitations

The RF (cable or antenna) input of a television set—even a top-grade model—is limited by NTSC broadcast parameters to no more than 330 lines of horizontal (and vertical) resolution. If you hook up a standard VHS recorder to the RF input, you will get no more than 240 lines, because of the limits of the recorder itself (see Chapter 6 for details). If you hook up a disc player, Super VHS, or satellite input to the RF input, you will still get no more than 330 lines, because of the NTSC RF-input parameters, even though such signal-delivering devices are capable of better performance. However, the *video* input (or inputs) of a monitor-type set will allow you to bypass the RF-decoding circuitry and get all the resolution your source can deliver.

Remember that hooking a standard VHS VCR into the video inputs will still net you no more than 240 lines, because of the recorder circuitry limits. Therefore, there is not a lot of visual difference between programs sent to the monitor via the video inputs and those sent to it via the RF inputs. There will probably be a bit more color noise with the RF input, but turning down the color control a smidgen will usually offset this problem. Consequently, with standard tape- and antenna-signal sources, a good-quality "non-monitor" TV set can nearly match one with monitor inputs in terms of subjective picture quality. This is good news for enthusiasts on a budget who watch only cable TV and taped programs and cannot afford a monitor-type set and such exotic items as large- or small-dish satellite systems and/or videodisc players.

GETTING READY TO WATCH

Optimum Viewing Distance

In the 1940s, Otto Schade's research proved that a person with normal vision should be able to delineate about 320 lines of vertical or horizontal resolution at a viewing distance of roughly seven times the vertical height of the screen containing the image. This is one reason (another was the technology then available) that those who developed the earliest video systems chose 330 lines as acceptable resolution. Given the expanse of the typical home living rooms of that era and the size limitations of the first TV picture tubes, it was felt that most people would view their sets from about that distance. The 330-line limit was therefore deemed adequate.

viewing distance

However, modern program watchers view their sets from about 5 or 6 times the vertical screen height—or even less. (Your author watches videodiscs on his big monitor from 4.5 times the vertical.) Modern TV picture tubes are much larger than they used to be, but the living areas of typical American homes are only slightly larger. Under such conditions, a sharp-eyed viewer may see the difference between a really good set and an average one or between a top-quality program source and a lesser one. This is one reason that high-quality equipment is important if a theater-like effect is to be approximated. If you wish to judge the optimum viewing distance of your present set or one that you plan to purchase, use a tape measure and make the necessary measurements. The Schade experiments partially explain why it is normally easy to see the difference between a 330-line cable program and its 240-line VHS copy. (Increased color noise in the taped program will be another reason.)

Traditional vs. Wide-Screen Monitors

Traditional NTSC sets have a screen width-to-height ratio of 4 to 3 (often referred to as 1.33 to 1 and sometimes called 12 to 9). One reason for this was economic. It closely mimicked the ratio of theater screens used until 1953 (the actual theater ratio was 1.37:1), allowing television sets to decently reproduce old movies. This made it possible for studios and networks to recycle old films for additional income. Another reason, of course, is that going to a wider ratio was technically difficult at the time. Getting a sharp picture on a wide-screen tube has never been easy.

wide screen

Consequently, the NTSC wide-screen format with a width-to-height ratio of 16 to 9 (1.77 to 1) has been a notable development. This shape dovetails much better with modern wide-screen films that have been difficult to adapt to video. The 34-inch, direct-view monitor is probably the most popular size available (the larger sets are front- or rear-projection models). It has about the same screen height as a 27-inch conventional (4:3) ratio set, but its width is over an inch more than that of a conventional-ratio 35-inch set. This new format is *not* digital "high-definition" television and has no more picture-resolving power than current NTSC sets, although a few versions incorporate line-doubling (simulated progressive scan) IDTV circuitry. However, even the cheaper, interlaced-scan-line

Sharp Direct-View Wide-Screen TV

(Photo courtesy of Sharp Electronics Corporation)

A wide-screen 34-inch set like this NTSC-spec model by Sharp is not true "high-definition" TV, but it offers much of the latter's impact. It features the same wide-screen feature as the much more expensive HDTV, which for most viewers is its most important feature. Given most home-viewing situations, wide-screen, NTSC-spec television sets (with and without line-doubling) offer satisfaction close to that of HDTV.

A 34-inch diagonal 16:9-ratio wide-screen set (HDTV or standard, NTSC resolution) will have a screen *height* of about 16½ inches, about the same as a conventional 27-inch 4:3 "standard" ratio set. Thus, a 34-inch wide-screen set will display individual *images* within the picture that are about the same size as those displayed by physically smaller, 27-inch, standard-ratio sets—but will place them on a screen surface that is about *8 inches wider*. Note that when a "letterboxed," wide-screen image fills out the screen on a 34-inch wide-screen model, it will only be slightly less than 3 inches narrower than what can be achieved when the same-ratio image is reproduced by a physically much larger 40-inch 4:3-ratio set (which displays blacked-out bands above and below the letterboxed image) and will be more than 1 inch wider than what can be presented on a still substantial, 4:3-ratio 35-inch model.[1]

An important feature of any wide-screen TV is its ability to "zoom" or "expand" the picture so that standard, 4:3-ratio NTSC images, letterboxed, wide-screen movies, and wide-screen, 1.77:1 broadcasts and digital videodiscs can be size-adjusted to suit viewer tastes. This feature will be of particular interest to enthusiasts with substantial letter-boxed LV disc collections or expanding DVD collections, provided they are willing to shell out the cash for sets of this caliber.

However, while the line-doubling circuitry available with some wide-screen sets will "clean up" most full-screen images quite well, the inherent loss of vertical resolution in any letterboxed program will limit what this circuitry can do. When it is zoomed to vertically

[1]See "Screen and Image Size," page 177, "Wide-Screen Zooming," page 178, and "Letterboxing," page 231, for more information on screen-size ratios and wide-screen video letterboxing.

(continued)

Sharp Direct-View Wide-Screen TV *(continued)*

fill out a 1.77:1 screen, a motion-picture ratio 2.35:1 image (sometimes called 21:9) will lose sharpness—not to mention the extreme left and right edges of the projected image—no matter what kind of digital, picture-improving manipulations are applied. (The vertical resolution losses are in the software itself and not the television monitor. With letterboxing, this is a result of not using all the scan lines available.) Consequently, the best-quality images for wide-screen TV will be those that are presented in a *full-frame* 16:9 ratio, via digital tape, digital satellite, or digital videodisc. These images will have none of the vertical resolution losses that result from having a large percentage of the available screen area unused. A full-frame input allows a wide-screen television monitor like this one to live up to its potential.

One interesting feature of this particular Sharp model is its ability to "scroll" an image—that is, move it up or down. This comes in handy when viewing letterboxed images that have subtitles below the picture area, and can also improve 4:3-ratio images that have been expanded to fill out the entire wide screen, by moving the picture down slightly to eliminate truncated foreheads.

jective standpoint than the 4:3 system we now use (at least for wide-screen movies), and some models are no more expensive than good conventional-ratio large-screen sets, making it possible for individuals with less than plutocratic incomes to enjoy wide-screen video in the home.

High-Definition Television

For more on MPEG, see page 234.

High-definition, "advanced," or "advanced-digital" sets resolve up to 1,080 horizontal lines (making them compatible with computer inputs). U.S. versions should be able to handle several display modes, including both interlaced- and progressive-scan formats at rates of 24, 30, and 60 frames per second. Subjectively, HDTV vertical resolution is much better than the 525-line NTSC format, particularly with larger sets. While a number of delivery systems were proposed for HDTV in the past, the one that has tentatively been chosen involves advanced MPEG (Motion Picture Experts Group), data-reduced digital video and audio delivered by any number of proposed sources, including tape, disc, cable, phone lines, and satellite. MPEG-1 digital video delivered VHS-quality pictures and two channels of audio. It was the format used with early 5¼-inch CD-videodiscs and early small-dish digital

Screen and Image Size

How "big" is a standard 4:3-ratio TV screen, and how reduced in size is a letterboxed movie image when viewed on it? You rarely see these meaningful measurements when reading test reports. The chart below will give you an idea of what you see when you view four different movie ratios on a 4:3-ratio set: the standard 1.33:1 picture, plus 1.85:1, 2:1, and 2.35:1 letterboxed images. Note that when viewing material having different ratios, the screen *height* changes, because the picture width will always reach to the edges of the 4:3 set's screen. It is the height of the picture that determines the size of individual images (people, cars, terminators, etc.); with extreme letterboxing, those images can shrink dramatically from what they would be in the same film when viewed in a non-letterboxed format. A 1.33:1 pan-and-scan presentation of a movie shown on a 35-inch set will have individual images identical in size to those of a 1.85:1 letterboxed version of the same film displayed on a same-ratio 55-inch set. However, the letterboxed picture on the larger set would be a foot wider, and there would be blacked-out bands above and below it. The height/width dimensions here are approximate; all measurements are in inches.

Screen (diagonal)	4:3 Screen Size	1.85:1 Picture Size	2:1 Picture Size	2.35:1 Picture Size
27	16 × 21.3	11.5 × 21.3	10.6 × 21.3	9.1 × 21.3
30	18 × 24	13 × 24	12 × 24	10.2 × 24
32	19 × 25.3	13.7 × 25.3	12.6 × 25.3	10.8 × 25.3
35	21 × 28	15 × 28	14 × 28	11.9 × 28
40	24 × 32	17.3 × 32	16 × 32	13.6 × 32
45	27 × 36	19.5 × 36	18 × 36	15.3 × 36
50	30 × 40	21.6 × 40	20 × 40	17 × 40
55	33 × 44	23.8 × 44	22 × 44	18.7 × 44
60	36 × 48	26 × 48	24 × 48	20.4 × 48
70	42 × 56	30 × 56	28 × 56	23.8 × 56

satellite transmissions. MPEG-2 brings the picture up to better-than-analog LaserVideo quality for both satellite-transmitted and DVD material. In addition to the video, MPEG-2 carries six (OK, 5.1) channels of audio, conceptually similar to Dolby AC-3 surround sound (see Chapter 8). Data reduction allows video systems to transmit full-bandwidth signals over conventional NTSC 6-mHz broadcast channels.

For many people, HDTV's most attractive feature is its wide-screen image and not its superior resolution, particularly compared with wide-screen NTSC sets that make use of good line-doubling circuitry. This is

Wide-Screen "Zooming"

Wide-screen, 16:9-ratio sets are typically measured diagonally, just like "standard" 4:3 ones. However, because of the different aspect ratios, these measurements can be misleading. A 55-inch wide-screen set has fewer square inches of picture area than a 55-inch 4:3-ratio set. In addition, the zoom, or multiple-magnification, feature of wide-screen sets will crop off a certain amount of image area with some program material. The chart below gives you an idea of several *wide-ratio*, screen sizes, along with the unused screen areas at the top and bottom that result from viewing noncropped letterboxing. Note that filling out the screen with additional image enlargement only affects visible picture *height*. The additional expansion to the sides extends the picture edges off the screen. The dimensions are approximate, and all measurements are in inches.

Compare the screen sizes here with those of the conventional sets outlined in "Screen and Image Size," page 177. A wide-screen 34-inch-diagonal set has a picture a smidgen taller than a conventional 27-inch set, but the screen is 8 inches wider. A wide-screen 55-inch-diagonal picture is the same height as that of a conventional 45-inch set, but the presentation is a *foot* wider.

Note that in the noncropped mode, there is only 0.7 inch of vertical screen area unused on a 34-inch set when playing 1.85:1-ratio letterboxed movies—the most common ratio used for modern films. On a 55-inch set, the top and bottom blacked-out areas are only one-half inch each. This is much better than the oversized blacked out areas that exist when this material is shown on standard-ratio sets. Wide-screen TV is nearly ideal for reproducing this format. However, it is clear that the 2:1 and 2.35:1 ratios would probably look better (in other words, bigger) if bumped up to fill out the top and bottom of the screen, provided that the side-edge losses incurred as the picture was further enlarged were not excessive and there were no major loss in sharpness. Note that all wide-screen sets have the ability to zoom, but most do so only in discrete steps and not continually.

How much image is lost with a strongly zoomed picture? On a 34-inch wide-screen set, a 2:1 picture that is zoomed to fill the screen clear to the top and bottom will have 3.6 inches of its width lopped off as it extends past the edges of the screen. A 55-inch wide-screen set will lose 6 inches. (In both cases, this is a loss of about 10 percent of the total width of the original wide-screen image.) With a letterboxed movie shot at the 2.35:1 ratio and shown on a 34-inch set, zooming so that the picture fills the screen top to bottom will result in a loss of 8.5 inches of picture from the sides (4.25 inches on each edge), and a 55-inch set will have 15.5 inches cropped from

Screen (diagonal)	16:9 Screen Size	1.85:1 Picture Size	2:1 Picture Size	2.35:1 Picture Size
34	16.5 × 29.3	15.8 × 29.3	14.6 × 29.3	12.5 × 29.3
50	24.5 × 43.5	23.5 × 43.5	21.7 × 43.5	18.5 × 43.5
55	27 × 48	26 × 48	24 × 48	20.4 × 48

(continued)

Wide-Screen "Zooming" *(continued)*

the screen (7.75 inches from each edge). (This is a loss of about 25 percent of the total width of the original wide-screen picture.)

While some presentations will not suffer from this kind of side cropping, others (particularly those old CinemaScope and Panavision 2.35:1 epics that exhibit action right out to the edges) may not look so good, as peripheral characters are edged off the screen. It will be up to you to uncover any compromises. However, under no conditions will the cropping be as bad as with a standard set that is working with material that has been panned and scanned. When viewed on a conventional-ratio set, a "pan-and-scan" full-screen picture of an original 1.85:1 film gives up about 25 percent of its picture width, a 2:1 film loses about 33 percent, and a 2.35:1 film loses a whopping 45 percent!

because the typical viewer will still only be able to afford a set of modest size (no larger than the already available 34-inch wide-screen NTSC sets) and will watch it from the standard 5 to 7 screen heights distance. This is not close enough (remember the findings of Otto Schade) to easily detect the difference between a 330-line picture and one resolving upwards of 800 lines. Most people will not feel comfortable snuggling up close enough to their TV sets to clearly illuminate the advantages of HDTV over simple wide-screen standard-resolution versions.

SHOPPING: NEW VS. USED

Open any newspaper and you will find scads of retail ads for TV sets from department, specialty, and discount stores; open the same paper to the classified section and you'll see a similar diversity of ads for used TVs. TV sets are easy to find, although prices will vary from near-giveaway to super-expensive, depending on the model, vendor, and age of the unit.

Shopping for a New Set

INSTORE STRATEGIES

The neat thing about shopping for a new TV is that—most of the time—you can easily tell how well a given model works. This is particularly true in a large store where the merchandise is lined up in rows making unit-to-unit comparisons easy. Well, almost easy. The store environment (the lighting, mainly) may not mimic that found in your home, and the signal source in the showroom may be better (or worse) than your own, further complicating your shopping expedition. Here are some important new-set shopping pointers.

Compare only sets that are physically close to each other. Trying to pair off a set in one section of a store with another two aisles over is chancy. The lighting may be different, the program being shown may keep changing, and your optical memory may play tricks on you. Side by side is best, even if it means having to move things around.

Often, good sets are maladjusted to serve as "straw men."

Check the input source going to each set in a store demo. A multiset cable feed may be assisted by a signal "booster," supposedly to enhance the set's performance. However, this can overload the input circuitry of an otherwise good set, and it will suffer in comparison to one that is not as good but better handles the extra power. Look over several other sets to see how they are reacting to the multiple feed. Sometimes store owners mistakenly create a setup that actually promotes poor performance in their demo models! I have been in stores that fed all their sets from a LV player, only to have the advantage of the high-resolution video source mostly nullified by the management's use of the player's resolution-limited RF output (routed through a full-throttle $15 signal booster!). Much better picture results would be possible by employing the composite-video or S-Video connections, not to mention the discrete audio outputs to give stereo "hi-fi" sound.

CHECKING COLOR, BRIGHTNESS, AND CONTRAST

Check and—if possible—adjust the color and brightness-control settings of any sets being compared to get them into the same performance ballpark. As long as those controls do not have to be set to extreme or near-extreme positions to get identical brightness and richness from the sets being compared, it is not necessary to fret about either being shifted from its factory-set, "normal" position. Rather than improperly adjusting the sets to compensate for excessive showroom lighting levels, try to get the salesperson to turn the lights down to more living-room-like, or even home-theater-like, levels. Note that sales personnel may purposely skew the adjustments of a top-grade set and situate it next to a visually "tweaked" lesser model that the management wants to move out of inventory.

Check for sharpness at all four edges and corners of the screen as well as the center. This will require a high-quality input, so try to use a laser disc or a really pristine antenna feed for this evaluation. If you cannot use a top-notch source, at least pay attention to the sharpness of the background noise, or "grain," that will exist in any NTSC image. Check to see if the set has a "sharpness" control, and note how well it affects the picture as it is adjusted. It is important to remember that the extreme sharpness of a set will not be

sharpness

critically important if you plan to only watch standard cable and homemade and/or rental videotapes. Indeed, a super-sharp set may actually show source-material deficiencies better than a lesser set and be less enjoyable to watch under some conditions. However, the finer a set is in terms of picture resolution, the more likely it will be prepared for future super-grade input sources.

A set with a lot of "contrast" may look more impressive than one with a flatter picture, and it may very well be a better set. However, make certain

Video Signal Boosters

The three-way splitter shown in Figure 5.5 routed the signal from the cable or antenna system to three components. Under most conditions (especially if the signal was from a cable feed, which is ordinarily "hot" enough to easily supply a number of receivers) this would not cause any signal-loss problems. However, if the signal was from a roof or indoor antenna or a bit weak to begin with and/or even more video receivers were tied in (say as a result of using multiple splitters daisy-chained together as well as multiposition switches controlling several additional television sets and VCRs at the main system or "networked" throughout the house), it might be desirable to use a video *signal booster* (also called a *distribution amplifier*) in place of the splitter. A booster looks like an oversized, odd-shaped splitter and plugs into an electrical wall outlet to get its power. It may also contain a knob or screwdriver-adjustable control to accommodate its output to the number of video receivers in the system.

While a booster can provide a whole house full of A/V gear with the required video signals, cheaper models, particularly those that do not have adequate shielding, may behave like miniature, maladjusted TV stations and end up transmitting spurious interference to a number of A/V components in the vicinity.[1] To determine if the sample you have is working correctly, do an impromptu A/B test by viewing a TV that is tied into the complete booster-driven network and that same set when plugged into the antenna or cable feed by itself. A quick-disconnect hook-up is mandatory for doing this kind of test with reasonable efficiency and, of course, you must leave the booster plugged into the wall outlet during the procedure. When adjusting any level control on the device, make sure to carefully monitor the effect on several receivers (both TV sets and VCRs) throughout the network, as well as on TV sets that are not attached to the cable and using indoor antennas.

Boosters are readily available through assorted A/V mail-order outlets, local electronics-parts houses, and, needless to say, Radio Shack.

[1]Cable companies must obey strict RFI regulations that are determined by the FCC, and some regularly cruise neighborhoods while using special radio-frequency-detecting equipment to "sniff out" offenders. If you once wondered how your cable company discovered that you had a booster hooked up to your system, you now know: you wired in a poorly shielded booster.

that the snappy picture has decent shadow detail: the dark areas should have adequate shade differentiation (i.e., you should be able to see the difference between dark gray and black). A good set should be able to handle the shadow detail in dark "film noire" movies like *Citizen Kane* and *The Third Man* and in horror and/or action films like *Aliens, The Abyss, The Hunt for Red October*, and the three Batman movies.

projection sets

If you are shopping for a rear-projection monitor, make certain that the three-lens convergence control is properly adjusted before judging the set's picture. These sets have an internally produced and projected test pattern you can use to make this adjustment; make certain that all three colored crosshairs in the test pattern are overlaid uniformly from top to bottom and left to right. When comparing sets, observe each from somewhat off to the side (corresponding to where the farthest-off-axis viewing position will be in your home), and note which maintains brightness best and has the least color shift. Remember to view projection sets *from a height equal to the center height of the screen*, because their vertical radiation is almost always factory-adjusted to be quite narrow and limited to what would be experienced from a seated position.

VIDEO INPUTS

When shopping for a monitor, try out its separate video inputs. That will show you the best side of a set, particularly if a videodisc source is auditioned.[10] Some sets have two or even three A/V inputs, a nice touch. Remember that a set with only one input may require that you add an external switchbox to attach multiple VCRs or LV players to it. (External switches of all kinds can be had at Radio Shack and other parts houses.) A set with one A/V input will work fine if you integrate it into an A/V system with a receiver (or integrated amp or preamp) employing video-switching capabilities of its own, in addition to the usual audio amenities (see Chapter 3 for details). If you plan to use a camcorder and want to feed its signal directly to the set instead of from a separate VCR (this may be the case if you use an 8-millimeter or Hi-8 model), try to get a monitor that also has front-panel inputs. (Some A/V receivers, integrated amps, and preamps also have this feature.)

See page 208 for more on the S-Video feature.

If you plan on getting a Super VHS or digital-tape recorder—these are often available used, and, who knows, maybe the price will continue to fall to where those on a budget can afford new ones—you owe it to yourself to get a monitor/receiver with separate video inputs. The RF input of a standard TV set will not do justice to these formats. Better yet, you should shop for a set with at least one of what is called an "S-Video" or Y/C video input for the very best in high-resolution performance, particularly if digital satellite pro-

[10]As noted previously, however, there may little visible difference between the RF and video inputs if a conventional (not S-VHS) videotape source is used as source material.

grams or digital videodiscs are also going to be watched. (The Y is the black-and-white part of the video signal, and the C is the color or chroma part.)[11] We'll more fully involve ourselves with this in Chapter 6.

STEREO/SURROUND SOUND

A TV equipped with a stereo tuner is nice, but remember that if you have a hi-fi VCR (which by definition will be stereo) you can use the deck's tuner to handle (and, if necessary, to record) stereo programs. The nonstereo tuner in the TV can still be used for monophonic program reception. The main advantage with an *MTS*-equipped set is that the manufacturer probably included some extra video features (and consequently better picture quality) in the more expensive package.

MTS

A number of high-quality sets contain surround-sound circuitry. (A few VCRs are also offering this feature.) While this is nice if the set is going to be a stand-alone model (possibly also driving some external surround speakers), it is pointless if you are going to integrate it into an audio system that already has a surround-sound receiver or surround processor.

REMOTE CONTROL

Look over the remote control on the set to see if it offers operational convenience. Is it too big? Too small? Some controls are more complex than a 747's cockpit panel and have illogically laid-out buttons that are impossible to work with in dim light. Some are not particularly powerful and will require careful aiming at the set's photoreceptor to get it to respond to a command. A good remote will be powerful enough to operate the set even if its controlling infrared beam is bounced off a white ceiling, an important feature if you are squinting at the thing in dim light while slumped in your favorite chair. See if the set itself has easy-to-operate front-panel controls that duplicate most of the remote's features. While most of us rarely even bother to fool with the controls on the set's chassis, this can be important if the remote's batteries fail during a program or if the remote breaks down. (Most remotes are electrically and mechanically pretty durable, but even the best of them can be terminated if they are dropped on a hard floor.)

REVIEWS

Read reviews in the video (and sometimes even audio) hobby magazines and *Consumer Reports.* Even if the sets they analyze are beyond your means, con-

[11]Some sets also have a "component video" input, which gives them the best possible picture from digital videodisc players with special outputs.

Sharp LCD Projector

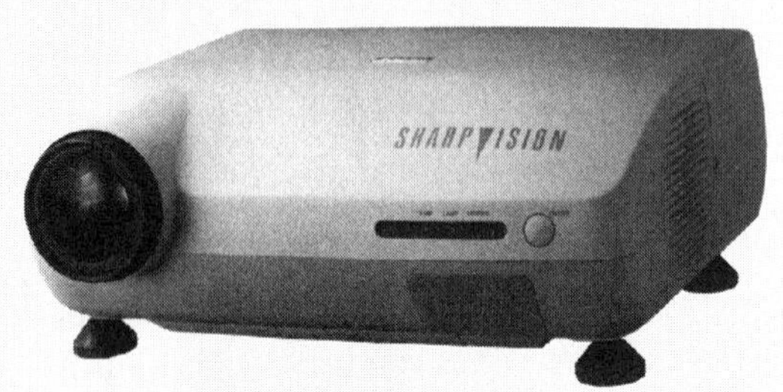

(Photo courtesy of Sharp Electronics Coproration)

Sharp is a pioneer in the production of front-projection liquid-crystal-display video monitors. Unlike standard, CRT (cathode-ray tube), triple-gun front-projection sets, which often weigh in like a load of bricks and can also be perplexing to set up correctly, LCD projectors are lightweight, simple to use, and free of operational quirks. LCD sets are also cooler running, easier to service (the projection bulb in this model is user-replaceable), and roughly equal in cost to comparable CRT sets. Perhaps the biggest advantage to front-projector sets like this one is their ability to *continously* zoom images, making them adaptable for wide-screen software and picture-size tweaking.

This particular model can be set on a table, mounted on a wall, or hung from a ceiling. It has about 500 lines of resolution, and while the individual "pixels" are visible when extreme magnifications are used (the set can zoom images from 25 to 200 inches, via a back-lit remote control), the picture does not exhibit as much scan-line flicker as conventional sets. If pixels are visible, they can be subdued by a slight fuzzing up of the set's sharpness control with minimal effect on image quality. The set also has a masking feature that can eliminate extraneous artifacts when displaying letterboxed material. While an LCD monitor does not ordinarily display a picture as bright as that produced by most direct-view sets, this particular model is brighter than most of the LCD sets that came before it.

Brightness notwithstanding, all front-projection models—LCD or CRT—have one big disadvantage. Because they use a highly reflective screen, they will not be able to reproduce acceptable dark areas if the room has even a small amount of ambient light. Dark picture areas can be no darker than the screen itself, and a white screen can only look "black" in a room with no extraneous illumination. These are not sets you can watch while reading the newspaper or having a cocktail party. However, they can deliver an impressive image in a room with subdued, movie-theater-level, background light, which is the way a home-theater image should be viewed for peak effectiveness anyway.

This set is not cheap, but it is lower in cost than many CRT models and does include a decent line-doubling circuit to further clean up the picture—a feature that has traditionally been very expensive. It is nice to think that after watching a really large-screen home-theater presentation you can roll up the screen, transport the projector to a closet (if it is not permanently mounted on the ceiling or back wall), and end up with a living area that does not look like the control room of the Starship *Enterprise*.

While flat-screen, plasma, digital-light-processor, or laser-display sets may become the TV presentation formats of the future, I believe the highest-video-tech way to go at this time, unless you have a small fortune to spend on a CRT projector and elaborate line-doubling processor, is with a liquid-crystal projector.

sider them as benchmarks, and try to compare any low-priced or newer models that you discover while shopping with models that are given good reviews. Go to your local public library for recent issues of those magazines. Read at least *some* test data before going on a shopping spree. TV sets are expensive.

Shopping for Used Sets

Good TV sets last a long time. Some, short of abuse, may last for more than a dozen years. A projection set used strictly for special-program viewing may still be looking good long after it is technically obsolete in the eyes of video aficionados. Because of this, a used set may be a good buy. While the rules outlined above are helpful whether you are shopping for a new set or a used one, it is important to cover some additional areas if you are buying someone else's set. Here are a few additional used-set shopping pointers.

Stay away from off-brands. It may be difficult to get parts for a "house-brand" or no-name set that is several years old if it prematurely breaks down—which it probably will.

Get it in writing.

Get some kind of warranty. Most stores will offer a buyer-protection warranty for at least 30 days. However, individuals often don't like to guarantee a set's performance; some even sell items "as is," take it or leave it. If the seller is willing, I advise making up a dated receipt, noting how long the seller warrants that it will operate. Again, 30 days should be long enough for you to determine if the set has any operational deficiencies.

Offer the lowest price possible. The seller, at least if a private party, may be desperate to unload the set, particularly if a new set was just installed. The old one takes up space and ruins the decor; the owner may be selling all sorts of furnishings and be in a hurry to get moved out of a house or apartment. You are probably doing the seller a favor by taking the set, even if you give next to nothing for it. However, *really* low-balling may require giving up the written warranty. "Fifty bucks, as is," the seller says.

Remember the "slob factor"; never buy a used set if there is evidence that the owner mistreated it.

PROTECT YOUR TV

Surge Protectors

Although many individuals go out of their way to protect a television's power line with a surge protector (even whole-house models are now available), the single biggest enemy of your set is the *cable or antenna* lightning strike. You have two defenses against this mishap. You can get an antenna surge protector. It will protect the sensitive RF-input section of your TV (or VCR) from moderately strong electrical surges produced by distant strikes within a cable system, but will do little for a near miss. The ultimate protection is quick-disconnects for your antenna/cable input. Most cable hookups (or 75-ohm, roof-antenna leads) have screw-type attachments called "F-connectors." They fasten very securely to your set but can be annoying if you need to rapidly "pull the plug" during a sudden series of energetic lightning strikes. A quick-disconnect plug can be securely attached to the standard F-connector and will allow you to rapidly decouple the set from the cable. A number of ready-made RF cables come standard with quick disconnects, a real convenience when hooking up a more elaborate A/V system. Radio Shack has some very fine versions, as will any decent electronic parts supply house. Whatever type of quick disconnect is selected, it should always be adjusted to achieve a snug fit for proper signal transfer.

quick disconnect

Incidentally, screw-in-type F-connectors should always be secured snugly to either the set or the quick-disconnect plug, since a wobbly fit will have adverse effects on signal quality, not only with your video receiver but possibly with other nearby receivers. Use a short 7⁄16-inch wrench to tighten it, not your fingers. Be careful not to over-tighten the connection, however. This is an antenna connection, not a cylinder head.

POWER-LINE SURGE PROTECTION

Power-line surges can also be a problem, although minor ones will cause no damage if your hardware is turned off. Power-line surge protectors range in price from a few bucks to upwards of a hundred. The better ones are electronic and have very-high-quality low-level filtering networks that are helpful mainly to computer users. Whether the typical television or audio system benefits from a power line RF filter is debatable, but with the increasing number of digital products (CD players, DVD players, computers, MiniDisc players, DBS, IDTV circuits, etc.), these filters cannot hurt.

Budget-level surge protectors found in computer stores and discount houses usually contain nothing more than a capacitor and a varistor for spike blocking but still offer far more protection than nothing at all. Most medium-priced models may have extra features but may offer nothing more than the budget models. A decent power-line surge protector will have protection for both the "hot" and "ground" leads; always go for this "full-protection" model. Also, make sure the unit has a warning light to let you know if it is still operational. A surge protector can only absorb a limited number of "hits," and it usually fails in the "open" mode (which means that it passes current but offers *no* surge protection) without any indication or warning,

full protection

unless it is equipped with a warning light. A reasonably good surge protector should cost no more than about $15 to $20.

No surge protector designed for home use will defend you against a near-miss lightning strike (let alone a direct hit), so it is a good idea to turn your system off if the weather gets violent. It is also a good idea to unplug everything (even if it is just a small TV in the back bedroom) if a big, dangerous storm is imminent. If you have your system plugged into a power-line strip (surge-protected or not), remember that merely switching off the strip is *not* equal to unplugging it. A near-miss jolt might be able to jump the open (but still physically close together) contacts within the switch. Goodbye TV set; farewell home-entertainment center!

Protecting the Picture Tube

A TV set's weakest component is its picture tube. While it may be tempting to crank up the brightness control to offset the daylight streaming through the windows or to overpower an array of lamps in a well-lit living room, your TV will last longer and work better if you close the curtains, dim the lights, and then run the set at a normal or even a reduced brightness level; the picture will also look better. Keeping the brightness adjustment turned low is even more important when using a projection set.

factory warranty

Most picture tubes last a long time if treated with reasonable care. However, it might not be a bad idea to get an extended factory warranty on the picture tube if you intend to keep the set running for long periods of time.

If you plan on taking a viewing break while watching your set and intend to come back within a half hour or so, leave the set on. The on/off cycle hurts a set more than leaving it on for a short while. If you inadvertently turn the set off before you intended (which is surprisingly easy to do with some remote controls, especially when viewing in a darkened room), let the set sit a minute before turning it back on.

Clean it!

If you think the picture is getting a bit dim on your "couple-of-years-old" set, try cleaning the screen surface before you increase the brightness setting or call for service. Picture tubes have tremendous static pull and attract dust, cooking-oil fumes, and tobacco particles from the air. (If you want to experience the static firsthand, just touch the screen while the set is running; it won't hurt you, but it feels funny.) The surface may simply be coated with grime. (I once cleaned the cover glass and screen of an old set owned by my in-laws, and the screen brightness subjectively doubled.) To prevent any cleaner (Windex or the like) from dribbling into the set, spray the cleaning fluid directly on your cleaning cloth, and then wipe it onto the glass surface.

You cannot safely clean the semitranslucent plastic screen of a rear-pro-

Tweaking a Set: Manuals and Test Discs

Early television receivers were blessed with a multitude of adjustments and alignment controls. Many of those could be found on the rear of the chassis, but a surprising number were up front, where you could manipulate them easily. In the old days, many stations transmitted a test pattern for the first 15 to 30 minutes of the day's broadcast, allowing set owners to fine-tune their machines for the best reception.

Things have changed. Few sets these days—even expensive ones—have the kind of broad-scale, user-friendly, picture-adjustment flexibility necessary to achieve top-grade performance. While some sets come from the factory fairly well adjusted, all will drift out of alignment over a period of time, as internal components age. The alignment controls I am discussing are not just the simple "tint," "picture," or "color" controls found on most contemporary sets. I am referring to the adjustments that allow you to precisely regulate the vertical and horizontal size of a set's picture, as well as its linearity (shape) and position on the screen. Getting to these controls on some sets requires that the chassis be removed from the case.

My projection monitor is a good one, and in addition to the standard, color "crosshairs" adjustment, it has over two dozen electrical and mechanical controls for picture linearity behind a cover under the screen that also mounts the speakers and protects the projection guns. With careful use of the factory service manual, I was able to "tweak" it to a performance level considerably beyond what came from the factory. The result was a picture that was razor sharp and almost three-dimensional.

A "dedicated" service manual for your set is an item that every owner should consider purchasing, even if you have no intention of ever working on the equipment. Typical repair shops cannot have manuals for every set they service, and if they have to send off for one to fix yours, it will increase the repair wait and possibly the cost. If you have a manual of your own, you can bring it in with the set when repairs are necessary (or have it at your home if the device is too big to carry to the shop and requires a service visit), earning the gratitude and respect of the technician. A typical service manual costs about $20 to $30.

Of course, most stations no longer broadcast test patterns, so I use Reference Recordings' professional-grade, laser-video monitor-alignment disc *A Video Standard* (LD-101). A newer, more "consumer-oriented" version, *Video Essentials*, is also available from the company. Each disc has a multitude of picture shape and color test patterns, as well as other recorded signals, which simulate what could only be achieved by thousands of dollars worth of precision hardware in a good laboratory. In addition, they contain fairly good audio signals that will assist the serious enthusiast in adjusting surround-sound hardware and in determining the broad-band tonal capabilities of their speakers. (*Essentials* even has a Dolby Digital AC-3 soundtrack, plus standard Dolby Surround.) Each is a worthwhile investment and educational tool, even if you never intend to do repair or alignment work on your audio or video gear.

(continued)

Tweaking a Set: Manuals and Test Discs *(continued)*

By making use of the discs and the manuals that come with them (written by Joe Kane, the *Video Standard* manual is a tutorial in video theory by itself), I was able to significantly improve the video performance of my monitors and also learned more about the surround-sound processors of both my A/V systems. While not everyone will—or should—open up a TV set and perform major adjustments, buying this disc (or the later, simpler, consumer version) and reading through the manual will help any owner determine if it is worthwhile to call in a service technician to do the job.

jection set the way you would rub down the glass screen of a direct-view set. See your owner's manual for instructions. Fortunately, a projection set's screen should not get covered with as much static-charge dust, but the "guns" and the mirror down in the innards may eventually need cleaning. However, this is not a home handyperson's job; definitely call a qualified repair shop, unless you are quite mechanically and electrically capable and have a copy of the set's shop manual.

Commonsense Rules: Keep It Dry and Cool

Keep your set dry. A TV set—like all electronic hardware—is sensitive to water, including the humidity in the air. Any piece of electrical gear will last longer if the wetness level in the viewing room is kept reasonably low. Forty to sixty percent relative humidity is about right. If necessary, get a dehumidifier and use it to keep the room dry when you are not using the set. (Unfortunately, a running dehumidifier is noisy as hell and makes listening to music or dialogue difficult, so it will have to be shut down during most programs.)

Keep your set cool. More than one A/V buff has discovered that the monitor that they "built in" to a custom cabinet overheated in the oven-like environment. Sets need to breathe. The electronics may be mostly solid state (transistorized), but they can still get pretty warm, and the back of the picture tube will heat up. A set needs several inches of air space around it (particularly in the rear), and there needs to be an air-flow route from top to bottom. Also, do not place a set where it can be heated up by other components or a home-heating vent. If you must build a TV set into a cabinet, install

cooling fan

some kind of internal cooling fan with appropriate inlets *and* exits for air flow. You should be able to find a quiet-running fan at any number of electronic-parts houses or mail-order outlets, including, needless to say, Radio Shack.

Do not place a set where sunlight can hit the screen surface. There may not be heat damage, but over time the solar radiation can damage the screen's surface phosphors.

Electromechanical Interference

Keep electrical motors and magnets away from the picture tube of a *direct-view set*, especially if it is a large-screen model. It will be OK to run your vacuum cleaner near the set, but do not turn the motor on or off when it is close to it. The collapsing magnetic field of the decelerating motor can temporarily affect the picture color. Needless to say, it is not a good idea to place loudspeakers (which are also motors) near a set unless their often very powerful magnetic assemblies are "shielded."[12] All built-in direct-view TV speakers are shielded.

shielded speakers

Try to position any non-magnetically shielded speakers so that their driver-magnet structures are at least a foot from the picture tube. (Some multiple-driver speakers may need more than a foot and a half.) If you manage to magnetize a tube, there will be discolorations within the image, usually toward the magnetized side. Once the offending magnet is removed from the set's vicinity (and over a short period of operational time), a well-designed set should be able to correct this, because all models partially "degauss" (demagnetize) the tube every time they are turned on. However, if a tube is severely affected it may have to be professionally degaussed.

Rear-projection TV monitors are less easily affected by magnetic interference, because their three compact tubes are farther from the set's cabinet edges than would be the case with a much larger direct-view tube. Those tubes, because of their narrow shapes and limited electron-beam deviation requirements, are also more resistant to close-up, external magnetic fields than a direct-view tube. It should not be necessary to use magnetically shielded speakers with these monitors. Needless to say, a *front-projection set*—by virtue of the projector's location—will not be influenced by the positioning of any of the front speakers of an audio system.

[12]When video speakers are shielded, it usually means that additional speaker magnets are reverse-mounted to the main magnets to counter their external magnetic fields.

6 Video Recorders and Players

NO HOME THEATER would be complete without some means of recording and playing video material. The videotape recorder is the biggest thing to hit home entertainment since television itself was introduced. The growth of video rental facilities has made it possible for us to enjoy first-run theatrical releases in the comfort on our home, often within months of their presentation in theaters. And now that videotape machines have become almost as common as TV sets, there's a new generation of laser-read equipment that promises to make viewing even more exciting.

In this chapter, we'll first discuss the ubiquitous videotape machine, including popular camcorders, and then the laser-video revolution, including both LaserVideo (LV) and Digital Videodisc (DVD) formats.

VIDEOTAPE

Originally pioneered by Ampex as a very expensive and often cumbersome process for studio use, videotape became an American institution when a Japanese company, Sony, released the first "Betamax" recorder for home use over two decades ago. Not too long after Sony began the revolution, the Japan Victor Corporation (JVC) joined the fray by producing a competitive system of its own, VHS ("Video Home System"). This variant allowed for longer recording times, thereby increasing user convenience and reducing tape costs. This performance edge, coupled with a somewhat less complex drive system (reducing production costs and possibly increasing reliability), helped JVC obtain a foothold in a market that should have been dominated by the Sony Betamax system. The VHS format lacked the visual qual-

Figure 6.1 Hi-Fi VCR

(Photo courtesy of Toshiba)

It pays to go the "hi-fi" route when shopping for a new or used VCR. Models like this Toshiba VCR cost only a bit more than decent-quality non-hi-fi versions, and the improvement in sound will be worth the difference, even if you only play the recorder's audio through the built-in speakers on a decent stereo TV monitor/receiver. If you watch TV programs and tapes on a standard set—one without any external A/V inputs—having a hi-fi VCR still insures that any tapes you make now will have good sound when you finally replace your current set with one having a set of monitor inputs and stereo speakers, or combine the old set with a good stereo hi-fi. Besides the hardware upgrade, the advanced hookup will only cost you a few bucks for a pair of audio cables to connect the hi-fi outputs of your VCR to your audio receiver.

ity of the Beta design (because of the slightly slower tape-writing speed), but the deficiency was simply not all that evident on the television sets of the era.

During this time, JVC not only edged out Sony in terms of perceived performance but also outmarketed its rival by allowing just about any other company to license its patent. Sony was far more protective of its design than JVC, and in the long run this cost the Beta developer plenty. By the early 1990s, VHS had essentially supplanted Beta as the video recording format for America. Although Sony continued to produce Beta systems for Betamax diehards, even it began manufacturing VHS recorders—many of them of excellent design.

Super Beta and VHS HQ

During the format war between Beta and VHS, each side made attempts to gain an advantage over its rival. The "Super" Beta system was an improvement over basic Beta and offered improved resolution and detail, as well as a reduction in background picture noise. JVC responded to this with VHS "HQ" (high quality), which slightly improved edge detail and also reduced video noise. Nearly all VHS recorders now include the HQ circuitry.

The HQ upgrade actually includes *four* enhancements. One, a 20-percent increase in white-clip level (WCL) that improves edge detail is added to all HQ decks. In addition, at least one of three additional improvements also is included in any deck sporting the HQ logo. These include a detail-enhancement circuit that works only in the record mode, a luminance vertical processor to improve noise canceling (luminance-noise reduction, or YNR), and a chroma-vertical processor to reduce color noise (chroma-noise reduction, or CNR). The latter two improvements are playback-only functions that reduce snow and color blotching. Because they are the most expensive HQ upgrades to implement, they are usually not packaged in cheaper decks. Premium decks often include all four enhancements, giving them a small quality edge over budget-grade models.

With both the Beta and VHS upgrades, there was full compatibility with previous Beta and VHS recorders. Many individuals felt that the Beta "Super" upgrade was more significant than the "HQ" modifications, putting Sony even further ahead in the picture-quality sweepstakes. Sadly for Sony, the mainstream viewing public, with its often run-of-the-mill-quality TV sets, continued to prefer VHS over Beta.

Beta Hi-Fi

The video format war heated up in earnest in 1983, when Sony introduced "Beta Hi-Fi" sound to its system. Prior to this time, most video recorders offered only monophonic sound, although a primitive stereo system was available. While early TV sets had internal sound systems that were "lo-fi," and thus could not take advantage of video recorders with good sound quality, the introduction of stereo television sets (initially by Sony, of course) ushered in a new era of video enjoyment. Suddenly, a video recorder could produce true high-fidelity sound that subjectively approached that of digital systems. At first, it looked like Sony had JVC checkmated, but it was only a matter of time (about a year) before the latter produced "VHS Hi-Fi."

VHS hi-fi

Not only was the VHS hi-fi system capable of stereophonic, high-fidelity reproduction, but many astute enthusiasts still feel that it had a slight fidelity edge over the Sony version. JVC had counterpunched Sony for a third time.

Early prerecorded videotapes of motion pictures were made in the "lo-fi" format. This was not such a bad thing with older films, which had inherent

Not Every "Stereo" VCR Is Hi-Fi

In the "pre-" hi-fi VCR era, all consumer-grade videocassette sound was monophonic and was laid down in a single track along the very narrow edge of the ½-inch-wide videotape. However, the sound of VCR audio at that time was worse than what was found on even budget-level audio recorders, because the VCR decks had inherently more wow and flutter, more background noise (they had no noise-reduction circuitry), and, most importantly, did not offer stereo sound.

For a while, in order to offer people the "sound of stereo" without having them invest in full-blown stereo units, a number of VCR manufacturers offered decks with *non*-hi-fi-stereo sound. Those units, now thankfully discontinued but which may still be available new in some stores or for sale used, made use of the same analog-track space that is found on standard decks but split it lengthwise to allow left and right channel information. Ordinarily, this would substantially increase background noise. The solution to this dilemma was to encode the stereo information (whether made with the deck itself or included on a few prerecorded tapes) with Dolby B noise reduction. This attenuated the noise to barely tolerable levels. The wow-and-flutter problem, unfortunately, remained.

This kind of stereo was *vastly* inferior to that possible with a true hi-fi VCR. If you are going to go to the trouble of integrating video into your audio system, *get a VCR that has the logo "Hi-Fi" on the faceplate*. Note that some prerecorded tapes, particularly those of older films or made-for-television programs, may be "Hi-Fi Mono," which means that they will deliver sound as good in terms of tonality as what was presented in the theater but that the performance will not be stereo, because the format cannot reproduce something that was not there to begin with.

Also note that Dolby noise reduction is *not* the same as Dolby Surround Sound. Dolby noise-reduction circuitry, in addition to being used in analog taping systems and non-hi-fi stereo-video soundtracks, is also used (in a variant of the Dolby B process) to help attenuate background noise artifacts in just the *surround* channel produced by nondigital Dolby Surround both in theaters and in Pro Logic decoded hi-fi video programs—including laser discs. However, the noise-reduction feature has nothing to do with the surround process itself.

fidelity limitations anyway, but it certainly did not make for a theater-like experience in viewing sonic blockbusters. Gradually, prerecorded tapes began to appear with stereo hi-fi soundtracks, and individuals with stereo television sets that had A/V monitor inputs for stereophonic audio as well as video discovered that TV viewing could sometimes be pretty exciting. Indeed, those who had decent separate audio systems discovered that it was possible to integrate all of their A/V hardware—with the hi-fi VCR as the focal point—and have "home theater."

Figure 6.2 Stereo TV-VCR Hookup

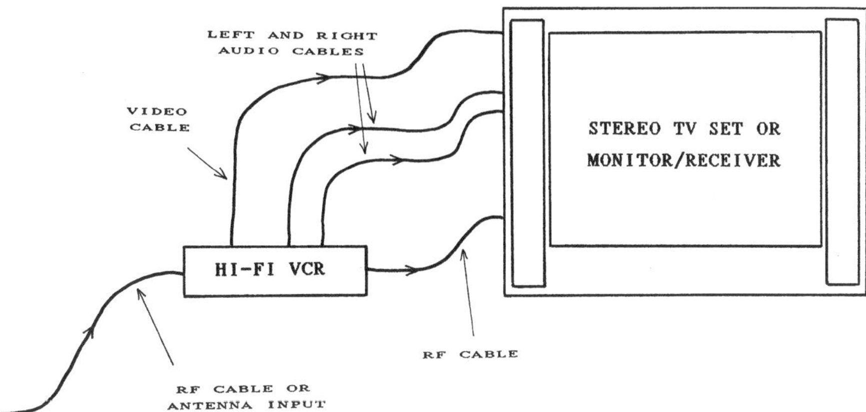

This is about as basic as a hi-fi VCR hookup can get. The hi-fi VCR has a built-in cable splitter just like a standard VCR: the cable/antenna input enters the recorder where it is made available to the "record" circuitry, and the same signal is also routed through the recorder to the antenna input of the television where it will be treated by the TV's tuner in a normal manner. The set can tune stations independently of the VCR, and the latter can record one program while you watch another. However, the VCR has three additional cables running to the set. These are the single "video" and left/right channel "audio" outputs of the VCR. They bypass the channel 3- and 4-encoding circuitry of the VCR and allow it to deliver high-quality video and audio directly to the monitor. The latter will have a video-input feature, allowing it to switch off its antenna input and use the direct signals from the VCR. Most monitor sets have more than one A/V input, allowing the use of several high-quality program sources.

Compare this setup with the one shown in Figure 5.3. That setup will probably not have as good a picture, because of its lack of a direct-video input (the video is supplied to the TV via the standard RF cable), but it will have better sound than the system illustrated here, because it uses a hi-fi audio receiver and larger, free-standing, hi-fi speakers.

Super VHS

In spite of its limitations (mostly involving color-noise artifacts and resolution—particularly color resolution), the triumphant VHS system was able to satisfy most viewers. Indeed, a well-made VHS tape produced on a good recorder can be quite enjoyable. For enthusiasts who demand more, JVC eventually came up with the "Super" VHS format. It increased horizontal resolution from the roughly 240-line maximum of standard VHS to about 400 lines or even a bit more. Unfortunately, S-VHS does not improve color resolution or reduce background color noise. Because of this, it is occasionally possible to see annoying artifacts that do not show up on conventional VHS recorders. Consequently, the increased sharpness of S-VHS has been a somewhat lopsided improvement.

Super VHS recorders also remain annoyingly expensive, and the special tape required is also costly. Prerecorded software producers are acutely aware of this and have released few S-VHS copies of films.[1] Because few S-VHS software releases are available, there has been little incentive for manufacturers to produce moderately priced recorders. Fortunately, all Super-VHS recorders are backward-compatible and will record and play back standard VHS tapes. This makes it appealing to those who want the potential for the best but also want the option of using VHS when necessary.

Other Formats

Sony has made several attempts to regain lost ground by concentrating on other aspects of the home-video market. They have developed four other recording formats. ED Beta is a super-high-resolution tape system (500+ horizontal lines) which is expensive and not backward-compatible to standard Beta (as S-VHS is with VHS). It is obviously aimed at pros and serious amateurs and is not going to be of interest to most home-video enthusiasts. Sony's 8-millimeter format is primarily aimed at camcorder users. It has an obvious advantage in terms of convenience and portability but has no edge in video performance over what can be obtained from decent VHS camcorders. Its audio performance is better than that found in non-hi-fi VHS machines, but most people will not appreciate the difference. Hi-8 is a high-resolution version of 8 mm and is comparable to S-VHS in performance. Like the latter, it is expensive to purchase and use. Finally, Sony introduced digital videotape, a product it hopes will compete successfully with JVC's equally expensive digital-tape system

None of these four formats should be seriously considered by individuals who want inexpensive, basic home-taping systems, particularly if they are just getting started in video and/or plan to view mainly rental tapes. No significant number of prerecorded items are available in any of these alternative formats, and their shelf-type recorders are prohibitively expensive. Those opting for serious picture and sound quality with prerecorded material should consider videodiscs.

[1]Most mainstream prerecorded software producers feel that well-heeled types will opt for videodiscs if they want really high-quality prerecorded video—even in preference to digital-video tape.

Camcorders

A reasonably priced camcorder is a particularly nice item if you have a growing family and/or take a lot of vacations. There are at least eight different camcorder formats to choose from, with the basic designs available from two manufacturing groups.

For the reasonably near future, the typical camcorder shopper wanting to spend as little money as possible has two basic choices: VHS if you want home-recorder compatibility and rock-bottom costs and do not care about the rather large size of the unit, or 8 mm if you want small size, reasonable cost, and convenience and do not mind always having to play your tapes back from the camcorder instead of from a rather costly 8-mm table-model deck.

Sony

8 mm	Small size, reasonably low price, lightweight; picture quality similar to that of VHS	Unless expensive tabletop unit also owned, tapes must be played back through recorder— increasing wear and tear
Hi 8	Similar to 8 mm, with picture quality similar to that of S-VHS	Same as 8 mm, with tabletop unit even more expensive
ED Beta	Very high-quality pro/amateur format	Expensive; not compatible with other Sony systems
DV	Better than ED Beta; possibly wave of the future for home users as price drops	Very expensive; not compatible with other Sony systems

JVC

VHS	Well-established format; compatible with common tabletop recorders; some models reasonably priced	Large cassette makes for bulky recorder
S-VHS	Compatible with tabletop recorders; picture better than VHS and equal to Hi 8	Same as VHS; expensive
VHS-C	Small size; with adapter, compatible with common, VHS tabletop recorders; picture quality at fast tape speed equal to VHS and to 8 mm	Short record and play time at fast speed; meager picture at slower speed—worse than 8 mm
Super VHS-C	Similar to VHS-C, with quality equal to Hi 8 at fast tape speed; decent picture at slow tape speed	Expensive; short record and play time at normal speed; picture worse than Hi 8 at slow tape speed
DV	Better than S-VHS; possibly wave of digital future for home users as price drops	Very expensive

HOW DO THEY WORK?

In order to properly work and maintain your VCR, it is important to understand a little bit of the technical functioning of the equipment. We'll focus on the VHS and S-VHS formats, because Beta is effectively finished, and digital-videotape is so expensive that is not yet a practical format for most viewers.

Heads

Unlike an analog audio-tape recorder, a modern home-video recorder does not use stationary heads, at least for the video signals. The required slow tape speeds make such a design impractical. Instead, by using a rotating "helical-scan" head spinning at about 1,800 rpm, linear tape "movement" speeds can be kept low while the "writing" speed can still be fairly high. This gives us long recording times, even though the tape is still being scanned rapidly by the moving heads.

Current VHS units have a *linear* tape movement speed of 33.35 millimeters (about 1.31 inches) per second at the "standard play" setting, with a *write* speed of 5.8 meters (5,800 millimeters, or about 228.5 inches) per second. This is done by "wrapping" the tape about halfway around the drum at an angle and scanning it with the rapidly spinning drum-mounted heads in a series of diagonal, "helical" strokes (tracks). Each drum half-revolution produces one head pass over the tape; thus, a full revolution produces two passes. Each pass produces one complete frame on the TV screen; one complete revolution of the drum gives us two passes and the full 525 scan lines needed to achieve one NTSC image. Because the drum is turning at 1,800 rpm, or 30 revolutions per second, the image produced by each head cycles on and off at 30 times per second. Each of these full-screen images is "interlaced" with the other, alternating at a combined rate of 60 times per second, matching the NTSC broadcast standard (see Chapter 5 for a review of interlacing).

The tape-head turn-on/turn-off electronic switching circuitry is critical, because any error in the 30-Hz trigger signals for each of them will result in serious picture distortion. The procedure is roughly analogous to the rotating distributor in a car that would not function properly if the electrical trigger from the coil did not arrive at the proper plug wire in time.

Linear tape speed drops to about 16.8 mm (0.66 inches) per second at the middle record and play speed (rarely used these days) and about 11.8 mm (0.44 inches) per second at the EP (SLP) setting. Interestingly, the "write" speed actually *increases* slightly at the slower record and playback speeds

Videotape Head Assembly

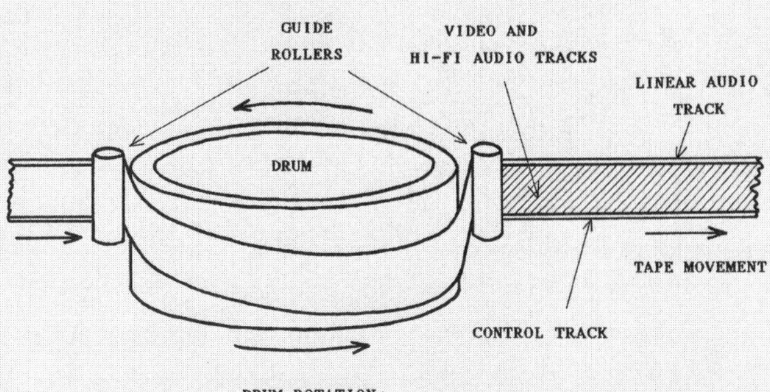

On a typical VCR, the video record and playback heads are located around the center of the shiny perimeter of the rapidly rotating tape drum. On two-head decks, the heads are spaced at opposite sides of the drum. As the drum spins, grooves along the edges create air pressure and actually cause the tape to float slightly above the drum surface, reducing tape wear and controlling tension. Tape touches only the tiny heads.

The "head-switching" network in the deck's electronic circuitry turns on each head when it comes into contact with the moving tape and switches it off after the record or playback pass is finished. On some VCRs, the "tracking-control" adjustment on the front panel fine-tunes this network, while other decks have automatic-tracking circuits.

On four-head decks, the heads can sometimes be at 90-degree intervals around the head and sometimes set in pairs closer together, but each *operating* pair (one for extended play and one for standard play or special effects) is still at a 180-degree spacing. In a hi-fi deck, the two audio heads are also placed opposite each other at other positions on the drum.

The recorded video signal, as seen on the tape at the right side of the drawing, consists of a series of hash-mark-like short scans (helical scanning) running diagonally across the tape. These are the successive swipes made by the heads as the tape passes at an *angle* across the rotating drum surface. At the top of the tape is the linear-audio track and at the bottom is the "control" track, containing the information essential for the deck to know how to handle a tape when it is played back. This linear-track information is encoded by fixed heads (not pictured here) located before and after the drum feed. (On a non-hi-fi "stereo" deck—see "Not Every 'Stereo' VCR Is Hi-Fi," page 194—the linear-audio track is split lengthwise into two even-narrower tracks.)

Also included in any VCR is the fixed "erase" head (not pictured here), which is located ahead of all other heads in the tape path. Its job, of course, is to eliminate all information before a tape is rerecorded. The space that exists between the fixed erase head and the recording heads creates the interference during playback at the beginning of a tape that has been reused. One way to correct this problem is to use a pair of "flying" erase heads, which are located on the drum and

(continued)

Videotape Head Assembly *(continued)*

completely erase any previous video signals just prior to new ones' being recorded.

A hi-fi VCR will also have diagonal, FM-encoded audio signals running across the tape in much the same way as the video ones, but they will be recorded deeper into the emulsion than the latter, and the heads themselves will also be at a different—and mutual-interference-defeating—record/playback *azimuth* than the video heads. Azimuth involves the angle at which a tape head gap intersects the scan movement.

Note that the drum rotates in the same direction as the tape movement. At first thought, this would seem to be detrimental to gaining the best performance at higher tape speeds, because the tape is actually moving more slowly in relation to the heads in the standard-play mode than in the extended-play mode. However, the actual difference is slight, given the rapid (1,800-rpm) drum speed in relation to any of the tape-transport speeds. What is different at dissimilar speeds is the *spacing* between tracks. In the standard-play mode, the gap between adjacent "swipes" is 28 microns wide; at the middle speed, the tracks nearly touch, and at the slowest speed they actually overlap by 11 microns. This closer track proximity is the reason for the increased noise and distortion at the slower record and playback speeds.

If it seems odd that the signal overlap at the slowest speed does not cause gross picture distortion, remember that—while any overlap should theoretically result in a very high level of interference between adjacent scans—the different azimuth adjustments of each head in relation to its partner prevent them from picking up too much unwanted information. One head gap will be situated at plus 6 degrees from the vertical, and the other will be at minus 6 degrees, resulting in a 12-degree difference. This is more than enough to limit really obnoxious cross-talk.

(about 229.1 inches per second at the middle speed and about 229.3 inches per second at the slowest speed), because of the drum rotating in the same direction as the tape. The picture gets worse in spite of this, because of the narrower track spacing at those lower recording speeds.

It is vital that the capstan-control circuitry keep the tape moving at the proper speed to allow the heads to exactly retrace the recorded data. The linear-speed-control circuitry must insure that the 30-Hz trigger signal sent to the heads dovetails correctly with the moving tape. All decks contain speed-detection circuitry that pinpoints the speed at which a given prerecorded program was made. In most modern analog-video decks, that circuitry is digital for reasons of economy, space, and longevity.

Even the most basic recorder has at least two heads on its drum, and others have as many as seven: five for first-class video and two additional ones for the hi-fi audio. While two heads are enough to produce a decent picture on a VCR and can offer full slow-motion and freeze-frame operation at the slowest tape speed, additional video heads (usually totaling four) allow a deck to produce decent-quality slow motion and excellent freezes with tapes recorded at the higher tape speeds. Four-video-head decks can also be designed to have each head pair optimized for standard- and extended-play viewing, rather than special-effects implementation. This permits better overall picture quality.

Audio

In addition to the video heads, VHS hi-fi decks have a pair of audio heads mounted on the rapidly rotating drum. Because of the higher writing speed and the way the heads are configured, an AFM-modulated audio signal can be laid down first and then the video signal laid over it. The lower frequencies that are inherent in the audio signal, plus their great magnetic strength—due mainly to the large record-head gaps—allow the audio signals to be buried deeper into the tape coating than the video signals. Thus, impossible as it may appear, the video and audio signals will not interfere with each other. The higher writing speed of the precisely controlled rotating drum also makes it possible for the audio to be subjectively free of wow-and-flutter speed errors.

Hi-fi VCR audio signals are not as inherently noise-free as those produced by digital systems. Consequently, it is necessary for the AFM signals to be processed by proprietary compression/expansion circuitry located within the deck to achieve true high-fidelity signal-to-noise ratios. The operation is distantly akin to systems developed by dbx and Dolby.

modulation noise

A potential problem with video hi-fi sound is modulation noise. This is a result of a slight mismatch between the recording speed and the playback speed. A common symptom is a slight "fluttery" or "bubbly-like" quality to simple tones (voice, for instance), which rises and falls with the level of the source. There is no way to correct this effect with any of the adjustments found on a typical playback unit. Modulation noise was common on some of the early-generation *prerecorded* tapes but has all but disappeared on all the recent samples I have auditioned. Usually, the effect is only noticeable with test tones or white noise in work with home-produced tapes. For the most part, hi-fi video sound is exemplary—be it homemade or prerecorded.

Tuners and Hookups

A typical VCR has a tuner just like your TV set's. Because of the way the recorder is hooked up to the television, it is possible with basic cable or ordinary antenna inputs to record one program while watching a completely different one. The VCR's tuner handles the program being recorded, and the TV set's tuner handles the one being watched. This is achieved because the recorder has both a signal "splitter" and an antenna-to-tape (TV/VCR) switch

Figure 6.3 VCR-to-TV and Amp-Speaker Hookup

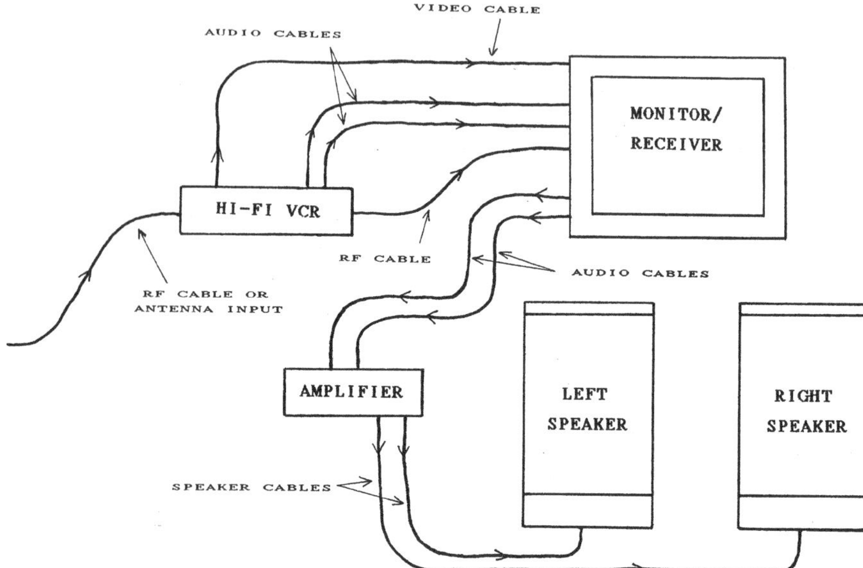

When a VCR is hooked up this way, it is using the full control facilities of a television monitor. Instead of sending its video (via the RF cable) to a nonmonitor set and its audio to a receiver for amplification for speakers, as illustrated in Figures 5.3 and 5.5, the recorder is sending *all* of its outputs to the set. In this case, as with Figure 6.2, the TV monitor is being used as a control center for all of the audio in this hi-fi "video-only" system.

Most monitors have built-in amplifiers to handle either internal (see Figure 6.2) or even external speakers, although most are low powered. The external power amp here can be a modest one and may be easily stashed out of sight. Note that this kind of hookup will not function with an amplifier that has no level controls unless the television has *variable* audio outputs. If the TV has no variable-audio-out feature, then either a power amplifier with level controls, an audio receiver, or an integrated amp (a receiver minus a tuner) will have to be used.

built into it. The internal splitter makes the antenna input available to both the VCR tuner and the TV.[2]

On the front panel of the unit, you will always find a switch or button that says something like "VCR/TV" or "Tape/Antenna." In one switch position, the signal available will be coming from the VCR's tuner, and in the

[2]Home satellite systems may not be able to handle simultaneous multiple program sources, depending on the complexity of the dish and receiver; some systems require multiple receivers for multiple program reception. Scrambled cable feeds may also be at a disadvantage here unless loopthroughs, external hookups, descramblers, and switches are employed.

Figure 6.4 High-Quality VCR-to-Audio-System Hookup

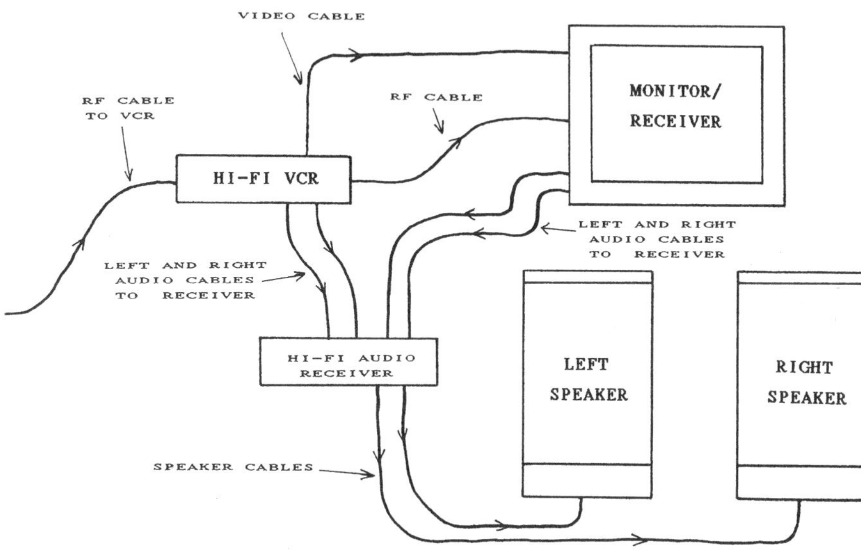

Unlike the one shown in Figure 6.3, this hookup has the audio outputs from the video recorder going directly to a hi-fi audio receiver instead of the TV monitor. (The video and pass-through RF signals still go to the TV.) The advantage of this hookup is that the audio control and switching circuitry of the receiver will almost certainly be superior to that of the television. Consequently, the sound will probably have more "oomph" and less background noise. Another advantage, obviously, is that this configuration can also accept additional inputs (it's a receiver, remember), and so the whole affair can be part of a true A/V system. The disadvantage is that you will have to use the receiver's controls to manipulate the audio while using the TV set's separate controls to handle the video. Note that the audio output of the TV monitor is also run to the receiver, allowing the latter to play back hi-fi sound from the set's antenna feed and audio circuitry.

Good planning requires obtaining a receiver with lots of inputs to handle any A/V eventuality. These days, true A/V receivers also accept audio *and* video inputs, allowing them to control video and audio switching simultaneously, further simplifying control. With this kind of flexibility, a TV monitor only needs one video input, because all audio/video source material (multiple VCRs, videodisc players, etc.) is switched by the A/V receiver, with the video part of the input routed to the monitor via the receiver's "video" output.

other it will be coming from the VCR's tape-playback circuits. In either case, the signal from the VCR will be available on channel 3 or 4 of the TV, depending on how the VCR is adjusted. Most recorders have a rear-panel switch to allow you to select either channel 3 or 4 as an output.

With a monitor-type set, the direct video and audio outputs of a VCR (including stereo outputs) can be run to the set's audio inputs. Then, when you want to watch a taped program, the "Source"/"TV/Video"/"Monitor"/"External Input"/"Antenna/External"/"External Source" control (the nomenclature varies) on the monitor or its remote control is engaged, and a better picture will result. (See Figure 6.2 for an example of this hookup.) Some sets have as many as three inputs available—for example, External 1, External 2, External 3—in addition to the antenna input, and maybe even an external input on the front panel for camcorder use. The monitor-type hookup has no effect on the dual-tuner nature of the combination, but it will no longer be necessary (or even desirable) to use the channel 3/4 feature to watch a program generated by the VCR.

Dubbing

All decent VCRs allow you to switch-select a "line"-source input in addition to that of the antenna, and most mid- and upper-level models have an "audio-only" feature. Both of these features bypass the antenna-decoding circuitry and allow the operator to record directly from a playback source to the VCR. This makes it possible to duplicate videotapes (a handy feature if you have a camcorder) or copy a program from a videodisc player; this also makes it possible to use the VCR as a high-grade audio tape deck, as discussed in Chapter 4.

When duplicating from a LaserVideo disc by use of the "line" (audio-video copying) feature, or recording a non-copy-protected program from a digital-satellite feed, a good hi-fi VCR can do a remarkable job of replicating a program. Indeed, it is likely that the quality of the dub will be superior to that of any commercially prerecorded tape version of the program, especially if an S-VHS recorder is used (and most assuredly if a digital-video recorder, which can download the digital source directly, is involved), because disc and digital satellite sources are themselves so high in quality. (Note that DVD's have an anti-copying feature.) As an experiment, I once copied one of my LV discs to good-quality VHS tape, and the results were both visually and audibly nearly equal to the original when viewed on a reasonably good-quality medium-sized monitor! If a Super-VHS recorder had done the job, the results would be hard to tell from the original with any monitor. On a still higher plane, digital-tape, dubs from digital-video sources look like clones.

However, barring the use of digital equipment, dubbing a program from one *tape* to another is not usually going to be so successful, unless the original is a fine S-VHS (or Hi-8) camcorder program or a standard VHS or 8-mm camcorder tape produced by a high-quality machine. Most enthusiasts feel that standard (purchased or rented) prerecorded VHS tapes are just adequate when viewed on a large monitor as it is. When these are copied, the result is a flat-looking, fuzzy, and visually noisy (grainy) program. In addition, most commercially produced tapes are copy-protected with a "Macrovision" or similar signal that makes the visual results of copying them even worse. Even if Macrovision did not exist, however, I would be reluctant to pirate any store-bought or rental tapes for serious use, because of the abysmal quality of dubs made on even the best machines, including S-VHS models.

Macrovision

Macrovision is a jamming signal designed to confuse the record circuitry of any machine trying to copy ("pirate") a prerecorded tape from a second recorder. If you try to dub a commercially produced Macrovision-encoded tape to another VHS-type VCR, the copy will usually have a picture that alternates between bright and dim, because the jamming signals will make the VCR think that a perfectly normal picture is suddenly too bright or dim. (The signals are inserted in an irregular manner to foil simple eliminators that would just amplify the dark signals back to near normal.) The picture may also have a loss of color, color noise, a loss of vertical stabilization (the picture will roll), or picture bending at the top of the screen. The effect will depend on the VCR models used (the one copying as well as the one being copied from) and the TV set used to play back the dub. You can spot a Macro-encoded tape by shifting the picture up or down with the vertical-hold control of your set (some sets do not have this control feature). Normally a black bar, the VBI, will be between screens. Macrovision-encoded tapes have a lot of blinking white squares or lines enclosed within that area.

I have mixed feelings about Macrovision. While I am aware that most film producers wish to protect their products from piracy, when the jamming code is improperly (or even in some cases properly) added to a prerecorded signal by the producer, the resulting taped *original* program can be unwatchable on certain otherwise-normal TV sets, because the new signals resemble false synchronization pulses. It's happened to a number of innocent people, including me.

Macrovision "eliminators" or "scrubbers" are advertised in all the video journals. Better-made versions of these devices can usually overcome the jamming signals at the cost of a small loss in picture quality. (The Macrovision Corporation has tried for years to get such hardware outlawed.) If your TV set has problems with some Macrovision tapes, you may have to obtain an eliminator to watch them. However, my experience with recent tapes indicates that Macrovision production techniques have been improved, so a scrubber is unnecessary unless you plan to dub from prerecorded material.

You will be happy to know that Macrovision has no effect on copies made with Beta or 8-mm machines. However, any VHS copies made from the 8-mm or Beta copies will be jammed, because the dubbing deck records the codes as well as the picture. Conventional laser-video (LV) discs cannot be encoded with Macrovision signals; consequently, excellent tape dubs can be made from these. However, digital videodiscs (DVD) have their own digital blocking code that prevents copies from being made.

The "audio-only" copying feature found on some decks shuts down the video recording circuitry completely. If a deck lacks this feature, an attempt to record an audio-only program in the video-oriented "line" mode may cause the video circuits to produce audible artifacts as they electronically search for the nonexistent signal.

Another feature not usually found on contemporary decks—even fairly pricey hi-fi models that otherwise perform well—is a manually operated audio-input-level control. All VCRs must have automatic video and audio gain controls (AGCs) to keep the source signals from overloading the recorder's input circuitry during an "unattended" recording session. If a deck did not have an audio AGC, shifting loudness levels in both channels caused by extreme variations in program audio level or poorly operated cable systems might cause problems. Improper level settings could lead to distortion during playback if the input gain was set too high or to excessive background noise if the input gain was inadvertently set too low and the playback level had to be turned up to compensate.

video AGC

Unfortunately, many audio AGCs overcompensate and reduce the input level so that the optimum signal-to-noise ratio is compromised. While proper audio-input settings will result in very good, clean, and dynamic sound when the level meters are running just a smidgen "into the red," the AGC systems prevent the input levels from ever approaching the red zone—even during the loudest passages. In some models, the AGC may even compress the working dynamic range of a program, limiting its high-fidelity impact. Therefore, if serious audio recording is to be attempted, a deck should have a manual override feature that allows you to monitor and adjust the input levels in much the same way as with an audio-cassette deck.

The main problem with having the option of using a manual control for "attended" recording (and one reason that some manufacturers omit this feature) is that some users may forget to switch back to automatic during unattended recording and end up with audibly substandard tapes.

Super-VHS

Super-VHS decks still cost more than most enthusiasts are willing to spend, particularly because there is little prerecorded software for the format. Their sole advantage over standard VHS is increased luminance (intensity and resolution, or "Y") signal bandwidth, which is increased from between 3.4 and 4.4 megahertz for standard VHS to between 5.4 and 7 megahertz. (Think of it as being comparable to increasing the highway speed limit from 55 to 88 mph.) Super-VHS tapes are capable of handling the higher frequencies because of their complex and quality-critical formulations. Consequently, they cost considerably more than standard VHS tapes—even premium ones. As noted previously, because Super VHS makes no change in the chrominance (color or "C") bandwidth response, color quality, color resolution, and background noise are the same in those decks as they are in standard models. Ironically, because of the improved, flaw-revealing luminance sharpness, some VHS tapes may look worse when played on an S-VHS system.

Another advantage S-VHS offers, in addition to the standard RF and composite hookups, is a special Y/C or "S-Video" interface for television monitor/receivers that can accept it. This also helps improve color and sharp-

Digital Videotape

The wave of the videotape future, if tape can be said to have any kind of serious future at all, is consumer-grade digital videotape, as exemplified in JVC's D-VHS system.[1] D-VHS—the *D* stands for data and not digital—is a system that can record data-reduced digital transmissions directly, be they from satellite, disc, or cable. This includes both video and audio material, which means that the format is compatible with 5.1-channel surround decoders.

Note that we are *not* talking here about "professional" digital video recorders or expensive consumer-grade versions—camcorders and table-top decks—that can encode and decode digitally compressed signals. The latter two items, great though they may be, are too expensive for all but the most serious and well-heeled enthusiasts—and will probably remain that way for some time. The professional digital format has been around for years and is used for, among other things, mastering the prerecorded tapes and discs we already enjoy.

In any case, don't get too excited about consumer-grade "data" video recorders. DVT's only salient feature is its direct-dubbing ability: during a download, the unaltered digital information is simply copied into the recording system. Unlike true digital recorders, D-VHS recorders will not make digital copies from analog sources, because the analog-to-digital (A/D) and digital reduction circuitry required to do so is much more complex and expensive than the digital-to-analog conversion hardware found in consumer-oriented products. (This feature is built in with digital video camcorders, however—but remember that they are very expensive.) For those wanting to record from analog sources, these recorders offer conventional S-VHS abilities. Indeed, D-VHS decks are actually S-VHS decks with an added digital-dubbing feature.

High cost aside, the only other drawback this format has in relation to disc-playback systems is the usual one: an inability to match any disc player's access time. However, no disc players in the near future will be able to record, at least from analog sources, so those wanting the very best in video recording from satellites and DVD will have to opt for a recorder that offers DVT dubbing.

[1]Of course, the real wave will be a recordable rotating disc, followed in due time by a fully solid-state format.

ness quality. The color and black-and-white resolution signals on a videotape are recorded separately from each other. When sent to a typical TV monitor, via the standard composite video cable, they remain mixed and are electronically separated within the TV to form the distinct color and sharpness parts of the picture. Modern sets utilize sophisticated, often digital, color "comb filter" circuits to keep these signals from interfering with each other during processing. The result is a picture that is superior to what is typically sent over the RF-cable output of a VCR to the antenna inputs of a set.

S-Video hookup

The S-Video connector is a dual-wire special-plug hookup that keeps the color and luminance signals separate all the way to the set, eliminating the need for the comb filters. The result is a picture with fewer annoying picture artifacts such as "dot crawl" and the line shimmer that happens when multiple horizontal lines are slowly moving up or down on the screen. While the improvement is subtle, it can be significant under some conditions. It is unfortunate that the option is not available on standard VHS decks.

The first S-VHS recorders were very expensive, and they remain premium priced. Does it pay to buy one? Probably not, if all you do is record cable or roof-antenna TV programs for later watching and then erasure. (You may not even require hi-fi audio for this.) Remember, the limit for NTSC antenna inputs is 330 horizontal lines of resolution, so the 400-line capability of S-VHS will not be fully utilized. Under most such conditions, this kind of recorder is only going to be marginally better than one that delivers the 240 lines of the standard VHS format, particularly with TV monitors of less than 30 inches. With anything but a top-quality monitor (one that has S-Video inputs, for sure), a deck like this is overkill. However, if you have a Super VHS camcorder or digital-satellite system or want to copy videodiscs for whatever reason, a Super-VHS deck (or D-VHS variant) is worth the expense if you can afford it.

VIDEOTAPE USE

VHS videotape comes in a variety of lengths, but the two most common ones are the T-120 and the T-160. At the highest speed, a T-120 will give you two hours of recording/playback time, and a T-160 will give you 40 minutes more than that. At the lowest speed, the T-120 grants you six hours, and the T-160 increases that to eight hours. Even longer tapes are available, but they are expensive.

T-160 tapes have two disadvantages over the T-120. *First*, the thinner (15- vs. 20-micron) material used on the longer tape means that picture quality will be compromised. T-160s have a bit more noise and also have more tape "dropouts," although most people will not notice the difference from a T-120, at least on a typical TV set. Thankfully, name-brand T-160s of the current generation are superior to many fine T-120s of only a few years back. Note that hi-fi sound quality should be the same with either tape.

The *second* disadvantage is that the thinner base material employed by the T-160 makes it more prone to damage from rough handling. The solution is to avoid rapid back-and-forth shuttling of the tape. It is also a good idea to avoid removing the tape cassette from a recorder unless it is fully wound; failure to do so may eventually cause the fragile material to snag or break. Fortunately, most modern decks handle all kinds of name-brand tape well, and it is unlikely that a decent T-160 will break or snag if it is handled properly. It's best to use T-160 tape mainly for recording single, long programs, rather than a group of short ones.

T-160 tapes have one obvious advantage over the T-120: time. Quite a few current movies exceed the 120-minute limit of the T-120 at the highest (SP) speed, and a T-160 may be just long enough to do the job. Even a "middling-level" T-160 at the high-speed-record setting will give a better picture than a top-grade T-120 tape at the slowest speed. I know a number of "time

Tape Dropouts

One big difference between cheap and expensive tapes (and even to an extent between good medium-grade and *very* expensive tapes) is the dropout rate. If the emulsion on a tape is not uniform and clean, the signals may not always be encoded uniformly on the tape surface. For some reason, most dropouts show up at the beginning or end of a tape, with the lowest count coming in the middle. When they want the best in record and playback quality, some enthusiasts leave the first minute or so of a tape blank.

In spite of the very real existence of blemishes on most tapes, visible dropouts are not a problem if the tape is reasonably good and the VCR is of decent quality, because a deck will have a *dropout-compensation circuit* to handle minor glitches. In a typical circuit of this kind, each scan line, in addition to being used to immediately make one interlaced picture line on the screen, is momentarily stored in a small delay circuit and is available to replace the next one if a tape dropout blocks it. The dropout circuit monitors the scanned signal, and if the signal strength of any given line drops below a preset level, the briefly "stored" line is substituted for it. Usually, the data on the substituted line is so similar and fleeting that you never notice the switch. The result of a single uncompensated dropout (which would appear if two or more adjacent dropouts overloaded the compensation circuit) will be a fine white line on the screen where the signal should have appeared during tape playback.

Tape Life

The Sony Corporation itself has stated that typical good-quality VCR program material will begin to deteriorate after about 15 years. However, don't panic! They didn't say that the tape will vaporize after 15 years. All they have stated is that the tape will begin to show visible signs of deterioration after that period of time. This means that those wedding tapes, children-growing-up tapes, and your collection of James Bond, Terminator, and Cary Grant movies will begin to lose picture quality sooner than you may have thought.

To avoid the heartache of lost memories, try following a few simple recommendations. Use only premium-grade tape when what you are recording is important enough to save. Premium tape holds up better than the cheap stuff. (Unfortunately, most prerecorded movies—even good ones—are not reproduced on premium tape.) If you are planning on taping an important occasion like a wedding, a bar mitzvah, or your child's birthday party, make certain to also take, or have somebody else take, regular color-film photographs. Kodacolor prints—but *not* the negatives—will last considerably longer than 15 years if they are well cared for, and Kodachrome (not Ektachrome) slides will probably last longer than you will.

Copying your aging tapes will not do the job, because the loss due to copying is worse than time-generated losses, and the copy tape will begin to deteriorate at the same rate as the original as soon as the copying is complete. Tape deterioration is continuous and the *results* of that deterioration are what begin to manifest themselves after 15 years. If you have tapes of films, enjoy them while you can, and make plans to either make heavier use of your local tape-rental facility (let them worry about replacement tapes) or invest in videodisc versions.

Another worry some tape buffs have is magnetic deterioration caused by placing tapes near powerful running motors or loudspeakers. However, my experience shows that a tape must be placed *very* close to a strong magnetic field for perceptible damage to occur. This is not to say that *storing* your tapes right next to a color TV set (with its degaussing electromagnet wrapped around the edge of the picture tube), a loudspeaker system, or a running motor will never cause problems. Magnetic-field effects on tape will be the result of both the strength of the magnet and the exposure time.

Just for fun, I once left a homemade test tape on top of a *very* potent speaker for several hours. Playing back the tape showed no visible signal degradation. I then put the same tape in a dust-tight plastic bag on top of my vacuum cleaner and ran the thing all over the house (cleaning the rug, of course). Even though the tape was only a couple of inches from the vacuum's unshielded motor and the motor was stopped and started a number of times, I could detect no visual deterioration when playing the tape.

My conclusion: be careful, but not paranoid, when it comes to keeping tapes out of magnetic fields.

shifters" (those who record a program at one time in order to view it later on) who use both T-120 and T-160 tapes at the slowest (SLP or EP) record speeds as a matter of course. While this may seem economical, I find that the picture quality is uniformly poor for serious viewing by anyone on a big-screen set. The colors and textures are usually one-dimensional, and the sharpness is several cuts below what is achievable at the higher speed. The only way I could justify using the slow speed would be if I were recording a documen-

Off-Brand Tape

Go into nearly any place that sells *blank* videotape, and you may find a rack or two of budget-priced or house-brand cassettes sandwiched in with the name-brand stuff. Usually, these tapes cost no more than about half of what the medium- or even basic-grade material made by Scotch, Maxell, TDK, BASF, and other reputable outfits sells for. Is this generic-grade stuff worth purchasing?

When video magazines have tested these tapes, they usually flunk in a big way. Among other video defects, off-brand tapes usually have huge dropout rates due to irregular emulsion coating, but you (well, some of you) could almost live with that. The more serious problem is that cheap tapes may contain no head-cleaning agents, and improperly applied oxides may shed and gum up your VCR's heads. Even if the shedding is not catastrophic, the buildup of oxide junk on the heads over a period of time will happen much more rapidly than if you use good tape.

While I do not feel that it is essential that you always purchase "premium-grade" tapes, it is important that you at least get your VCR tapes from reputable, high-profile companies. These days, name-brand tapes, be they premium, medium, or even standard grade in quality, are all quite good, although the premium stuff is marginally better than the standard material for camcorder use (especially in terms of analog-audio performance), and it is mandatory to use Super-VHS grade tape for best results in an S-VHS recorder.

Other budget products on the scene these days are long-play *prerecorded* tapes. Often, those will contain recordings of blockbuster-grade films, but they will be recorded at the middle—or even the slowest—playback speed. (These often will be advertised as "Recorded in the LP—or EP—Mode to save you money.") This is *supposed* to work in your favor, because the reduced quantity of tape needed for the film results in cheaper production costs—which are then passed on to you. However, the amount of money saved by producing a tape in the LP or EP mode is slight. The real savings for the producers comes from using cheap tape. When you play these tapes, not only do you get a poor picture and usually substandard sound, but you also risk gumming up your recorder's heads the same way you would if you used cheap blank tapes.

tary or something similar for one-time only use and needed all the tape space I could get. If I were recording the same item as a "keeper," I would use the faster speed and invest in additional tape. If it were a "special" keeper, I would not even use T-160 tape and would get as many T-120s as I needed.

wear and tear

In any case, use the slow speed only when it is absolutely required to fit a long program on a tape. Nothing is saved by using a tape at the slow speed to record programs that will fit on tape when run at the higher speed. Indeed, while tape-recorder *head* wear is pretty much the same at any speed, *tape* wear itself will be much greater when short programs are recorded over and over at the same place on a tape at slow speeds than if the same programs were spread over the much larger area employed at the faster speeds.

A number of decks have the ability to record at a "middle" (LP) speed. Indeed, Panasonic and Magnavox once made this a design feature. While the four-hour T-120 and five-hour-and-twenty-minute T-160 times may seem like a good compromise over the standard and long-play modes, the middle speed's picture-quality improvement compared with the slowest speed is almost nil. Most current decks have given up on the middle record speed, because it was never authorized by JVC in the first place, although many decks can still play back at that setting. Playing back a tape in the middle speed will not offer clear slow- and fast-motion effects, because the tape heads are not optimized for anything but the fast and/or slow speeds.

SOME PROGRAMMING NIGHTMARES EXPLAINED

In the dark days before the introduction of easy-programming standards including VCR Plus, VideoGuide, TV Guide Plus, and StarSight, the happy video-recorder owner was faced with the daunting task of programming the machine. Owners had the most difficulty with the "time shift" operation; setting the clock in advance to record a specific program on a specific channel to be viewed at a later time. This procedure almost certainly required that the VCR programmer open the owner's manual and then read—and digest—some of the often cryptic instructions concerning this maneuver.

It is unsettling enough for easily intimidated owners to analyze the basic hookup instructions and plug everything together after the initial purchase. It is also certainly no fun for the average TV-program junkie to have to remember to switch the TV set to "channel 3" when using the VCR, and then try to intelligently switch over from "TV" to "VCR" whenever a tape has to be watched. The further agony of deciphering the rococo *record* instructions

Easy Programming

Here are four of the most notable VCR programming or on-screen menu options available today, along with some of their advantages, disadvantages, and operational quirks.

The operation of these systems is continually evolving, so you may need to check with the provider to see if features have changed. Newer systems may also appear, and the ones listed may change radically or even disappear. However, the advantages of one system in comparison with the others as they continue to develop will be marginal at best.

All four of these systems work quite well, and choosing from among them is something that you will have to decide for yourself. While VCR Plus requires no subscription-service payments and no outboard-mounted control box, StarSight, TV Guide Plus, and VideoGuide offer you a menu system, along with additional features. Recording with any of these systems is easy, once you get the hang of their operation.

System	Advantages	Disadvantages	Basic Operation
VCR Plus	Found in many new TVs and VCRs; can work with older models by means of a programmable handset	May not be compatible with certain cable boxes or satellite systems; requires elaborate initial setup procedures; TV must be on to use; will not aid in finding current or future programs to watch	Following prompt on TV screen, enter code number from TV listing; system handles all operational functions, including channel selection and start/stop times
StarSight	Available on some VCRs, TVs, and satellite receivers; also available as outboard decoder; can work with all cable/satellite delivery systems; offers onscreen guide to programs; some versions can even set a VCR's clock	Subscription service requires fee; initial setup may be complex; TV must be turned on to program VCR ahead	Select an item from onscreen menu to watch or record now or later; inputs can be made via remote control; system handles all functions, including channel selection and start/stop times
TV Guide Plus	Similar to StarSight, although cheaper and easier to use, because of less-elaborate design	Similar to StarSight, although early version did not require subscription	Similar to StarSight
VideoGuide	Available as black-box add-on that functions similarly to StarSight, but menu is easier to read and provides more information; box can work with a variety of other gear	Similar to StarSight, although interface can be more difficult to use; black box takes up additional shelf space	Similar to StarSight

that come with recorders is often the final blow to a cowed owner's self-esteem. Well, I assure you that at least some aspects of setting up a VCR to record are not really all that big a deal. Trust me on this one; read on.

Get help!

Perhaps the best way to learn to program any VCR is to get help. Not just the kind of superbly written help you are experiencing here and certainly not the kind of help offered by the owner's manual (which may offer more confusion than clarification), but *real* help in the form of someone who has already programmed videocassette recorders and knows how to do the job. You need a competent teacher for this task: an "audio/video nerd." When you find one, have him or her methodically give you a firsthand demonstration of what you are about to read.

OTR

First, some good news: for most recording, it will not be necessary to use the "record" button on your VCR at all! The majority of recording jobs can be handled with the one-time record (OTR) or instant-record feature found on nearly all models.[3] The *OTR* procedure is designed to allow the user to *immediately* record a program for a specific time interval without going through the arcane machinations required even on VCR Plus–equipped models to set recording parameters. You just press the OTR "button" on the front panel (or remote control), and the VCR immediately springs into action to record the program indicated on its channel dial. (Of course, before doing this you will have installed a rewound tape of the required length and set the recorder to the required speed.) Just keep pressing the OTR button to extend the recording time in 15- or 30-minute increments as indicated on the VCR's front panel or on the TV screen if on-screen readouts are available.

standby function

On some models, a standby function can be used in conjunction with OTR to allow the user to "OTR" *several hours into the future*, eliminating those nettling, menu-controlled programming operations. Again, a single, easy-to-see front panel or remote-control "button" sets things up. *One push* will switch the system to "standby," and *each successive push* extends the time on the screen a half (or quarter) hour into the future. When the desired start-up time is displayed, the operator then has about 5 seconds to use the OTR button to set the program length to be recorded. The whole procedure takes only a few moments. In most manuals, this operation is explained fairly well. If you fail to properly program a deck on the first try, many decks allow you to void the commands by pressing the OTR and standby buttons simultaneously. Then you can give it another try. For me, the old-fashioned OTR-standby functions are vastly easier to use than any other programming procedures—including VCR Plus, VideoGuide, and StarSight.

[3]This function, depending on who made the VCR, may also be referred to as XPR, one-touch record, quick timer, instant timer record, express record, IRT, or something similar. On some models this operation may also be used in combination with some kind of *standby* function (see text).

Note however that with most VCRs, the OTR-standby mode will only work a few hours into the future and can be set to only record *one program at a time.* (OTR must complete its job before it can be reprogrammed to add more to a tape.) If you wish to program further ahead than this or desire to record a number of programs in a row on different channels (say a group of half- or one-hour soap operas, sitcoms, children's cartoons, or talk shows running each day or several times each day during that annual two-week vacation in Memphis, Miami Beach, or Marseilles), it will be necessary to use either the more elaborate "programming" mode of the recorder or VCR Plus/StarSight/Videoguide (if your machine is so equipped).

It's impossible for me to guide you through the assorted procedures involved in programming the large number of VCR brands available. However, a few helpful hints can clarify the procedure for the neophyte programmer. Once experienced, it will be second nature to you.

on-screen display

Be aware of the VCR's on-screen-display option (often listed as "OSD" on the control panel or remote). If your VCR offers this option, you can switch on the TV set and use the big screen to advantage while programming, rather than trying to interpret your VCR's often-cryptic front-panel display. However, once you are adept at programming, I advise you to skip the TV-screen prompt, because the operation can damage the set's circuitry and will shorten its life span. (More and more VCRs have no self-contained programming displays whatsoever, and all programming procedures—even setting the clock, for Pete's sake—*must* be done on the TV screen; I *do not* like this.)

Use the remote control to program your VCR, whether or not your set has on-screen programming. It has been my experience with most modern VCRs that the remote has a better layout than the recorder's front panel, and the markings are usually more legible.

Usually, when a VCR is set up to tape a program, the parameters that are ready to be set will blink or be highlighted on its front panel or on the TV screen. This will assist you in setting them one at a time. On many models, you can program up to eight (or on some machines, more) programs to be recorded at different times, on different days, and on different channels. On screen, these will be listed individually in order; if your VCR has only a self-contained panel display, the program number should appear before the channel/time/date information.

With most remote controls, it should be easy to program the first available slot by following the instructions in the manual. After setting up that particular program, all that will usually be required is a press on a button labeled something like "record," "timer," or "memory." (This button may sometimes be confused with the regular record button.) Then the readout should indicate that selection number two is ready to be programmed. If you only care to record one item, you will then have to press an escape or setup button to put the deck in the "programmed record" mode, locking in the VCR and making it ready to come on and do its job at the time you have selected. If you wish to continue programming (to nab all those soap operas while you are out of town), you just repeat the same procedures for programs number 2, number 3, etc., and then wrap things up with the escape or setup button. Once the VCR is put into the programmed-record mode, you will *not* be able to use it in any other way until it is done recording or you take it out of that mode.

If you are stringing a number of programs together, it is important to make sure that the *channel* for each one is correctly chosen. This is probably the second most common error made by most of us (after failing to rewind a tape far enough to make room for a new program). VCR Plus, VideoGuide, and StarSight have eliminated this snag.

Many VCRs have a "memory backup" that allows them to retain programming instructions if the power goes off for a moment or two. However, many pay-TV cable decoders have no such memory. If you have set your recorder to get a program on a given decoder-selected station and the power fails for even a second, chances are that the decoder will come back on at a different channel from what it was on when you set up the operation (when I used a decoder, a power glitch always caused it to default to channel 2). Thus, you may end up recording 2 hours of a 1950s sitcom or *Star Trek* reruns instead of *Rambo VIII* or that X-rated "jiggle movie" that you were hoping to pick up.

Check the tape length.

Remember to carefully check how much tape will be needed to record a given program. Your VCR doesn't "know" how long the installed tape is, and it may accept your programming instructions even if the total or remaining tape length and/or the selected tape speed are inadequate, even if you have VCR Plus. If you do not have enough time on a tape, either switch to a slower record speed or get a longer tape. This tape-length-check procedure is particularly important when stringing a series of short programs together. A few models have an "auto-speed" feature available, which, if SP has been selected, will automatically shift to EP (SLP) if the recorder determines that there is not enough tape to record the selected program, a nice feature but one that requires that the tape be installed rewound and that the tape counter be reset.

Use as much of the tape as you can. Some individuals always use the slow (6-hour) speed to record 1- or 2-hour programs over and over at the beginning of the tape. However, the tape will erode faster as you concentrate the rapid, rotating head movement repeatedly over a small area at the beginning of the slow-moving tape. (When that small area wears out, you still have to discard the whole cassette.) You also sacrifice picture quality. If you need to record a long program, see if a T-160 will do the job before using the slower speed.

SHOPPING: NEW VS. USED

The same rules that applied to the purchase of used audio-only tape recorders and CD players apply here. VCRs are complex items, and purchasing one of them secondhand can be a mistake unless you have a good idea of the history of the unit under consideration. If you purchase one that is on its last legs—even if it is a real bargain—you may discover that repairs are so expensive that purchasing a new unit, complete with warranty, would have been a better idea.

Get it in writing.

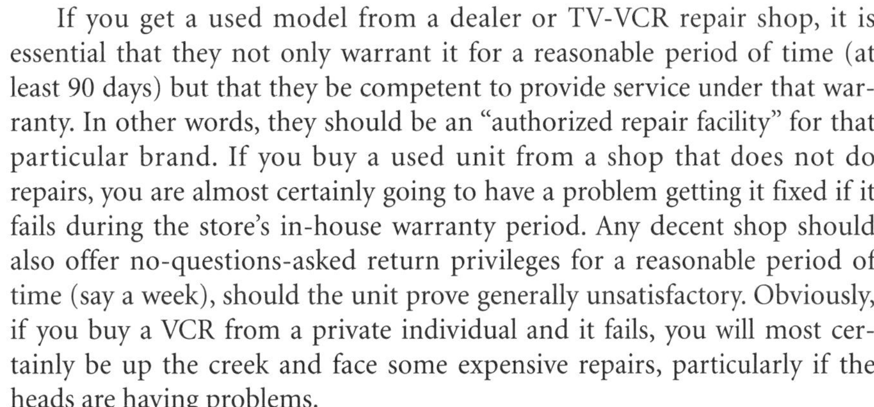

If you get a used model from a dealer or TV-VCR repair shop, it is essential that they not only warrant it for a reasonable period of time (at least 90 days) but that they be competent to provide service under that warranty. In other words, they should be an "authorized repair facility" for that particular brand. If you buy a used unit from a shop that does not do repairs, you are almost certainly going to have a problem getting it fixed if it fails during the store's in-house warranty period. Any decent shop should also offer no-questions-asked return privileges for a reasonable period of time (say a week), should the unit prove generally unsatisfactory. Obviously, if you buy a VCR from a private individual and it fails, you will most certainly be up the creek and face some expensive repairs, particularly if the heads are having problems.

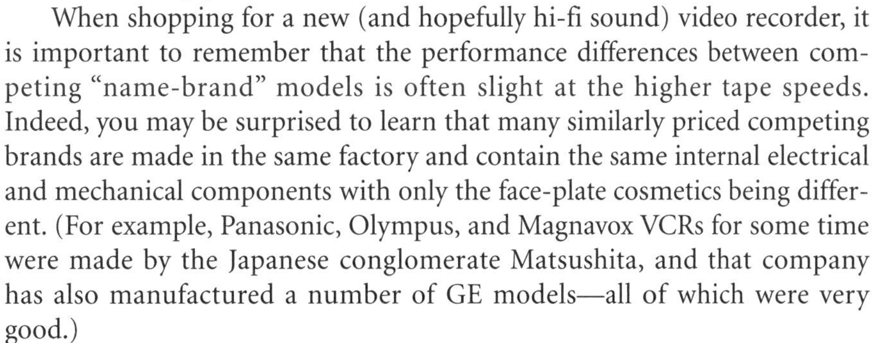

When shopping for a new (and hopefully hi-fi sound) video recorder, it is important to remember that the performance differences between competing "name-brand" models is often slight at the higher tape speeds. Indeed, you may be surprised to learn that many similarly priced competing brands are made in the same factory and contain the same internal electrical and mechanical components with only the face-plate cosmetics being different. (For example, Panasonic, Olympus, and Magnavox VCRs for some time were made by the Japanese conglomerate Matsushita, and that company has also manufactured a number of GE models—all of which were very good.)

However, some of the more expensive multihead models do offer a better picture at slow speeds. If you time-shift a lot and/or like to pile a great deal of material onto a tape, then it is important that you closely evaluate the slow-speed performance of a deck. If slow-speed video performance is important to you, the purchase of an S-VHS recorder should be seriously considered. Ditto if "ultimate picture quality" is high on your priority list, because at the highest speed a Super-VHS machine easily outclasses any standard VHS model.

Integrating a Decoder

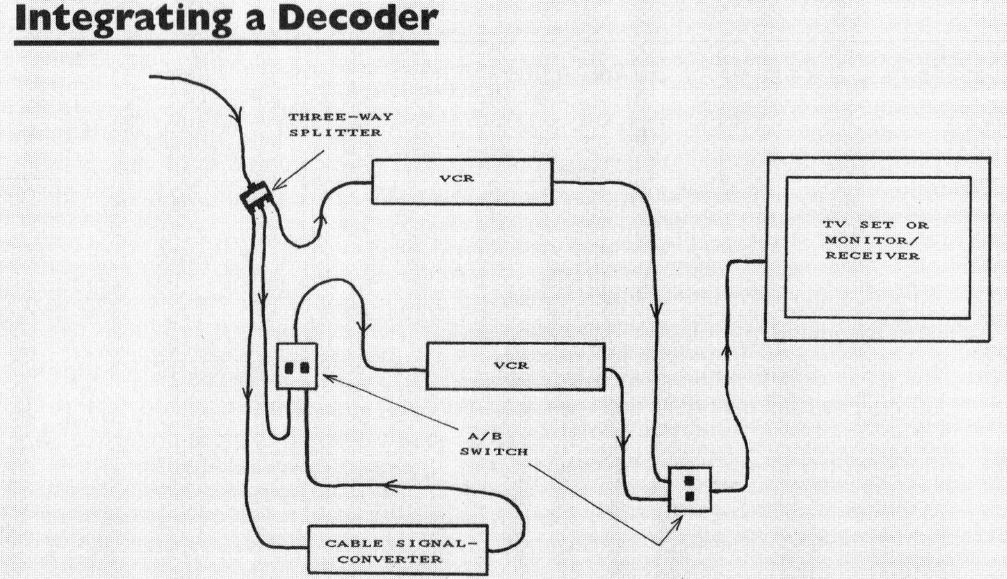

If you use a cable "converter" box to get pay-TV programs, you can't watch one show while recording another with your VCR. The converter is a single-source device, and the signal it puts out will allow you to record and/or watch one program only. Well, not if you improvise!

The hookup shown here allows you to tape or watch a program coming from the converter (or bypassing it) while taping or watching a second "unconverted" program. This is a two-deck setup, but eliminating one VCR will still allow you to watch an unconverted program while taping a second converted or unconverted one. The secret is the creative use of A/B or A/B/C switches and splitters (available at Radio Shack and other electronic supply stores). Let this diagram give you some ideas about creative hookups for your system. Remember that hookups like this can be dovetailed with configurations like those pictured in Figures 5.3, 5.5, 6.2, 6.3, and 6.4 to further expand their capabilities.

Videocassette Recorder Buyer's Checklist

Remote control:

- Easy to read? In dim light?
- Do the button labels, functions, and positions make sense?

Programming display:

- Front panel? Most convenient in the long haul.
- TV only? TV must be on to program—often awkward.

Is the owner's manual coherent?

Records and plays back at all three speeds (low, middle, and high)? Most users will not miss middle speed.

Important audio-only recording features:

Front-panel tuner/line/audio switch included?

Input-level control and audio-AGC override?

True hi-fi stereo (not linear-track, lo-fi "stereo")?

Perhaps the best thing for the neophyte and even the seasoned videophile to do first when shopping for a new model is to dig up a copy of the latest test that *Consumer Reports* has done on VCRs. (You could also use their older tests to check out the new-model quality of a used unit you might be looking at.) *Consumer Reports* usually compares a large group of units at a time and is not afraid to trash the dogs. Even if the model you are considering was not tested by the magazine, you can get a passable idea of its quality by comparing other models made by the same company with the competition. *Consumer Reports* also does research on "frequency of repair." Whenever they do an extensive test of VCRs (or TV sets—or automobiles, too, for that matter), they will usually include the results of an extensive "owner's survey" on reliability. This is important. By intelligently using these data, prospective shoppers will save themselves a lot of headaches.

returns

As with the purchase of any other new or used component, investigate the return privileges offered by the store. A week-long "grace" period would be minimum, but many local discount-type stores allow as much as a month. (Oddly enough, many high-priced specialty shops will not give you this much time.) When you get the new unit home, run it a lot to flush out any bugs. Remember, most audio-video components that are going to malfunction will do so either during the first few hours of use or after many years of use. *Get over those first few hours before the grace period expires.* Once that time period is up, any recorder with defects will be ineligible for an exchange and will require warranty action.

The same admonition regarding "authorized repair facility" status applies to new units as to used ones. Discount houses and audio-video sales emporiums are nice places to save money, as are mail-order businesses. However, if you end up having to send a unit off for a fix at a repair facility because your local store does not provide repair service, be prepared to wait a while to get it back. If your shop is an authorized facility, the chances are good that the fix will be much faster and more hassle free.

mail order

Buying a VCR through a mail-order house raises problems similar to those that we discussed earlier regarding other componentry. If the unit works fine and you have no problems with it, you are an automatic winner. You saved a bit of money. However, if the unit malfunctions right out of the box, you are going to have to go to the trouble of sending it back (once you contact the company and get authorization to do so) and then spend time worrying about whether it will get back to them in one piece. (Of course, you will already have gone through the process of worrying about whether it was going to get to *you* OK.) The most important thing to consider with a mail-order house is their short-term return privileges. At least one week is a fair period of time. On the whole, I would not purchase an item as expensive and persnickety as a VCR from a mail-order house until I had already had some experience with them and knew that they offer excellent customer service.

TAPE AND RECORDER CARE

Tape Care

Videotape is maintained in pretty much the same way as audio tape. Do not leave a cassette in a place that gets hot. Long-term storage of a tape above a "warm-running" component like a TV set, close to a heater duct, or in proximity to a fireplace is obviously not a good thing to do. Even low-level equipment heat (say in the neighborhood of 90 to 100 degrees) will gradually do subtle damage. The single biggest problem with rental tapes is short-term heat damage. Leaving that freshly rented "blockbuster" in your tightly sealed car on a hot day while you go shopping (particularly if it is in the sun on the rear shelf under the back window) is *not* a good idea. This can partially melt the emulsion and foul up the heads on your deck so badly that it may not be possible to clean them—even professionally; they will have to be replaced. When returning that rental tape, remember to continue to protect it as if it were your property. You do not need to screw up the next customer's VCR any more than you need to ruin yours.

storage

Do not store tapes in damp areas like a basement or out on the back porch. Tape emulsions are hygroscopic and will absorb moisture readily. If your home humidity levels are too high at certain times of the year, consider getting a dehumidifier. Also do not store recording tape near strong magnetic fields like a loudspeaker, TV set, or big motor (like the one in the dehumidifier!), in spite of the fact that we have shown that videotapes are surprisingly resistant to short-term exposure to such fields. Why take chances?

Like audio tape, videotape is best stored in the "played" position, because that leaves it in the most uniformly packed state. A fast rewind packs the tape in a less-than-regular manner, submitting it to nonlinear tension stresses. During a long period of storage, those stresses can distort the shape of the tape. Rewind just prior to using. (Rental stores have you rewind because the tape will almost certainly be used again in a short time, whereas your tapes may be in storage for several months or even years.) Many tape experts recommend that you fast forward and rewind a tape at least once a year to keep it "exercised." On paper this sounds good, but the result may be that it is stored "fast rewound" for still another year. Thus, there is no substitute for *watching* a tape at least once a year.

Tape Rewinders

Tape rewinders are generally available for under twenty-five bucks, although some of the more versatile models cost a bit more. At first glance, it is hard to see how one of them could reduce wear and tear on your recorder as the ads claim. After all, a VCR is designed to run for hours at a time. Just how much additional deterioration can happen to a recorder's innards during a few brief minutes of tape rewinding? Well, during the rewinding process the recorder is not being pushed hard, but the sudden jolt that happens at the end, when the tape-reel brakes are applied to the rapidly rotating drive mechanism, can affect the brake assemblies considerably. In addition, while most recorders pull the tape back into the cassette before rewinding, some keep the tape threaded within the serpentine playback mechanism to allow for rapid startups at the end of a fast shuttle. Decks like those may be fairly rough on tapes if they are the least bit out of alignment.

While it may save wear and tear on your recorder, a rewinder can be needlessly rough on tape. At the end of a rewind, a VCR is triggered to stop, thanks to either incandescent or infrared sensors that detect the clear mylar leader at either end of a tape. The brakes are applied before the end of the leader is reached. Most non-upscale rewinders depend on the sudden buildup of tension that occurs when the tape snaps tight at the end of a rewind to trigger the quick-stop action of its motor. Obviously, a rewinder will be considerably rougher on the leader connection at the end of a tape than a typical recorder. If it is too rough, it will cause damage to your (and/or your rental shop's) tapes. The most typical mishap is that the tape is jerked loose from the spool inside of the cassette.

The way to combat this problem is to get a "slow" rewinder. A number of models have fairly powerful motors to get the rewind job done fast. However, in the case of fragile videotapes, slower is better. Most 9-volt models are gentle enough on your tapes to not cause them any harm. Unfortunately, these units may not have sufficient starting torque to rewind a tape if it is only partially played through. Under such conditions, you will have to opt for the rewind feature of your recorder.

Some of the more elaborate (and expensive) rewinders on the market have tape counters and can fast forward as well as rewind. I do not find these features particularly helpful, because it is a pain to try to calculate the necessary counter numbers for specific program locations in either direction. I have also heard of a rewinder that cleans the tape as it works, but I strongly advise against using such a machine. Tapes do not clean gracefully at high speeds. Some of the better rewinders do have a photocell shutoff feature, and that is definitely a worthwhile option.

Hands off the video cassette!

Never try to open a tape cassette and "fix" things. If the cassette seems substandard or worn out, get a replacement. Some popular magazine articles show you how to fix tapes, but if you botch the job and the tape fouls up your VCR, you will end up spending a lot more on repairs than what a new tape would have cost. If an important tape (records of children at play, a wedding, etc.) is damaged, take it to a competent repair facility and have a qualified technician copy it to a new tape. There will be a quality loss due to the transfer, but at least the new tape will be workable.

Incidentally, if the shell of a tape cassette you are planning to rent looks like it has a lot of mileage on it, you may wish to avoid taking it home at all. A really worn rental cassette shell is a sign of heavy use, and heavily used tapes can actually be abrasive to heads. Always check the condition of a cassette you are renting (particularly if it is an older film that may have been rented many times) before you take it out of the store. If it is beat up, get another copy.

Recorder Maintenance

Video recorders, in spite of their complexity, are actually pretty tough items once they have passed through the initial trial period. If you use quality tape (name-brand medium or premium grades, which have cleaning compounds built in), it is unlikely that you will have to have yours serviced for a good 1,000 hours of use—or even considerably more. That's 500+ 2-hour T-120 tape passes at the standard speed. Five hundred rental-tape sessions also adds up to about the same amount of time, but rental tapes may be considerably less kind to the machine than quality blank tapes, so do not be surprised if your deck starts to give trouble at a shorter interval.

After about 1,000 hours of (sane) use, it will probably be a good idea to have the tape path cleaned, the belts checked, and the cassette and tape-loading mechanism lubricated. After about 1,000 additional hours, a well-made deck should get a complete lubrication, probably have the drive belts replaced, and have the assorted rollers checked and/or replaced. If you have been using quality tapes during the life of your VCR, a gradual loss in picture quality over the kind of time periods we are discussing here will mean that your heads are worn and in need of replacement.

dirty heads

If you habitually run rental tapes through your deck, you may occasionally notice a loss in picture quality. In addition to potential heat problems, rental tapes may encounter all sorts of unfriendly (read dirty) environments as they travel around and get played. Your deck's heads may pay the price. The usual symptom for *one* dirty head is snow across the entire TV screen, while the picture still remains. This is because each head operating during

Stick with good-quality tape and you may never have to clean the heads.

playback is responsible for one full interlaced field across the entire screen; the "clean" head will continue to produce its half of the interlace, even after the dirty one fails. *Two* dirty heads will give you nothing but snow—no picture at all. Under normal conditions, the chances of both heads clogging up at the same time are pretty slim. A complete loss of a picture may mean that professional service is required.

If a head gets fouled, hold off lugging your machine to the local repair shop until you take a stab at cleaning the unit yourself with a decent "head-cleaning tape." There are a lot of those available. Both wet- and dry-process cleaners seem to work well, but some are probably rougher on heads than others. Fabric cleaning tapes can become frayed and cause damage, and I suggest you avoid them. The best head-cleaning cassette I have seen is made by Scotch. It contains a section of pretty-much-normal videotape coated with more than the usual amount of head-cleaning compound and less picture-retaining emulsion than is found on typical tapes. It also has a video message encoded on it. When you can clearly read the words as they appear on your TV screen, the heads are clean. Unless you have experience and/or are willing to take a chance at ruining your VCR, leave more serious head cleaning to a good technician.

pets

Pet owners should note that cat hairs can give tape heads and other internal parts trouble. Because felines love to nap on warm surfaces (like the top of your VCR), invest in a good plastic or cloth cover if your unit is out in the open. Without a cover, the hair a cat leaves behind will eventually work itself into the recorder through the top vents and cause problems. Radio Shack has stocked some nice cloth covers (which vent off small levels of residual heat), with clear plastic fronts. *Do not* leave the cover on if you have programmed the unit, because the heat buildup during the record or play mode requires more ventilation than even a cloth cover can supply. If your cat sleeps on the VCR during its operation, you need to have a talk with him or her.

Finally, invest ten to twenty bucks in a full-protection surge protector to help keep your VCR from becoming toast in a thunderstorm. Remember that a VCR's clock circuits and infrared remote control receiver are still powered even when it's turned off. If storms are approaching or forecast the morning before you go to work, disconnect the roof antenna or cable from your system. If the sky looks really bad, unplug the system from the electrical power source, too. The only inconvenience will be that you will have to reset the clock if the system is left disconnected for a long period of time.[4] If you *must*

[4]Many decks have a memory feature that allows them to retain a clock setting for some time if the power goes off, and some upscale models will use the time and date signals from local PBS stations to automatically reset the clock.

tape programs during an occasional storm, invest in an antenna surge protector, which will block small surges. Then, when you tape, disconnect the VCR from everything else that has been separated from the power and antenna inputs, and the recorder alone will take the blow if a big electrical surge hits the area; no use blowing up the TV set, too.

THE LASER FRONTIER

Although videotape players are an essential part of any home-theater system, laser video—whether it be frequency-modulated analog videodiscs (LV) or CD-sized, MPEG-configured, digital videodiscs (DVD)—has yet to catch on. Both formats are more expensive than most prerecorded VCR tapes or CDs, and most videodisc players cost more than a hi-fi VCR and considerably more than a CD player. What's more, money-saving, disc-rental facilities are much less common than videotape rental stores. However, each offers an improvement in fidelity over videotape, and some feel that DVD is the wave of the future. Although the LV system is doomed to become an outmoded format, it still rates serious analysis, because many people have substantial money invested in hardware and software. What's more, both players and discs will end up being available used, as habitual audio-video "upgraders" convert to the all-digital DVD format and software companies liquidate stock.

LASERVIDEO: THE FM-ANALOG VIDEODISC

The current FM-analog system as developed by Pioneer, and known to video enthusiasts for years as "LaserVideo" or LV, employs an optical pickup to track a frequency-modulated signal embossed into an aluminized coating below a thin, transparent plastic surface. Frequency-modulated (FM) video and audio signals are encoded on one side of two extremely thin aluminum discs by means of an engraved, microscopic, outward-swirling pattern of dots, or "pits."[5] The two aluminum sheets are laminated back to back and encapsulated by a hard, transparent, plastic coating. During playback, the engraved signals are read by an infrared laser, decoded by appropriate circuitry within the player, and sent to a television monitor as a composite that is further processed into a picture.

[5]Microscopic is the word here; if the disc were enlarged so that each dot were about the size of a shoebox, the playing surface would be 30 miles across; one pit is 700 times smaller than a pinprick.

Figure 6.5 Laser-Video Player

(Photo reprinted with the permission of Pioneer Electronics [USA] Inc.)

Combi players like this Pioneer model cost much less than they used to, thanks to advances in LV technology and the market impact of the new digital videodisc players. A typical unit costs $300 to $800 (discounted) and should be able to deliver a picture and sound that are subjectively (although not always measurably) equal to that of the more expensive models—although not quite in a class with that of DVD units. One difference between budget-grade LV players and many upscale models is the inclusion of a digital-field-memory circuit that allows slow motion and freeze framing with CLV discs. Many expensive players also improve picture quality with digital time-based correction and field-noise reduction circuitry, as well as digital comb filtering and digital dropout compensation. Upscale units like the model pictured here also incorporate highly effective Dolby AC-3 Digital Surround, and some Pioneer models have the ability to play three formats: LV, CD, *and* DVD (see Chapter 7).

Unlike a CD, the LV disc's program is recorded on both sides to increase storage capacity. In addition, the LV disc has always been much larger than a CD, and it spins at a higher rate of speed. The large size and two-sided recording are necessary to contain the lengthy video programs the system was designed to reproduce, and the high rotational speed is necessary to handle the bandwidth requirements of the video part of the signal.

The laser-scanned *video* material on LV discs is not digital. Minute variations in the size of the pits change the quantity of light reflected, and these "modulations" are converted by the player into video; this is very different from the way digital video is handled. The AFM (analog frequency-modulated) audio on early LV discs was not comparable in quality to what was available on properly manufactured hi-fi VCR tapes. Fortunately, not too long after its inception, the "CX" compression/expansion noise-reduction process was added to the LV system. With CX processing, the format had the potential to subjectively equal a hi-fi VCR in audio quality—particularly in bandwidth and dynamic range—and, of course, it already surpassed both

Super Beta and VHS HQ in picture quality. (Super-VHS, ED Beta, and consumer digital tape formats had yet to appear.)

These days, all LV discs that do not have Dolby AC-3 Digital Surround audio (see Chapter 8) include both CX-encoded *and* PCM-digital sound; this upgrade allows many high-powered LV programs to deliver substantially better audio than even the best nondigital videotapes. It also makes them compatible with older-model players lacking digital-audio circuitry. In some cases, the CX tracks are utilized for SAP (second audio program) commentaries on the screen action by directors, film editors, animators, film critics, etc. The remote control allows the viewer to easily switch back and forth between the PCM-digital, AC-3 digital (if so equipped), and CX tracks.[6]

The final evolution of analog LV development involved the incorporation of Dolby AC-3 Digital Surround, a standard feature with the digital videodiscs that we will be discussing up ahead. This five- (plus subwoofer) channel audio format has ushered in a new age of high-fidelity realism in the home for both video and audio-only applications.

CAV vs. CLV

Two basic types of the common 12-inch discs are available for the enthusiast who opts for the LV format.[7] The most versatile is the *CAV* (constant angular velocity), or "Standard-Play," disc, which always rotates at the same 1,800-rpm speed. An NTSC picture "frame" takes up one full revolution of the disc, making high-quality, slow-motion, and freeze frame effects possible. When a frame is selected to be frozen, the player just keeps repeating the same full-circumference interlaced frame over and over. When fast or slow motion is required, the player just moves from frame to frame at the user-selected speed.

CAV

CAV's flexibility comes at a price: the disc surface area is not efficiently used. One frame requires a full rotation, whether the disc is being scanned over its inner or its much longer outer circumference (although as the laser tracks towards the outer circumferences, the added surface area may slightly improve the video signal-to-noise ratio). Consequently, the playing time per side cannot exceed much more than a *half an hour*.[8] A two-hour movie

[6]With players that have both PCM-digital and CX-analog circuits, it is possible (and interesting) to A/B between each during playback to hear the advantages of digital-audio over the CX-analog counterpart. With older films in particular, the effect is so subtle as to go unnoticed. At other times, especially when the extremely deep-bass found on many releases is coming on strong or there is a lot of dynamic range, the digital improvement can be significant—if your speakers are good enough.

[7]Less popular 8-inchers are usually reserved for music-only programs.

[8]There are 54,000 video "frames" on each side of a CAV disc; the 30-frames-per-second retrieval speed adds up to 30 minutes per side.

requires two discs! CAV discs are the most expensive LV format, and many "deluxe" or "special-edition" films have been released in this format to allow serious film buffs to do frame-by-frame and slow-motion analyses.

CLV

The answer to CAV's high disc cost and short playback time is the CLV (constant linear velocity), or "extended-play," disc, which has a variable rotational speed of from 1,800 to 600 rpm. The disc begins playing at the inner circumference and tracks outwards—just like a CD—with its speed gradually decreasing as the program continues. Mechanical-rotation noise also decreases, something that does not happen with CAV. With the CLV format, each frame takes up the same amount of linear disc *space* as the full-circle scan at the inside circumference of the disc. As a result, three full frames can be fitted into the circumference at the outermost edge, with gradually reduced quantities toward the inner circumference. This variable, constant "linear" speed system allows a *one-hour* program to be shoe-horned on to one side of a disc, allowing it to hold a two-hour (or shorter) movie.

Beyond potential frame cross-talk problems and the need for more refined time-base correction circuitry in the player, CLV sacrifices still- and high-speed frame manipulation, unless an expensive player with digitally controlled still-frame and slow-motion capabilities is used. For budget-oriented enthusiasts who end up with less-expensive, less-versatile machines or who do not purchase CAV discs, the loss of these special-effects features is a small price to pay for being able to spend more time watching and less time flipping and/or switching discs. Only real diehard special-effects fanatics, serious students of film, or music students studying fingering techniques on classical-music discs need the still and slow-motion capabilities of expensive players or CAV discs—and they will be better served in the long run by embracing DVD technology.

Some feel the picture quality of the CAV disc is marginally better than that of the CLV disc, and although this is true in theory, I have never been able to detect this with any of my good-quality monitors. The sound quality of both versions is identical. Needless to say, economic realities have caused the vast bulk of LV discs produced to be CLV.

LV-CD "COMBI" PLAYERS

Nearly all LV players these days are "combi" models. These play the 12- and 8-inch videodiscs (the latter format is nearly defunct), 5-inch "CDV" videodiscs (a bygone format, which displayed about 5 minutes of high-quality digital

Video Resolution

In reading advertising claims about the capabilities of different video components—be they monitors, VCRs, or laser-video players—it is important to remember that test results almost always deal with ideal situations. Most software, be it prerecorded tape or videodiscs, will not measure up to those standards, so the often-fretted-over hair-splitting test-report-discovered differences between assorted components, once a *minimum level of quality* is reached, are usually meaningless. All other things being equal, the subjective picture-quality difference between a TV set that can resolve 500 horizontal lines and one that can resolve 800 is nil; no consumer-available prerecorded source material, including DVD, would overwork either.

Nearly every disc player on the market can achieve 350+ lines of horizontal resolution (remember, vertical resolution is pegged at 330 lines by limits set by the FCC), which is more than enough to handle the usual 300 to 350 lines found on most commercially produced discs. Although this is not state-of-the art subjective performance (that is only usually achieved with test discs, laboratory test hardware, good DVD productions, or possibly some THX releases), it is substantially better than standard (not S-VHS) commercially produced VHS or Beta tapes. The owner of even the most rock-bottom-priced laser-video player, provided it is properly hooked up to a good monitor, will get a dramatically better picture than what is possible with even the best non-super-type videotape recorder.

One advantage that discs have over tape—even Super-VHS tape—involves color resolution. The NTSC maximum for color sharpness is only about *120 horizontal lines*, and the LV disc (and proper NTSC broadcast signals) can deliver this. However, videotape—even S-VHS—can only deliver about *30 to 50 lines* of horizontal color detail. This results in significant color fringing effects (color bleedover or smearing outside of the more highly resolved boundaries of the basic luminance image) and will always make these particular tape formats inferior to videodiscs.

audio and FM-analog video followed by about 20 more minutes of straight digital audio), and standard CDs and 3-inch "single" discs (also an economically dead format). The video performance of these machines is not compromised by the additional features, and their subjective audio-only competence equals that of even high-end CD players. A few combi players are also carousel-style CD changers.

The advent of the combi player put new life into the pre-DVD videodisc, and the medium, while never booming along in a way comparable to prerecorded hi-fi videotape programs, did fairly well for a number of years. More than 8,000 different film and concert discs were eventually available, and a

Videodisc Hookups

Hooking up an analog LV or DVD player is no more difficult than hooking up a hi-fi VCR and only a little more complex than hooking up a CD player. Connect the direct right and left, *audio outputs* (the red and white RCA plugs on the player) to the appropriate audio inputs of your preamp, integrated amplifier, A/V receiver, or TV monitor. If your LV player includes a Dolby AC-3 Digital Surround hookup, things are even less complex; you only have to route a special RF cable[1] from the player to the AC-3 fitting on your surround-sound processor, if it has one (see Chapter 7). As in the case of the hi-fi VCR, the direct, *video output* (the yellow RCA plug on the player) is then run to the appropriate video input on your monitor.

If your system has a receiver, integrated amp, or preamp with video-switching options, you can route the video output to it instead of the TV monitor, along with the audio outputs, connecting it to the same input "source" as the player's audio inputs. (An additional RCA-plug cable is then routed from the video output of the preamp, receiver, or integrated amp to the TV monitor's video input.) The ability of a receiver, preamp, or integrated amplifier to control both the video and audio inputs together is an important—but not essential—convenience if you plan to work with multiple video inputs: say from an LV player and one or more VCRs.

Many LV players also have an RF "antenna" output (F-connector) that allows you to hook them to a TV set via the standard antenna cable inputs. This means that the signal must be further processed within the player and sent off as a standard signal on channel 3 or 4. Given the visual performance limits of that output, you would do well to also obtain a TV monitor with at least one set of separate audio-video inputs. While the horizontal resolution limits of a quality television set's RF input should be 330 or so lines—only a seemingly negligible amount less than the typical 350 or so lines of most commercial discs—the overall quality of the picture, taking into consideration color noise and other artifacts, will be visually inferior to what will be obtained by using the direct-video input. Also, many discs have horizontal resolution levels of up to 400 lines—THX-certified versions, for instance.

More importantly, the audio sent out over the "RF" connection is not stereo or hi-fi, and it will be *substantially* inferior to what will be obtained by using the dedicated, stereophonic, audio output-input feature of the player in combination with a good sound system. No matter how the audio has been produced, not only will there be no stereo, there will be no surround sound either.

[1]The RF cable still has an RCA-type connector and is not the same thing as the F-connector-outfitted RF-video cable.

number of enthusiasts built up large collections. Music enthusiasts should be aware that at one time there were more than 1,000 music-only programs available—both classical and popular—and that many of these will be in stores for some time.

DIGITAL VIDEODISCS

The LV format, with its cumbersome size and shortcomings, will eventually board the slow train to oblivion. The wave of the future is the near-CD-sized sigital videodisc (sometimes called the digital "versatile" disc and/or given the SD—super density—prefix to acknowledge the format's computer-ROM capabilities), employing MPEG video and audio data reduction[9] analogous to what is found in the digital satellite system pioneered by Hughes and Thomson and more distantly to the audio MiniDisc and Digital Compact Cassette.

DVD's advantage

The DVD has two major advantages over analog-video-format, LV discs. *First*, and this will mean plenty to a lot of us, it is small: CD sized. *Second*, it holds vastly more video and audio data per side than the large-disc analog format. While the latter can contain no more than 60 minutes of video and audio on a side (and only 30 minutes of CAV video and audio), the DVD can hold over 2 hours of full-motion A/V program material per side with single-layer encoding and can double that with dual-layering[10]—and double it again if both sides of the disc are used.

This means that nearly every popular movie ever made can be played without an annoying flip-over. In some cases, two complete movies can be placed on a single disc, making the format a natural for those ever-popular "Part I" and "Part II" type films. (OK, Rambo, Rocky, and *Nightmare on Elm Street* fans will have to buy more than one disc.) Even when a program is so long that it requires three sides (something like Ken Burns's *Civil War*, for example), the ease of use of the DVD is self-evident. This is certainly good news to anybody who has had to manipulate a pair of 12-inch conventional laser videodiscs during a really lengthy presentation like *Gone With the Wind*, *Spartacus*, or *Lawrence of Arabia*.

selectable aspect ratios

DVD also offers selectable aspect ratios with wide-screen presentations. While conventional analog-format discs may be available in separate pan-and-scan and wide-screen "letterboxed" versions, wide-screen DVD discs contain a program code that allows multiple aspect ratios to be delivered to your set *from the same disc.*[11] If you have an appropriate 16:9-ratio set, you can watch a wide-screen presentation with no loss in vertical resolution, which is impossible with a letterboxed LV disc. With the DVD, there will be no vertical-resolution loss due to the existence of black bars above and below the screen, a loss made more apparent with wide-screen sets when the image

[9]Data reduction is sometimes called "lossy compression."

[10]Dual-layer technology allows the laser to focus on either a bottom layer of encoded bits or a semi-transparent intermediate layer embedded between the disc surface and the bottom layer.

[11]This feature is also available with some digital satellite transmissions and involves anamorphic compression.

Letterboxing

Letterboxing allows a full wide-screen motion-picture image to be presented on a conventional, NTSC-spec, 4:3-ratio TV set. When an anamorphically produced or hard-matted wide-screen film-to-video transfer is adjusted so that it fills the whole screen, the sides of the original picture are obviously going to be cut off. Therefore, images at those extremes often have to be "panned and scanned" to be seen. For example, when a character on the right side of the screen talks, the image shifts right; when the character on the left answers, the image moves to the left. A scene that was originally stabilized will now have the camera swinging back and forth. On films where the back-and-forth dialogue is rapid, this pan-and-scan technique would look idiotic, so the video editor may be forced to either focus on one character or leave the scene centered somewhere between the two, lopping off critical parts of their respective anatomies.

Letterboxing supposedly solves the problem by placing the entire wide-screen image on the 4:3-ratio TV screen. Of course, when this is done, the top and bottom of the TV picture are blacked out and the individual characters and objects in the assorted scenes are reduced in size compared with a pan-and-scan image. The smaller, more rectangular picture image resembles the mouth of the squat "letterbox" found in British post offices. Many buffs swear by this technique, and a lot of film directors love it, too. After all, it allows their entire artistic statement to be present for the viewer's enjoyment.

The technique does have drawbacks, however. *First*, as noted in Chapter 5, there are a finite number of horizontal-running scan lines available to form an image on a TV screen. With the top and bottom of the picture blacked out, the lines within those areas are not being used. Vertical resolution may suffer dramatically because of this, especially with films having an extreme aspect ratio—beyond 2:1. *Second*, the individual characters or objects within a scene may be too small to see easily. When viewed on a conventional set having a screen diameter of less than 30 inches, radically letterboxed epic films like *Spartacus*, *Ben Hur*, and *Lawrence of Arabia* (which were shot at an extreme 2.35:1 aspect ratio or wider) will sometimes look like stretched-out ant farms. When a really huge TV set is employed to offset this effect, the image may be adequately sized. However, when the same set is then used to watch 4:3-ratio programs, the images look elephantine.

Interestingly, a number of modern films have been shot in the 4:3 ratio and then "matted" to a wide-screen ratio for showing in theaters. When shown on TV, disc, or tape, the viewer may actually see *more* of the total image area than was seen in the theatrical release! Unfortunately, the additional uncropped area often reveals production artifacts (film splices, microphones, scenery edges, etc.) that the directors would not want the audience to see. In addition, many modern films are shot in a hard-matted or anamorphic wide-screen format for theater release but are controlled by their directors on the set so that visual information at the left and right extremes is not all that important. Such true wide-screen films can be adapted to 4:3-ratio video viewing with little need to pan and scan.

Not all letterboxing techniques put the total film area on the video screen. In some cases, presentations that were shot *very* wide for theater release are transferred to video with a mixture of pan-and-scan and moderate letterboxing. Although some critics do not like this idea, it seems like a great one to me, provided no important information is lost. For one thing, it allows a super-wide-screen epic to be

(continued)

Letterboxing *(continued)*

brought up to a more moderate, near-screen-filling size with 4:3 sets, with better vertical resolution. For another, it makes it possible for an expertly panned and scanned moderately wide-screen program to be transferred directly to wide-screen NTSC or HDTV monitors with no blacked-out areas above and below the picture.

A fascinating tidbit about wide-screen films involves the film classic *Gone With the Wind*. It was originally shot 4:3 and was shown that way in theaters for years. Later on, when wide-screen films were becoming popular, the motion picture was cropped to a wide-screen ratio and reissued as a "new, wide-screen version." Actually, this was done in both 1954 and 1967, with the former restoration, overseen by David Selznick, given a moderate wide-screen ratio and also given a surprisingly good simulated stereo soundtrack. The 1967 job was thoroughly ruined by inexpert hacks, given a poorly cropped, wider, 2:1 ratio and overdone pseudo-stereo. Both releases, especially the one done in 1967, actually showed *less* of what the original producers wanted than the 1939 version—like "Super 35" gone wrong. A few scenes were thoroughly truncated in the later version. Some time back, the film was cleaned up, re-edited, and released on video in the original 4:3 ratio. (The sound is still mono, unfortunately, which is a shame, since the 1954 simulated stereo tracks were decent.)

Aesthetic and technical limitations should not alienate 4:3-ratio TV set owners from letterboxing. While *taped*, letterboxed movies may have marginal quality, due to their limited vertical *and* horizontal resolution, videodiscs may have enough horizontal detail to substantially offset letterboxing's inherent vertical-resolution losses. In addition, when watching letterboxed presentations, one can mostly remedy the small-image-size problem by reducing the gap between viewer and set—either by moving the set or the viewing chair. (If a surround-sound system is optimized for a given listening position, moving the TV a tad is probably the best option.) After watching a letterboxed film, the set-to-viewer distance can be readjusted so that normal programs are not subjectively too large.

Although some may be alienated by such machinations, smart shoppers will still choose a letterboxed video over a more squarish presentation nearly every time, because it may be impossible to know if a production was matted from a Super-35 master. When faced with the prospect of selecting either a pan-and-scanned or letterboxed videodisc, remember what I said about NTSC wide-screen and HDTV sets: those items have the ability to "zoom" an image to fill out the whole screen. Letterboxed material, especially at the less radical 1.85:1 ratio, is ideal for viewing on these sets. However, when a panned-and-scanned 4:3-ratio program is "zoomed" to fill out a wide-screen set for more dramatic effect, the top and bottom of the image must be cut off. The result is a picture that is truncated on all four sides: the left and right edges being axed by the video editor and the top and bottom being erased by the zoom feature of the wide-screen set. Those who have a 4:3-ratio set and plan to eventually get a wide-screen model will opt for letterboxing even if right now the sometimes ant-farm-sized images look disconcertingly small.

Of course, the DVD format rescues you from this dilemma by having digitally controlled letterboxing, pan-and-scan, and full-vertical-resolution wide-screen options *built into a single pressing*. The presentation codes are in the recording itself, allowing you to select either pan-and-scan or letterboxing if you have a 4:3-ratio TV set. If you have a wide-screen set, you merely select the full-screen option and gain all the advantages of letterboxing with no loss in full-screen resolution.

is "zoomed" to fill out the screen. If you have a standard 4:3-ratio set, you can configure the player to deliver a "letterboxed" image, or you can set the player to fill the screen to get a panned-and-scanned presentation.

Current digital videodiscs make use of MPEG-2 data reduction, and the players employ shorter-wavelength (read: smaller-size pits to scan) red lasers instead of the infrared versions used with LV and even CD players, although DVD combi-players will automatically refocus their lasers to track CDs. This

	VHS	LV	DVD
Sharpness	Fair: 200–240 lines	Good: 350–400 lines	The best: 450+ lines
Visual noise	Fair	Good	Excellent
Lifetime	Visible deterioration after 15 years	Substantial; possibly longer than CD	Same as CD; possibly decades
User-friendliness	Must be rewound; jamming or breakage possible; magnetic field or heat damage possible; scene searching difficult	No rewinding or jamming; immune to magnetic fields; more immune to heat than tape; quick scene access; separate letterboxed and pan-and-scan versions	Same as LV, except that pan-and-scan, letterboxed, and wide-screen versions are on same disc; excellent still and slow-motion performance
Storage and handling	Relatively compact size; convenient to store if kept away from magnetic fields and heat; dirty-fingered handling OK	Large size demands big shelf, with substantial end bracing; easy to damage with rough handling	Lightweight; easy as CD to store; more resistant to rough handling than LV but not as robust as tape cassette
Availability	Zillions of titles available; rental facilities abound	Reasonable number of titles available; as DVD spreads, quantities may dwindle; rental facilities rare	Up-and-coming format has fewest titles available, but will expand
Cost (software)	Tape purchases often initially pricey, but generally prices fall rapidly; rentals are reasonable	Often cheaper than tape at beginning, but more expensive later on; cutouts are often very cheap; rentals usually more expensive than tape	Medium to high, depending on program
Cost (hardware)	Reasonable, although premium and S-VHS versions can be expensive	Medium to high; discontinued machines can be very cheap	Medium to high; switching to this format almost mandates AC-3 processor and wide-screen TV
Reliability	Good; easy to find repair facilities	Fair; difficult to find repair facilities	Good; equal to CD player; difficult to find repair facility
Obsolescence	Format will be around for years to come	Discs and players will decline sharply in numbers over the next decade	The new kid on the block should last for many years—even into the high-definition TV era

MPEG and Video

The MPEG video compression standard is a method of providing digital video for satellite and cable television, DVD disc, and even desk-top computers. Uncompressed/unreduced digital video requires well over 1,000 megabytes of data per minute to achieve 350+ lines of horizontal resolution. This means that a typical movie would require 100–150 thousand megabytes of storage space, something that no affordable system within the foreseeable future could accommodate. In contrast, MPEG digital video can produce acceptable video images with a data rate of less than a dozen megabytes per minute.

The reduction process is similar in concept to what is used to deliver DCC and MiniDisc audio—indeed, the PASC data-reduction system used by DCC is an MPEG variation. However, MPEG has a very high compression ratio—as much as 200:1—and uses an intraframe compression scheme based upon the relationship between adjacent video frames. By using motion estimation or compensation, all redundant information—nonmoving material—is eliminated from digital consideration, and only moving artifacts in the picture are continually processed. Early MPEG-1 satellite transmissions, with their fixed data rates, were at their best with talk shows, where the background tends to be steady, and at their worst during sport events or action movies, where both players and the background move with considerable vigor, overloading the processing system. MPEG-2 and more advanced versions, which have variable data rates, do not suffer from this problem.

gives them video performance somewhat superior to that of analog videodiscs. In addition, these new discs can also handle up to 5.1 channels of perceptual-coded digital audio and if necessary have multiple-language capability. Proposed versions may eventually employ shorter-wavelength blue lasers to further increase capacity with HDTV playback.

The DVD has nearly everything going for it. Unless you are constrained by serious budget limitations and must opt for standard videotape and tape-rental stores for your collections, anyone starting out on a pursuit of the best in video quality would be crazy not to make DVD the first choice.

The table on page 233 gives you a brief comparision of VHS, LV, and DVD formats.

HARDWARE: NEW VS. USED

Unless 5.1-channel performance is important to you, LV players produced by different companies have performance levels so similar—particularly audio performance—that in most cases film or music-video enthusiasts need only

Figure 6.6 Toshiba DVD Player

(Photo courtesy of Toshiba)

A DVD player like this reasonably priced Toshiba model has just about all the control, performance, and technical frills that any home-theater enthusiast would want. Most are "combi" players, which means that, in addition to digital-video discs, they can also play CDs with as much reproductive accuracy as the best CD-only players. The DVD is a superior video reproducer, with resolution and video-noise levels that have only previously been achieved by professional-grade digital-videotape recorders.

When hooked up to a top-quality, wide-screen monitor and an appropriate surround-sound decoder-processor, a player like this one will deliver the best picture and sound performance this side of high-definition television. Indeed, given current viewing habits, most individuals would not be able to see a substantial difference between DVD and what is available from HDTV. What's more, this technology, which is based on MPEG algorithms, is upgradable to HDTV levels. Serious audiophiles should also rejoice, because DVD technology has the potential to deliver superb, 5.1-channel, audio-only sound, either by means of linear-PCM techniques or data reduction. For enthusiasts on a reasonable budget—and who want real impact in their home theater—the DVD system is the way to go.

Note that while this particular DVD player has both composite and S-Video outputs, Toshiba also offers a more upscale model with these plus a "component" output that is compatible with the component inputs used on data-grade projectors, high-end video displays, and line doublers. A component output-input insures superior quality from a DVD player–video-monitor combination—even better than what is possible with an S-Video connection. Some of Toshiba's wide-screen sets feature component inputs, meaning that they should be able to deliver DVD pictures a cut above what most other models offer.

pay attention to price and operational features. With DVD players, the real-world performance differences between models are microscopic. However, there are a few quality advantages and operational features that come with the more expensive models that may be of interest to the prospective buyer. Before shopping, remember how many additional movie or musical-program recordings you can buy for the price difference between an expensive, "feature-laden" player and a no-frills, "budget-level" model—if we dare talk "budget" when discussing these formats.

LV and DVD players cover the price spectrum from cheap (okay, decent-hi-fi-VCR "cheap") to plenty expensive. In terms of playing back movies for the simple enjoyment of watching them, the "econo-grade" LV players are subjectively about as good as the far more expensive ones; even AC-3 audio is available on reasonably cheap models. I see no reason to opt for a really expensive FM-analog LV player to get absolutely top-quality performance, because—let's be realistic—there are good DVD players available for those demanding the ultimate in video.

While a really large, top-notch TV monitor might be able to highlight the subtle optical advantages of a super-grade LV player (one that sports digital time-base correction and/or digital comb-filtering), most people would not be able to appreciate the difference. This would be true even if you were looking for picture-degrading artifacts while sitting four feet from the screen. The horizontal resolution and subjective color-noise performance of a high-end LV player may at times be slightly better than what is possible with the cheaper units—but most discs are just not good enough to show up these differences. OK, if you specialize in watching test discs or expensive THX editions this may not always be the case. However, true fanatics would probably skip the older technology completely and opt for a DVD unit.

However, upscale LV players may have features that budget-level units do not. Many advertisers tout an S-Video connection as an important feature. As already mentioned, this hookup subtly improves the picture coming from a Super-VHS, ED Beta, or Hi-8 videotape recorder, because it keeps the chrominance (color) and luminance (picture) signals separated until they reach the TV monitor. However, unlike a videotape, the LV disc does not have those signals stored separately; they exist as a "composite" and are already mixed together. In most cases, if your TV monitor is a newer, premium-grade model (which will almost certainly have a fine-comb filter), you probably will not gain much if you spend extra for a player that has an S-Video output. I should note here that the S-Video output on DVD players will deliver better pictures than the composite output.

Another feature lovingly promoted by advertisers is a digital-audio output connection, enabling you to hook up your unit to an outboard two-channel DAC, supposedly for improved sound quality. While future receivers and amplifiers may employ built-in DACs for CD-player use, the advent of the DVD and AC-3 has made outboard DACS anachronisms for video.

A number of high-end LV players have built-in digital circuitry that allows you to do freeze-frame and slow- or fast-motion work with CLV discs. However, remember that even better results may be possible by using CAV discs in combination with a cheaper player—or by forgoing LV and going directly to DVD which has spectacularly good freeze-frame and slow-motion abilities. Special effects may seem like a wonderful option before you buy a machine; however, my experience has been that after a few slow- or fast-motion manipulations the feature loses its attractiveness. Note also that a good VCR will also be able to produce stable freezes, whereas a CLV disc of the same program, at least when played on a budget-grade player, will not. Consequently, the best poor-man's special-effects medium may be videotape.

two-sided play

Finally, some mid- and upper-strata LV models offer automatic two-sided play. This involves a specially designed tracking mechanism that moves from one side of the disc to the other after the first side has been played, while the disc comes to a stop and goes into reverse. The process only takes a few seconds and is certainly more convenient than having to get up and turn over a disc. However, the function may only be important if you watch a lot of short-playing CAV discs. This dilemma will be nearly meaningless with DVD, since most movies will fit on one disc side. The obvious drawback to this feature is that auto-flip players are more expensive and add additional complexity to a mechanism that is already pretty complex to begin with. One obvious advantage to this feature is less disc handling by people, meaning there will be less of a chance of damaging a disc by dropping it, scuffing it, or placing it in the disc drawer wrong. I must admit that once you get used to the auto-flip feature, it can make watching even CLV discs more enjoyable.

Used Equipment

I would never purchase a used LV/CD combi player unless it were nearly new and maybe still under warranty, or owned by a videohound who hardly ever used the thing and never abused it. Used discs? Now that DVD is among us, decent-condition FM-laser videodiscs may become extremely good things to shop for. Indeed, when your friends sell their analog combi player to you cheap, maybe they will also throw in their old disc collection as a bonus.

SOFTWARE

As of this time, I know of no book-compendium of DVD reviews. For those, I suggest subscribing to (or having your local library subscribe to) one of the numerous home-theater or video magazines available. Other review sources include the *Laser Disc Newsletter*, *Stereophile Guide to Home Theater*, *Video* magazine, *Laserviews*, and *Widescreen Review*. In addition to the DVD, these magazines continue to publish review material on analog LV and on video technology in general.

As for shopping for older LV discs, particularly those available used or sitting in clearance-sale bins, I suggest you get a copy of *The Laser Video Disc Companion*, by Douglas Pratt (Baseline Books), editor and publisher of the *Laser Disc Newsletter*. Pratt knows his laser discs and has probably evaluated more of them than anyone else. The book leads off with an excellent introductory overview of conventional laservideo technology. While it is occasionally off base when discussing the artistic merits of certain movies and video concerts (Pratt is not the only reviewer to let his film-school preconceptions occasionally run wild), the book is excellent at analyzing their technical qualities. This makes it doubly helpful when several versions (CAV, CLV, letterboxed, import, reissued, etc.) of the same film are available.

Another helpful book dealing with LV releases is *The Laserdisc Film Guide*, by Jeff Rovin (St. Martin's Press). The first version of this book was published in 1993, and Rovin says that subsequent editions will follow with additional reviews. Let's hope so, because the book deals at length with only slightly over 300 titles, meaning that it has far, far fewer of them than Pratt's book. Nevertheless, the discussion of video and audio *technical* quality is more comprehensive, readable, and to the point than in Pratt's book, and Rovin even has a Zero- to Five-Star numerical rating system to summarize his views of both technical (further subdivided into video and audio categories) and artistic merits.

When shopping for *musical* discs, it is important to ask yourself if you really need a picture. Unless you are a Madonna, Paula Abdul, or Jackson-family fan, there is precious little of substance in most video performances compared with their audio-only versions. Indeed, the visual part of a performance may best be left to the imagination. In addition, there are a number of classical-concert symphonic discs available that reveal nothing more than what the camera can see as it pans the orchestra players in wide-angle and close-up modes. Is this important to you? If you enjoy the visuals during the Boston Pops presentations on PBS or were captivated by Leonard Bernstein's

Young People's Concerts or watch MTV or VH-1 regularly, then maybe you will like these programs.

music productions

High-quality, visual-oriented performances are another matter. If you love opera and/or ballet, then you will be thrilled with the assorted video productions available. For important information on the quality of both motion-picture and musical material, see the books and journals already mentioned. You can also rummage through back issues of older journals like *Laserviews*, *Video* magazine (combined some time back with *Sound & Image*), and *Video* review (which was discontinued a few years ago). The record-review magazines *Fanfare* and *The American Record Guide* also publish a few classical-music video reviews. Given the cost of LV and DVD discs, it is important to obtain recordings that deliver a high-quality audio, as well as video, performance, and the reviews will probably critique the sound as well as the picture.

If you plan to add a movie to your collection, be sure to obtain the latest reasonably priced version of it. In many cases, a film may be initially transferred to disc without the necessary care; it will then be reissued in an improved version. The improvements may involve audio or video quality (often with THX certification), technological upgrades (including surround-sound advancements such as AC-3 or reformatting from LV to DVD), and/or improvements in content including extended length. Good examples of this are *Gone With The Wind*, *The Star Wars Trilogy*, *Aliens*, *The Abyss*, and *Lawrence of Arabia*. Some early releases had substandard sound (some were not even given CX encoding, let alone digital audio), and many wide-screen films were panned and scanned when they should have been letterboxed. Some of the technically fine early transfers, such as *Terminator 2*, were made even better by THX remastering. In addition, many later editions are "director's cuts" (*Terminator 2*, again, for example), which often not only have additional footage beyond what appeared in the theatrical release but may also have a running artistic or technical commentary on the parallel analog-audio tracks.

Remember that, unless you have a wide-screen TV monitor, letterboxing may not always work to your advantage (see "Letterboxing," page 231–32). For example, pan-and-scan films on LV discs will have better close-up detail than the wide-screen versions. Also, many wide-screen films are actually shot at a 4:3 ratio on 35-mm stock and masked (matted) at the top and bottom to fill the wide theater screen. *Air America*, *Black Rain*, *Ghost*, *The Mosquito Coast*, *Star Trek VI*, *Top Gun*, and *Total Recall*, among a fair number of others, were done this way. Video 4:3-ratio releases of these films often have the full width *and* the restored top and bottom. They are *not* panned and scanned, and you see more of the picture area than you did at the theater! Even many

Director's Cuts

One of the most annoying aspects of videodisc reality is the mixed blessing of the "director's cut." Director's cuts offer many advantages to the discerning videophile. They are usually not only better than the earlier video releases of a film but are often artistically superior to, or at least longer than, the theater releases. Anyone who compared the THX-certified *Terminator 2*, the director's cut of *The Abyss*, or the longer version of *Aliens* with any of the earlier LV versions will agree that they were vastly improved after they were restored to the length and/or artistic configuration envisioned by James Cameron before his work was truncated by the bean counters in the front office who demanded more compact versions.[1] In addition, few complained when very early LV releases were replaced by cleaner, better-sounding, occasionally longer-running wide-screen LV versions after being given better technical and artistic editing. This was certainly the case with the reissues of *The Wild Bunch* (which even had AC-3 digital surround sound), *The Lion in Winter, The Big Chill, The Star Wars Trilogy, Gone With the Wind, Fantasia, Casablanca, Lawrence of Arabia*, and the THX-certified versions of *The Wizard of Oz* and *Oklahoma!* Indeed, the older LV release of the latter had been copied from a 35-mm CinemaScope filmstrip, whereas the newer, THX-certified transcription was made from a 65-mm Todd-AO sample and not only was technically better but also featured earlier outtakes (the Todd-AO scenes were shot first, and the grainier 35-mm scenes were shot afterwards) that had the performers often performing better.[2]

However, what annoys a fair number of us is a producer's or manufacturer's habit of releasing a technically good disc and then following it a few months later with a supposedly "better" one (longer, re-edited, improved sound, improved picture, etc., etc.) at a substantially higher price—even though it utilizes the same technology. *Why was the shorter, inferior version released on videodisc in the first place?* Why not just issue the LV disc or DVD edition as a "director's cut" right off the bat and save us cash-poor video enthusiasts the agony of having to decide if it is worth replacing our original versions with the supposedly "better" ones? Give us the best right off—and if the bean counters insist on high profits, charge us a premium price. The director's cut is sometimes nothing more than a marketing ploy to milk the same cow twice!

[1]Why do the powers that be produce shortened theater versions at all? Because short presentations allow more showings during the day—earning more money.

[2]The reason that two different film-stock versions were shot is that most theaters did not have the superior Todd-AO playback gear and the producers wanted to make sure that the film would get the widest-possible exposure around the country—while still having a superior version for theaters that had the better playback gear.

hard-matted or anamorphic wide-screen films are composed in such a way that moderately panned-and-scanned video versions transfer to the 4:3 ratio with little in the way of important detail or scenic losses—or unsettling pans.

Always keep the sales receipt and play the disc as soon as possible after a

defects

purchase to make sure it has no defects. Those may manifest themselves in many ways. Numerous white speckles or horizontal white lines scattered about at regular intervals on an LV disc (mostly in dark scenes) will probably be dropouts, and if there are a lot of them, the disc should be taken back and exchanged. Every disc will have at least a few, however, and they may be caused either by embedded dust or a defective aluminum substratum. Black speckles or lines and dark, irregular-pattern artifacts could indicate a problem with the original film stock, and it may be that an exchange will not correct this. However, remember that newer *editions* (such as *Oklahoma!* discussed in "Director's Cuts," page 240) may be transferred from more pristine stock. If no new edition is available, you will have to decide if keeping a disc made from substandard film stock is worth it. If it looks little better than the tape copy you rented or still own, then it obviously is not.

If there is any kind of tracking or cuing problem, the disc should also be replaced with a better copy. Note, however, that I have seen a few cases where an LV disc would not cue up on an otherwise normal player—and neither would its replacement copy—although both samples worked well on players made by other companies. The first player was reacting to an encoding artifact. Your only options would be to choose a tape version, get a new LV player, or switch to DVD.

warpage

An occasional 12-inch LV disc may be excessively warped, and—even though it may present a decent picture, thanks to the player's hard-working tracking mechanism and/or its error-correcting circuitry—it may create a lot of rumbling noise within the player itself. In time, noisily warped discs will cause excessive wear to the mechanism, so exchange them.

Sometimes, the description of the disc on the box—either the timing, contents, or video material—may be inaccurate, as advertising copywriters "puff up" the product. You'll either have to live with these inconsistencies or exchange the disc for another product. However, sometimes companies do make mistakes when initially releasing movies on disc. In their books, Rovin and Pratt both point out numerous poor mastering jobs. For example, according to Pratt, the audio channels are backward on the early CAV version of *ET: The Extra-Terrestrial;* the CLV version was OK. The moral: More expensive is not always better.

DISC CARE

Take care of analog LV discs and DVDs as you would CDs, remembering that the former, being heavier, will be more susceptible to physical damage than the lightweights. Avoid scratching or smudging the surfaces. Handle discs by

cleaning

the edges and center; for larger LVs, make use of your old LP-record-handling abilities. If discs get dirty enough to affect the picture or sound, clean them with the same kind of local-area, radial-direction-rubbing technique discussed in Chapter 4. If a disc has a fingerprint or a bit of dirt on a playing surface and it does *not* cause any visual or audio problems, don't fool with it; cleaning can sometimes put microscratches on the surface which cannot be removed and which may eventually lead to unsolvable problems. For stubborn hard-to-remove glop, good, proprietary cleaners produced by some record-care companies may work quite well, but use them carefully until you know if they are benign. One safe bet is to simply use a bit of distilled water as a cleaning substance. (Just breathing on a disc will aid minor cleanups.) Keep discs away from excessive heat and sunlight, and avoid laying them unwrapped on hard, dusty surfaces.

packaging

If you buy an LV disc that comes with one of those flimsy, semicircular, all-plastic, transparent inner sleeves, go to a record store and get a pack of square, plastic-lined, paper-sleeve replacements—similar to what quality LP records used to use. (Good ones are available from Radio Shack.) The flimsy plastic sleeve is hard to insert into the cardboard jacket (unless it is a hinged, box-type design), and it is so floppy that a sleeved disc may actually roll out when you pick the package up. The plastic-lined paper sleeve is stiff and is easier to work with, and its square shape will not let the disc cartwheel out of the sleeve if it is inserted properly. Put it into the cardboard jacket with its open end facing *up*, against the closed top of the jacket. That way, the sideways-facing, open end of the cardboard jacket will be sealed with a closed, paper/plastic sleeve edge; the disc cannot roll out.

loading

Take care when inserting a disc into a player, particularly an LV disc. If the disc is inserted partially or crookedly, the closing drawer may crunch down and ruin it. (This can also happen with compact discs and DVDs, but the larger LV disc is often harder to center up in the larger tray.) I once saw a salesman in a brightly lit store stick a 12-inch disc into a player improperly. When the drawer closed, the disc was "ramped" right up to the top of the player. It just slid across the top and skidded to a dusty stop. The salesman was speechless, but your writer was mightily impressed.

storing

Store DVDs the same way as you'd keep CDs. Easy storage is one of this format's strong points. Store large LV discs the same way you once intelligently kept your LP records. *Do not* insert them in one of those old spiral-type record holders high-school kids use for grooved recordings. Put them on a shelf and stabilize the row with a sturdy bookend. Place a 10- to 12-inch-square flat spacer (thick cardboard is fine but wood may be better) between the discs and the bookend as a protection against denting or bending. Do not allow the row to grow so large that it causes excessive pressure on the individual discs. Vertically stacking discs is not a good idea, but if you must, do not put more than a half-dozen in the pile. Finally, keep *all* videodiscs (and audio discs and all kinds of video and audio tape, as well) away from dust and things like cooking fumes, solvents, and heater ducts. And whatever you do, do not lend your discs—LV or DVD—to fumble-fingered, greasy-pawed friends or let young children play with them.

7 Surround Sound

SURROUND SOUND—THE CAPABILITY to simulate a concert hall or theatrical experience—is the last important element of any home theater. For those who have lived happily with traditional two-speaker stereo systems, they may be shocked by—and even resistant to—the idea of additional speakers. Yet one listen to a decent surround system should convince anyone that for audio-only or audio-video recordings that will approximate life-like experience, surround sound is the way to go.

HOW IT WORKS

The heart of any surround-sound system is the processor. In some cases, this may be a sophisticated add-on device that couples an existing stereo system to additional amplifiers and surround and even center speakers. More commonly, a surround processor may be built right into a preamplifier or included as part of an integrated amplifier or receiver. The latter two configurations run the gamut from budget/basic units, with some movie-soundtrack decoding circuitry and no hall-simulating features at all, to state-of-the-art devices that have a variety of enhancement options and come close to holding their own with the better "separates."

While separate, outboard-mounted processors and surround preamplifiers may have a few built-in amplifiers to augment the main amps, integrated amplifiers and receivers will nearly always have the ability to drive both the main and surround speakers—including a center channel—without the aid of outboard amplifiers. This is their big attraction, because it results in a simplified hookup procedure. Basic surround-equipped integrated amplifiers or receivers will have four or five speaker connections, and more elaborate ones may have as many as seven—in addition to an unamplified output for a subwoofer. Some can even control two subwoofers.

Figure 7.1 Onkyo Surround-Sound Receiver

(Photo courtesy of Onkyo)

A surround-sound receiver is the way to go if you are building a home-theater system on a budget and/or do not want to clutter up your living room with extra pieces of hardware. While models like this moderately priced one by Onkyo do not have the ultrasophisticated DSP circuitry that the more expensive Lexicon, Sony, and Yamaha—as well as other Onkyo products—feature, they often have enough synthesizing and matrixing capabilities to make a substantial improvement in the sound of nearly any audio recording. While at one time a luxury item, these days even fairly low-priced A/V receivers feature Dolby Pro Logic circuitry and video switching. Happily, the Pro Logic function of many cheaper models is as good as what is found on those costing considerably more, insuring that individuals with a good video monitor and five or more good speakers located in a decent environment (or even only three up front, working in the "Dolby 3 Channel" mode) can have a satisfactory movie-theater experience.

Universal or Dedicated Surround-Sound Circuitry

However it is packaged and whatever it costs, the circuitry within a processor may be universal, enhancing nearly any kind of stereo (or mono) recording, and/or dedicated. The former is designed to work with a variety of mostly audio recordings (usually straight stereo), while the latter is designed to work in conjunction with either video or audio software. Almost all processors these days, unless included within a very-low-cost A/V receiver, contain both universal and dedicated circuitry.

Universal processors can be subdivided further into two types. As a rule, the least costly is the *ambience-extraction* device, which works by *extracting* the reverberation and stereo difference signals (left minus right channel [L – R]) that exist on any stereo recording. This type of signal manipulation is

ambience extraction

often categorized as a "matrix" or "hall" mode on some surround-sound processors. Its effectiveness depends not only on the nature of the circuitry employed but also upon the amount of stereo and out-of-phase information a recording contains.

ambience synthesis

The more costly, but popular with many enthusiasts, universal processor is the *ambience-synthesizing* device. There are many different variations of these available, most of which use DSP (digital signal processing), but all work by *adding* processor-generated reverberation to the ambience already on the recording. Like the extraction-type processors, this additional reverb will usually be steered, delayed, and frequency-contoured, and then sent to two or more surround speakers—and sometimes even the main speakers.

Usually, these two types of universal surround processors—ambience extraction and ambience synthesizing—are audio (as opposed to video) oriented. DSP ambience synthesizing, ambience extraction, and video surround-sound capabilities will often be included as a complete package within upscale audio-video processors, and some models make it possible for these functions to be combined in a variety of ways.

Dedicated processors are designed to work with specific surround-encoded material. These include three standards developed by Dolby, plus the far-less-common *Ambisonic* system. The two best-known Dolby decoding schemes for home-theater use have been around for years: the Dolby Surround matrix configuration, which is similar to the L – R extraction method noted above; and the more advanced Pro Logic version of this same system. The third Dolby system is a much more effective digital format employing Dolby AC-3 Digital Surround encoding for truly discrete separation between all channels. Home-theater processors will incorporate at least one of these Dolby systems, and many will also have audio-only enhancement capabilities, either of the extraction or the synthesizing type—or both.

Decent results can be obtained with surround speakers that are small, bandwidth-limited, and cheap. This is particularly true with "matrixed" Dolby motion-picture program material that does not (theoretically at least) require surround speakers with formidable bass capabilities or an extended high range. (It is also true if you employ a subwoofer that is tied into the

subwoofer

"dedicated" subwoofer output in your processor, because the bass from *all* channels will be routed to the big woofer.) In contrast, it is important for the center speaker (even if it is also small) to be of fairly high quality, particularly throughout the midrange, because it will often be called upon to handle serious program material, particularly with motion pictures.

It may also be necessary to have high-quality wide-bandwidth surround speakers for standout results if they are to be used with audio-oriented DSP-type or other "synthesizing" processors, or for 5.1-channel technology not involving a processor-dedicated outboard subwoofer. If no subwoofer is employed, these formats may send a wide-bandwidth signal to the surrounds. While bass in the surround channels is redundant with what is going to the main speakers or being sent to a subwoofer, some individuals believe that having it come from multiple sources around the room will smooth out the bass by controlling room standing-wave effects.

All of this makes it possible to have a complete home surround-sound system with four to seven (or more) speakers scattered about the listening room, and enough wire, amplification, and control sophistication to daunt all but the most dedicated audio enthusiast. Combine this with the seating, video-monitor placement, and surround-sound requirements of good home theater, and you can see that this facet of audio-video can be expensive, complex, and confusing—but also interesting and rewarding. Let's take a look at some of the basic universal and dedicated processor types—as well as some surround-sound recording and playback techniques—and see what we can learn about high-fidelity sound reproduction and the enjoyment of music and video.

FILM AND VIDEO

Viable surround sound for visual programming goes back to the early 1940s, when Disney Studios released *Fantasia*, an animated motion picture that made use of "stereophonic" sound reproduction. Until that time, commercial movie-theater sound was handled by a single channel picked up from an audio track on the film. While the action on the screen might have lots of left-to-right movement and be supported by large-scale symphonic music, movie makers depended upon visual clues to help the audience orient the sound with the action.

The Disney "Fantasound" system delivered the output of *three* separate channels to a trio of speaker systems spread out behind the movie screen. The sound was delivered by an optical sound track on a separate reel, which was synchronized with the visual material on the picture reel. A number of "surround" speakers were placed behind or to the sides of the listeners to enhance the sense of spaciousness and drama. While the audio tracks of *Fantasia* were meant to give an accurate reproduction of an orchestra and not to create sound effects or to synchronize character movements on the screen, the discrete-channel system used in this film added a special new dimension to motion pictures. Unfortunately, World War II and the costs of outfitting theaters with proper equipment put surround-sound motion pictures on hold for a number of years.

The concept was revived in the 1950s and 1960s in a number of "wide-screen extravaganza" film productions. Films like *The Ten Commandments*,

Ben Hur, *Around the World in 80 Days*, and *Spartacus* were made much more gripping when on-screen multichannel audio complemented the wide-expanse visual material. The principal impetus for this renaissance of advanced motion-picture audio (not to mention the wide-screen image) was television, which had cut into theater revenues considerably. In addition to having the screen sound follow the screen visuals (as well as having the music simulate a live-orchestra experience as in *Fantasia*), this technology also allowed movie makers to portray (in the sound program) action happening *off* the screen. This sense of envelopment required five behind-screen speakers (*left*, *near left*, *center*, *near right*, and *right*), as well as an array of additional units mounted to the sides and behind the audience that reproduced a sixth channel: the *surround* information. The six magnetically encoded audio channels took up a lot of space on the film strip, mandating a much larger format (often as wide as 70 mm) than that of previous motion-picture systems. (A byproduct of this film was a better picture, particularly on extremely large screens.)

The costs involved limited the use of large formats to theaters that could afford the playback equipment. Installations were rare outside of larger cities, and in the later 1960s and 1970s distribution and financial problems ultimately forced many motion-picture production companies to back off from elaborate discrete-track multichannel formats. However, the concept did not completely die off, and from the late 1970s to the early 1990s a few large theaters continued to use discrete-surround playback systems for large-format versions of contemporary films.

matrixing

While these early adventures in motion-picture surround sound were taking place, a new audio-oriented matrixing-dematrixing concept was being developed by Peter Scheiber. This 4:2:4 process (four-channel to two-channel to four-channel) involved "folding" *two additional surround* channels into the *two main stereo* channels during the recording process. The two matrix-encoded channels could play back without problems on any two-channel playback system, giving a somewhat heightened sense of spaciousness. However, with suitable decoding they could also be dematrixed (the encoded information being mathematically extracted from the primary channels) into four reasonably discrete signals: *left front*, *right front*, *left rear*, and *right rear*. The idea was to develop a surround-sound format that could easily be integrated into the existing two channels of a stereophonic LP record, analog cassette-tape player, and stereo FM radio. The "SQ" and "QS" systems of that era were two manifestations of the matrixing concept. Unfortunately, none of these systems caught on for a variety of marketing, technical, and practical reasons, and matrixing fell out of favor for several years.

However, in 1977, a new theater-sound process was dramatically introduced to moviegoers—one that delivered surround enhancements to the mass market. This system, a spinoff of "QS," was designed to be cheaper and easier to work with than the earlier, more elaborate discrete-track systems. Instead of independent and costly surround, main, and center channels, the new system involved the Scheiber-pioneered 4:2:4 concept but reoriented the four channels to deliver *left front*, *center front*, and *right-front* outputs, plus a single channel of *surround* data. The four inputs were folded into two optically read, stereo-compatible channels that still functioned well in theaters having 35-mm stereo or even mono playback systems. However, when the same matrixed 35-mm film was played in theaters having the necessary decoding hardware, a decent facsimile of four-discrete-channel surround sound would result. This new theater-sound process was called *Dolby Stereo*, and the film that introduced it to the general public was *Star Wars*.[1]

The "remastered" release uses a discrete, digital surround process.

Although its initial implementation was not as well executed as it would be in subsequent films (including the two later *Star Wars* epics), Dolby Stereo caught on rapidly, and within a few years the process was being used in a multitude of films, with over 21,000 theaters worldwide currently equipped to play back the matrix. While not every Dolby-encoded film displays exemplary sound, even middling-level efforts are more impressive than mono or straight stereo. Dolby-surround-equipped theaters have since become the norm, although they, like the films themselves, vary in quality.

The Dolby matrix's simple two-channel storage requirement makes it possible for the sound to be transferred directly to a stereo-compatible video format. Thus, any hi-fi videotape or videodisc that is a copy of a Dolby Stereo film will also contain the surround data. To differentiate the theater version from the home-video version, the latter is called Dolby Surround Sound. If a home playback system has the necessary decoding gear and is properly configured, the Dolby matrix-surround theater experience will be duplicated—well, almost duplicated. For the next few pages, our main thrust will be to analyze Dolby Surround Sound and advanced variants of this system, plus, further on down the line, the newer 5.1 formats.

DOLBY SURROUND

In principle, basic Dolby Surround allows the front stereo signals in the program to pass through to the left and right main speaker systems unaltered. The additional, matrixed information is not perceived as something "alien" during stereo-only listening—although the intensity of the surround information may impart a certain additional spaciousness or out-of-phase quality to some parts of the sound. This is one reason why a Dolby soundtrack will

[1]Actually, Dolby Stereo was first applied in *A Star Is Born* in 1976, but the space epic's need for flyovers and surround ambience highlighted the impact of this new process and caught the public's attention.

Hafler Circuit

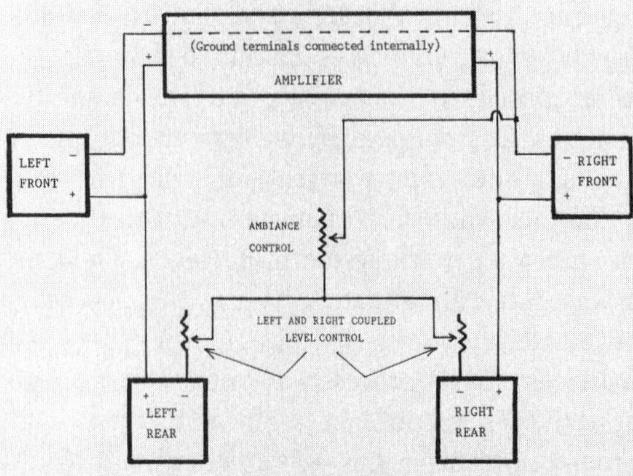

A spin-off of the basic Scheiber topology was introduced by David Hafler back in the 1970s. The "Hafler Circuit" diagrammed here is one way to make use of the ambient information found on any stereo recording.

Because the stereo effect requires that left minus right (L – R) data be present, it is possible to *extract* it and use it to feed sound to one or more rear- or side-mounted speakers for enhanced playback realism. The Hafler company and Dynaco have offered "black-box" devices as economical add-on components for some time to do this task, and in the past similar hookups were available in a few budget-grade Japanese receivers. A number of adept audio enthusiasts have also built similar devices from scratch. The latest Dynaco version even has a simple non-steered L + R center-channel output.

The circuit sends the extracted "difference signal" to two specially connected "surround" speakers. Note that this hookup does *not* require a separate surround-channel amplifier. The L – R information is extracted from the stereo signal by connecting the surround speakers across the "hot" leads of the amplifier or receiver. The ambience control (which comes with the factory-built jobs) allows you to adjust the relative amount of L – R information. By using this control, the perceived ambience can be varied from full L – R to a "mono" summation of the left and right main signals. The coupled level controls adjust the gain of the surround outputs relative to that of the main speakers. Typically, the surrounds should be adjusted to play about one-tenth as loud as the primary speakers (10 dB down). Also, for best results they should be somewhat farther from the listening position than the main speakers.[1]

Because it uses only the power provided by the two-channel amplifier, the impedance of the speakers and the internal design of the amplifier itself is important. If any of the speakers you use with this circuit are less than 8 ohms (particularly if all four are 4-ohm models), the extra current draw might overload the amplifier. Also, if the amp itself does not have internally connected ground terminals, a real electrical disaster could result from using this hookup. Before using current Hafler or Dynaco processors (or if you can purchase, used, an early Dynaco "Quadaptor") or before building a device yourself, contact your amplifier or receiver manufacturer to make sure that your amp has *internally connected grounds* and can handle the combined impedance loads.

Remember, this circuit is a bare-bones substitute for a surround system having 4 or 5 channels of discrete amplification for the main and ambience channels. It can make many recordings sound more open and spacious, but it cannot equal a major-league Dolby or other type of surround-sound setup.

[1]If the surrounds are closer to the listener than the mains, the result will be the curious effect of having the extracted ambience arrive before the primary signal that created it in the recording. More elaborate surround systems have an effects-channel delay circuit to prevent this from occurring.

usually sound fine on a conventional two-channel stereo playback system. Indeed, the "phasy-sound" of the surround-encoded material may produce surround-like effects if the listener is sitting near the central sweet spot, and the out-of-phase surround signals may even appear to be coming from off to the side—although sometimes the wrong side.

derived center channel

In my opinion, at least in terms of dramatic and artistic impact, the most important part of the Dolby-encoded signal is the "derived" *center* channel. This is because it contains the dialogue, as well as a substantial amount of music and sound effects. It is important to remember that this center channel is really nothing more than the *sum of the left and right signals* (L + R). Any left-*plus*-right information (identical, in-phase sound from the left and right main channels) that exists within a stereo film soundtrack is considered the "center" channel.

However, any hi-fi enthusiast knows that an audio recording or soundtrack's center channel can also be reproduced without a center speaker. Indeed, center-channel information exists in *all* stereophonic recordings, Dolby encoded or not. In a standard two-channel stereophonic set up, the L + R sound forms a "phantom" image for listeners sitting in the central-axis location between and somewhat in front of the main stereo pair: i.e., the "sweet spot." The phantom-center phenomenon allows a "Dolbyized" film to work when played back on just two front speakers.

extracted surround channel

Although less important from an on-screen drama standpoint than the center-channel information, the fourth or *surround* channel often heightens the impact of a Dolby-encoded program, particularly one with a lot of action. This "extracted" left-*minus*-right information contains the environment-simulating ambience that allows a sense of envelopment. In a conventional, stereo audio or video recording, the L – R part of the program will include front directional clues, stage-depth information, and a certain amount of hall or studio reverberation; this information is what makes stereo "stereo." A Dolby-film soundtrack also contains surround artifacts, and assorted—usually vaguely focused—"effects." Like the center-channel signals, specifically encoded surround signals in the left and right main channels are identical in spectral content. However, these identical signals are folded in to the main channels at a phase angle of +90° in the right channel and at a phase angle of –90° in the left channel—that is, *they are 180° out of phase with each other*. When heard on a standard stereophonic home-playback system, most—but not all—of those encoded artifacts will be heard as coming from the front area and will enhance the sound stage to some degree.

Film-sound engineers avoid putting surround effects that are critical to the story entirely into the surround channel. They leave a percentage of the surround material "in phase" so that it will also be reproduced up front.

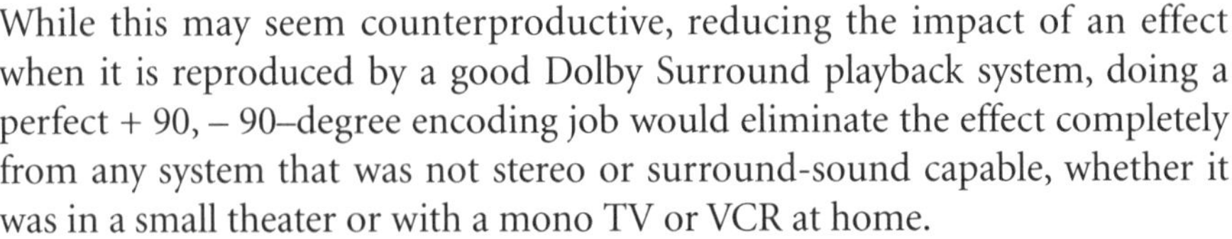

While this may seem counterproductive, reducing the impact of an effect when it is reproduced by a good Dolby Surround playback system, doing a perfect + 90, – 90–degree encoding job would eliminate the effect completely from any system that was not stereo or surround-sound capable, whether it was in a small theater or with a mono TV or VCR at home.

Because of the way these phase manipulations affect the ears (particularly if the viewer is situated at or near the "sweet spot"), some of those surround effects may be heard as if they were coming from well beyond the confines of the screen area, even when only two standard-stereo front speakers are employed. However, in a full-blown but basic Dolby Surround decoding system, after being picked off by means of a decoder, this identical L – R part of the front-stereo signal will also be reproduced by surround speakers to help put the viewer in the middle of the action.

Surround is mono.

It is important to remember that—no matter how many surround speakers are employed (and theaters and some elaborate home systems may employ quite a few)—the extracted surround signal in a standard Dolby Surround video or 35-mm theater installation is still *monophonic.* The surround speakers reproduce identical signals.[2] With the Dolby Stereo matrix process, it is *not* possible for the surround speakers to consistently place discrete sounds out to the sides or anywhere but directly behind the listeners.[3] In addition, no Dolby Surround system can impart vertical information, although, because of the impact of what is sometimes presented on the screen, it may seem that they can.

Under some conditions, it *is* possible for the effects speakers in a Dolby matrix playback system to display "flyby" and even directional clues, if they are located somewhat towards the rear rather than straight out at the sides. At some listening/viewing locations, this can be impressive (for example, the opening helicopter sequence of *Cliffhanger*), although there may also be a tendency for side images to shift erratically and be poorly focused. However, the Dolby matrix is mainly designed to simulate a sense of off-screen environmental space in the listening area—which requires that the surround speakers be placed more to the sides than toward the rear. This degree of envelopment is what sets it apart from simple front-speaker stereo.

[2]Even some of the magnetic 70-mm "discrete" six-channel systems have only a monophonic surround channel—although the Dolby 70-mm magnetic print of *Apocalypse Now* (with three, instead of five, channels up front) had a split-surround option; that option has been available on Dolby 70-mm theater releases ever since, and is also available with AC-3 surround-encoded material.

[3]Only the effects channels presenting two-channel surround information—in Dolby AC-3 or competing DTS or SDDS theater releases or AC-3 and DTS video- and audio-only material—can deliver specific images to the rear or sides of the listening area.

Setting Dolby Surround Delay Time

When working with surround speakers for video, the trickiest thing is figuring out where to place them. Most experts say that if only two "surround" speakers are being used, they should be placed to the *sides* of the main listening area, rather than towards the rear. If a six- or seven-speaker setup is employed, this can be modified with the two additional surround units placed toward the rear. Some manufacturers—notably Yamaha—locate all four surrounds near the front and rear corners of the room.

Once the positioning of the surround speakers is determined, it is necessary to set their delay times. The Dolby rule of thumb for video sound says that there should be a minimum delay of 20 ms between the sound arriving from the front speakers and the sound arriving from the surround speakers. This is the usual "default" setting with most Dolby processors. If the closest surround speaker is about the same distance from the listener as the front speakers in a simple four-speaker system (or the center speaker in a more elaborate Pro Logic setup), things should be fine, once all the speaker playback levels are adjusted.

If not, the delay time will have to be regulated. Sound moves at a bit over 1,100 feet per second at sea level, but for practical "audio-system" purposes we can fix the speed at 1,000 fps. With this in mind, it is easy to see that for every 1-foot difference in distance between the surrounds and the front speakers (or center speaker in a Pro Logic arrangement) we must adjust the delay by 1 ms to keep the subjective total at 20 ms. If the surround speakers are farther from the prime listening position than the main speakers are, you may want to shorten the delay time somewhat. If they are closer (a more typical—and potentially serious—situation with most systems), you should extend the delay time as required.

For example, if your main speakers (or center speaker) are 10 feet from you and the surrounds are 15 feet away, set the delay time to 15 ms (10 – 15 = –5; –5 + 20 = 15 ms) to achieve exactly 20 ms. If your main speakers are 12 feet away and your surrounds are 7 feet away, set the delay time to 25 ms (12 – 7 = 5; 5 + 20 = 25 ms). Note that if the surround speakers are *unequal* distances from the listening area, you should use the distance to the nearest one.

If the main speakers are unequal distances from the listening area, I suggest you upgrade your system to Pro Logic, add a center speaker, and then measure to it. If the surrounds are somewhat farther from the listening position than the mains or center, the increased delay will not be a problem (because the 20-ms delay is a minimum, not a hard-and-fast requirement as long as you do not exceed about 30 ms), but you may have to increase the surround gain somewhat to offset the increased distance.

Note: DSP-type systems designed for audio, as well as video, use delay times that may be widely variable, depending on the effects required; check the owner's manual for setup instructions.

Requirements for Playback Equipment

One characteristic of any high-quality surround-playback system designed for home use is that separate amplification is used for the effects speakers. This quality sets the better configurations apart from the really bare-bones basic designs occasionally found in rock-bottom budget-grade receivers. However, there are other parameters required of the playback equipment to satisfy "basic" *Dolby Surround Sound* licensing requirements.

1. The extracted signal sent to the surround speakers is given a moderate time delay to help to separate it from the sound of the front speakers. If this is not done, listeners may be able to localize front-channel artifacts that leak into the surround channel, skewing the on-screen sound image toward the rear. The delay also slightly increases the sense of spaciousness, although that is a very minor function. The optimal subjective delay is 20 milliseconds (20 thousandths of a second). Most systems have a variable delay, allowing variations of from 10 or 15 ms on up to 30 ms. This lets you compensate for situations when the surround speakers must be placed at less than optimum listening distances.

2. Frequencies above 7 kHz (during encoding and decoding) are attenuated to keep vocal sibilance and other high-frequency artifacts from leaking into the surround channels. High-frequency leakage could call attention to the surround channels and, especially with substandard software, tend to partially scatter the sound that was supposed to remain stabilized up front between the main speakers. The high-frequency attenuation works in conjunction with the 20-ms delay to isolate the surround channel sound from the front.

3. Because mild (only 5-dB) Dolby B noise reduction is applied to the surround signals during the recording-encoding process, complementary decoding is used to re-equalize the signal. This not only quiets prerecorded out-of-phase noise that may be picked up during the L – R extraction process but also helps to silence audible artifacts generated by the delay circuitry within the processor itself. (The rise of quieter digital recording techniques in video and motion-picture audio production has reduced the need for this feature.)

Limitations of Basic Dolby Surround

There are serious limitations to the basic Dolby Surround Sound process as it was originally configured for home use. For the most part, these handicaps have been overcome by modern technology, and an enthusiast who is willing to spend a bit extra for the necessary refinements should have no serious problems.

A phantom-center channel works decently only if the listener is sitting at, or at least fairly near, the central sweet spot. This keeps full-left, full-right, and center (left-plus-right) images stabilized. However, sitting off to the side results in the center image shifting toward the nearer left or right speaker. This can happen for two reasons. If you sit far enough off center for the sound from the nearer speaker to be 6 or more dB louder than that from the more distant one, the center stage may shift radically toward the nearer speaker. More importantly, sitting just a few feet to one side will also delay the sound from the more distant speaker, and if the delay exceeds 2 milliseconds (a shift to the side of just two or three feet can cause this to occur with some speaker arrangements), the whole center stage may skew into the nearer speaker. Whether or not surround speakers are involved, in working with two speakers up front, particularly if they are widely spaced, a central seating position is mandatory for decent imaging.

See page 93 for a review of image shifting.

The center-shift problem can be partially alleviated by placing the speakers closer together, but this may dilute the wide-stage effect (and magnetically affect the TV picture if the speakers are not shielded). Radically toeing in the speakers (even going so far as to aim them at each other) may also partially stabilize the central image. This technique allows intensity differences created by the inherently narrow, cardioid radiation pattern of the speakers at mid and high frequencies to partially offset the timing differences. This trick has been practiced by audio buffs for years to stabilize the center when listening from off-axis. However, both spacing and toe-in angle are critical for this to work.

See page 310 for more on speaker aiming.

center clarity

There can be problems with using only two front speakers even when sitting out along our preferred central axis between both speakers. The phantom-center image is a poor substitute for a good center speaker. A phantom center will result in ear-spacing effects that can cause a frequency-response notch in the midrange—at roughly 2 kHz—even when the two speakers involved are top-grade models.[4] In theory, this is not a good thing for dialogue clarity, although I believe that the reverberant field produced by a good pair of wide-dispersion speakers should tend to smooth out any notch, even one induced by ear spacing. Film-sound editors equalize the center sound for center-*speaker* playback, which means that a phantom center will not sound as intended.[5] In addition, the multiple asymmetrical reflections in most reverberant home-listening rooms make it impossible for two speakers to have the central focus and imaging precision of a discrete centrally positioned speaker system.

Unfortunately, this problem is only partially solved by adding a center

[4]According to the late Ralph Hodges, who, in the September 1990 issue of *Stereo Review*, summarized a paper delivered by George Augspurger.

[5]Also according to Hodges, summarizing the findings of Tomlinson Holman, the man chiefly responsible for the LucasFilm THX parameters.

speaker to a Dolby Surround setup. Because of limitations in the system design, the center channel can never be more than 3 dB different from either the left or right channels, which are still reproducing L + R center information themselves (in addition to fully left and fully right sound). This limits the *width* of the left-to-right spread as the center speaker's sound subjectively pulls full-left or full-right images away from the left and right main speakers. For example, when the left speaker is full on and the right speaker is full off, the center speaker is also reproducing the same left-side material at 3 dB below the standard L + R level (it is getting half its standard center input, and a half-loudness signal is minus 3 dB). Although attenuated 3 dB, the output from the center speaker still pulls the left-channel image towards the middle. The same effect happens in reverse with full-right-channel material.

center spread

In other words, while a derived center channel can tighten up the center image, it also pinches the enveloping frontal sound-stage spread that would be presented by a simple two-speaker phantom-center configuration. Yes, you can employ level controls to vary the output of the center speaker, but decreasing its level also reduces the amount of center focus. In a simple, three-front speaker, nonsteered Dolby matrixing system you can have a wide sound-stage spread *or* central focus—but not both simultaneously.

The separation between the L + R center-channel signal and the L – R surround signal is theoretically infinite.[6] However, the separation between the full-left or full-right signals and the surround channel will, as with the main speakers and the center, only be 3 dB. The L – R extraction, in addition to routing the identical but out-of-phase "surround" signals to the surround channel, also sends it copies of any independent left and right main-channel signals. This lack of separation allows the front speakers to partially pull the surround envelopment toward the front of the room, and also allows on-screen signals that are supposed to be stabilized at stage left or stage right to be skewed toward the rear. The surround channel's 20-ms delay and rolloff above 7 kHz reduces these effects, but not entirely.

DOLBY PRO LOGIC

These inherent limitations in Dolby Surround mostly can be solved by the use of what is called *steering logic.* This capability has been employed for some time in Dolby-equipped theaters and does a good job of delivering subjectively discrete left-, right-, center-, and surround-channel performance. What's more, moderately priced (and, of course, high-end), home-oriented surround-sound

[6]In practice, because of the vagaries of the program material and hardware, it will be in the neighborhood of 20 to 40 dB: subjectively as good as infinite.

systems have had their own brands of steering for some time, the most popular being *Dolby Pro Logic*, often called DPL.[7] The primary function of steering circuitry is to electrically (and sometimes digitally) keep the derived center (L + R) and extracted surround (L – R) signals from bleeding back into the left and right main speakers. Good steering circuitry behaves like automatic fast-responding level and signal-filtering controls for each of the four channels.

centered dialogue

A good Dolby-type steering system will isolate dialogue and central-image sound artifacts from the left and right main speakers, with anywhere from 20 to 40 dB of program-dependent separation, while at the same time keeping left and right signals where they belong. Keeping the center channel "centered" insures that dialogue remains clear and subjectively on the screen for viewers sitting both at the center and off to the side. At the same time, keeping the left- and right-channel signals isolated from the center insures that a wide, enveloping sound stage will exist up front. Interestingly, while many stereo blockbuster films of old had dialogue panned across the screen to correspond to the position of performers, modern Dolby-matrixed (but not AC-3) films keep the dialogue pretty much centered, because off-center dialogue has a tendency to bleed into the surround channel.

Pro Logic steering is also used to help isolate the surround signal going to the effects speakers, but the result may not be as immediately impressive as what occurs with the center channel. The blended nature of the sound going to all of the effects speakers, combined with the built-in vagueness of surround signals in general (and the effectiveness of the 7-kHz rolloff and 20-ms delay), tends to reduce the need for rear isolation exceeding what is delivered by a simple L – R matrix process. Still, under some conditions rear steering is important, particularly with poor source material.

The steering will also suppress most of the up-front "phasiness" of the two out-of-phase surround signals that constitute the mono-surround information, routing them to the surround channel as a reconstituted mono signal. However, only during the occasional rear-placed effect will a DPL decoder almost totally eliminate surround sounds from the front channels. Most of the time, at least some of the surround signal is allowed to bleed through to the front to help diffuse the effect. The steering also prevents full-left and full-right signals from being sent to the surround channel, a problem with nonsteered decoding. However, odd as it may seem, some listeners feel that the effects of steering on the surround signals may occasionally appear to detract from the swimming—somewhat phasy, sometimes pleasant—spaciousness delivered by a straight, nonsteered Dolby system.

[7]Even the non-Dolby or modified-Dolby options offered at one time or another by Shure, Fosgate, Yamaha, Lexicon, and Proton—as well as items given THX certification—build upon the basic Dolby parameters.

The "Dolby 3 Channel" mode available in most Dolby Surround processors and A/V receivers, although often thought of as a compromise, takes advantage of this phenomenon. Designed for those who cannot employ surround speakers, it allows the main speakers to reproduce those 180-degree-out-of-phase signals, slightly delayed, while still steering center-channel sounds to a center speaker. Dolby 3 can sound very good with Dolby-encoded material, even if the "surround" sounds display a phasy characteristic and offer odd localization cues, and is also a viable playback option for listening to certain stereophonic audio-only material that is not Dolby encoded. With the latter, if you sit somewhat away from the center axis and the center speaker is a good one (and you turn down the center volume a bit to reduce the occasional left-right squeeze), Dolby 3 can be very beneficial.

A number of companies have proprietary steering circuitry of their own, but they all are fairly similar to Pro Logic and usually differ mainly in the high-frequency rolloff curves chosen for the effects channels or surround delay times. In some cases, a manufacturer will feature proprietary steering circuitry within their hardware, in addition to the Pro Logic configuration. The competing design may even exceed Dolby standards (particularly in terms of surround-channel separation and center-channel dialogue stability with problem software), but the processor will nearly always still include the user-selectable Dolby option for marketing and legal reasons.

Incidentally, *Fantasia* has been reissued in a Dolby Surround video format on both tape and laser videodisc—with the latter being superior in every way to the former. The three discrete front channels were mixed down to two, with additional, synthesized reverb matrixed in for surround impact. With home-based Pro Logic, the new version decodes quite well back to the standard left, center, right, and surround outputs and is certainly something that you—and your children—may wish to rent or purchase. Do not expect the very latest in up-to-date, high-fidelity tonal quality—although the sound is certainly impressive for a 1940-era effort—but be prepared for often stimulating visuals.

Dolby Matrix Problems and Solutions

The Dolby matrix-dematrix structure is a compromise, but it still works surprisingly well—particularly when steering is employed. However, there are some problems that can develop with the system; fortunately, there are also some workable solutions.

Dolby Pro Logic Center-Channel Bass

When an "all-channels" subwoofer is employed in a surround-sound setup, the bass from both the main and center speakers is routed to this special speaker, saving them from being overly strained by the occasional high-bass output. However, without a subwoofer, the left and right main channels are operated at full bandwidth. Under these conditions, the center-channel function has two options. In the "normal"[1] or "small" bandwidth mode, the processor routes the center bass (below about 100 Hz) to the left and right main speakers—effectively "biamping" the center channel. This keeps a typical small center speaker from being pulverized by center-channel bass pyrotechnics. However, employing a larger, more bass-able, speaker for the center and switching the processor to the "large," "wide," or "full bandwidth" setting should, theoretically at least, result in better central focus, at least for individuals sitting a substantial distance from the ideal "sweet spot" between the speakers.

However, at the "wide" setting, the center may get more bass than either of the two main speakers. After auditioning a number of videodisc movies, I have concluded that, in the "wide" mode, the center woofer will ordinarily receive at least two-thirds of the total bass power at any given time (and often may have to deal with an even higher percentage), with the main speakers sharing the job of delivering the remaining third. As a result, while the "normal" setting allows a *pair* of full-range main woofers to handle the bass range, the "wide" setting will dump one-third greater bass energy (or even more) into the *single* woofer of the center system than either main woofer receives in the normal mode.

One solution to this dilemma—and possibly the best one—is to add a separate subwoofer hooked directly to a processor's subwoofer output, with three matched front speakers handling nothing but the upper-bass, mid, and treble range. Another is to add center woofers that are more robust than either of the main woofers, with the center-channel bass section possibly as powerful as that of the left and right main systems combined.

The latter solution can be implemented (although somewhat expensively) by adding an outboard subwoofer to handle, or at least augment, the low-bass range of the center-speaker woofers. This subwoofer will ordinarily handle very nondirectional frequencies (below 80 Hz if it has a decent crossover), so it can be located away from the center, indeed practically anywhere in the room. A good one will have a volume control to allow for proper adjustment of its output. If the main systems are already potent bass producers (or have a subwoofer or subwoofers of their own that are not connected to the surround processor's dedicated subwoofer output), this "triple whammy" could result in formidable performance.

While using the subwoofer output of a

[1]Some manufacturers use this term in a different way than I am using it here; check your owner's manual.

(continued)

Dolby Pro Logic Center-Channel Bass *(continued)*

surround processor should free all three front main speakers from having to work hard in the low-bass range (and also provide smoother, stronger, and more artifact-free low bass), under some conditions this option may have drawbacks. Some processors begin to roll off the bass energy to the three front mains as high up as 150 or even 200 Hz, gradually phasing in the subwoofer(s) at that point. (This is *not* the case with THX-certified or AC-3-equipped processors, which fix the crossover point at 80 to 120 Hz.) While this certainly takes the heat off the woofer sections of the three front speakers (indeed, it allows very small or even mini-speakers to be used exclusively for that job), it can cause image shifting if the subwoofer is not positioned up front and reasonably close to the center. While it is true that bass is still fairly difficult to localize over the 100- to 200-Hz range (and in some cases even up to 300 Hz), the crossover slopes of some processors allow a substantial amount of midrange energy to leak to the subwoofer(s). This arrangement almost demands that two subwoofers be used—with each symmetrically placed and near the left and right main speakers, even though the spaced woofers may exhibit minor low-frequency response irregularities due to interference effects.*

*When spaced subwoofers are employed, the nature of sound at low frequencies allows the midpoint exactly between them to behave as a boundary. This can trigger cancellation problems—suck out—similiar to the midbass Allison effect, but at lower frequencies (see page 55, and "The Boundary Effect," page 56).

Broadband Center-Channel Problems

Problem	Solution
The center-channel speaker may have to be smaller than the left and right mains for space or aesthetic reasons, and yet the center channel usually delivers the bulk of the sound in a program; if your center speaker is small and your listening room is big (and/or you habitually listen at high levels), there is little doubt that your main speakers are sometimes cruising while your diminutive center-channel speaker is getting pummeled.	Set the processor's center-function switch to "small speaker" or "normal" (which routes center bass to the main speakers) instead of the "full-bandwidth" or "wide" mode; (2) get a more robust center speaker; or (3), if you like the sound of your present one, get a duplicate to double the center-channel's output capability (see comments on "bracketing" below).

(continued)

Broadband Center-Channel Problems *(continued)*

Problem	Solution
With some processors it may be possible—particularly with action-adventure film sound effects—to hear midbass center-sound artifacts bleeding over to the left or right speakers when the center channel is set to "normal." This will be most noticable if one is sitting well off-center and close to one of the left or right speakers.	Either avoid sitting too close to the left or right speakers or get a subwoofer for the center channel and switch the processor to the "wide" mode; if the subwoofer-to-center-speaker crossover is low enough (80–100 Hz or below), the center-channel subwoofer will not call attention to itself and can be placed just about any place in the room—preferably in a corner, even one behind the listener; if you already have a mono subwoofer for the entire system connected to the subwoofer output of a surround processor—and it has a crossover point of 100 Hz or lower—you should have no problem with image shifting with any of the three front channels.
Unless you have a front-projection monitor and an acoustically transparent screen to place your center speaker behind (both expensive), you will have to place the center speaker above or below your TV screen. Unless your main speakers are positioned at the same height as the center unit (not always visually pleasing or physically possible) this results in left-to-right pans that are not level.	Bracket the monitor with *two* vertically oriented, magnetically shielded center speakers on either side of the screen; this will keep the center image similar in height to that produced by the main speakers, while still allowing the steered center material to be decently focused—although if you sit off to the side somewhat there will be some minor center-shifting; bracketing will also create some comb-filtering and 2-kHz notching effects even when you sit in the sweet spot, although these will not be as onerous as what is experienced in the "phantom-center" mode with widely spaced main speakers; you can also place the speakers above and below the picture, but this will result in substantial floor-reflection-induced frequency-response difficulties, and I do not believe it is a good idea.[1]
Fairly deep (40–50 Hz) bass sometimes "leaks" into the surround channel; while the levels involved should cause no problems with even the smallest decent surround speakers, there is always the potential for trouble; this is particularly true if you also listen to music run through a DSP-type or AC-3 processor, which often will route substantial bass to the surround speakers.	Use bigger surround speakers, or more of them, or add a small subwoofer to the surround channel; while bass leakage is usually accidental, good surround speakers will often add considerable impact to a program, particularly if the main speakers are not really bass potent; bass-capable surround speakers are said by some to also smooth out room-generated standing waves, although this is unlikely.[2]

[1]If you go the double-speaker route, they should be wired in series if they are 4-ohm models and in parallel if they are 8-ohm. Some surround processors have built-in provisions for hooking up two center-channel speakers.

[2]If your system has a big subwoofer connected to a subwoofer output on your Dolby-equipped receiver or processor, *all* the deep bass produced will be routed to the subwoofer, and none will get to the main, center, or surround speakers.

Center Polarity Test

The left and right main speakers in a stereo system must be "in phase" if you want them to sound right. But how can you be sure if the *center* speaker in an A/V set up is in phase with the main speakers? If all three front speakers are identical or at least made by the same (hopefully competent) company, simply getting the connections right (red to red, black to black, etc.) should do the trick. But if you have an outboard amplifier for the center channel (which may invert polarity), or your center speaker is made by a different company than the main speakers, or you suspect that your same-brand center speaker may be internally wired backward from the mains, you need to do a front- and center-speaker phase check.

If your stereo-main (left and right) and mono-center amplifiers are outboard from your processor or are built in to an upscale receiver or integrated amplifier that has them connected together by "jumper straps" on the back, a center-channel polarity test is easy.

To check the center *amplifier*, temporarily substitute it for one of the stereo amplifier's left or right channels (letting it drive one of the main speakers) and, after adjusting the balance control for close-to-equal outputs from each speaker, do a standard "stereo" speaker phase check as discussed in Chapter 2 with *mono* source material.[8] This will determine if the center amplifier is out of phase with the main stereo amplifier. Note that if this test pinpoints a problem, the center amp may still be OK: the main-channel stereo amplifier may have *its* polarity reversed. However, it doesn't really matter which component is reversed, because what you are after is to get all three operating *in phase* with each other, reversed or otherwise. (Polarity reversal is not uncommon with stereo amplifiers and will not cause problems with standard program material in a simple stereo system, because *both* channels would be reversed.)

See why a blanket reversal is not critical, page 83.

To check if a different brand or model center *speaker* is phase-compatible with the main speakers, temporarily replace either of the main speakers with the surround speaker. Then do a conventional "stereo" polarity test as before. If the center speaker is wired the way it should be, the image with a mono input should be fairly well centered up. (Use the balance control to get the image decently centered if the two speakers have a substantial difference in electrical efficiency.) If they produce a phasy-sounding effect, reverse the leads on the center speaker and check for an improvement.[9] Then, put things

[8]With built-in amps you will have to reposition the jumper straps or possibly substitute RCA-plug-equipped cables.

[9]You should not expect the tight central image that happens when doing this test with identical main speakers, even if the somewhat different central speaker is made by the same manufacturer; instead, what you are after is the most coherent phantom image between the speakers.

back as they were and connect the center speaker normally if it and the center amp have the same polarity orientation as the main-stereo amp and main speakers; reverse the center hookup leads if either the amp or the speaker has polarity that is reversed from the main amp and main speakers.

If you have a receiver or integrated surround processor that does not allow you to disconnect its internal amplifiers from the preamp/processor circuits, here is an alternate procedure that is a bit less systematic but works nearly as well as these others. If your system has an outboard subwoofer connected to the subwoofer output of the surround processor or receiver, temporarily disconnect it and readjust the processor controls so that the left and right front satellites operate as full-range speakers. Set your surround processor to the "normal" center-channel mode (which has the main-speaker woofers handle center-channel bass and low bass, while the center speaker reproduces only the upper bass, midrange, and treble) and play some mono pink noise with the bass tone control turned up a bit above normal. Listen to the result with the center speaker wired normally and again with the leads reversed.[10] If the center is wired out of phase from the mains, there should be a substantial dip in output at the 100–125-Hz transition point between the center and the main-speaker woofers.

If you cannot readily hear any null with either wiring orientation, purchase or rent a sound-level meter or even an RTA (real-time analyzer) and use it to measure the null. Use the center-speaker hookup that delivers the most uniform, dip-free response in the 100–125-Hz range. If you use an RTA it will probably have a pink-noise generator built in. A sound-level meter will be trickier to use, because it measures broadband noise. If you test by ear, you will almost certainly have to obtain a test disc with pink noise on it. There are a number of these available, but one of the best, and one that also has special tests for center-channel phasing, is the Delos *Surround Spectacular* (DE-3179).

Most test discs work best if you also use a sound-level meter.

As a final check, particularly if your center speaker is a full-range model, switch your system to the DPL "wideband" mode and play a stereo program with substantial bass content with all three front speakers hooked up and positioned normally. After noting the level of bass (and possibly measuring it with an RTA), reverse the center-channel speaker leads and listen for a difference in bass level. If the center is wired out of phase with the others, there should be a slight loss in bass power.

While a bass polarity check will not always insure that the midrange and

[10]Great care should be exercised when doing any kind of speaker-wire "switcharounds" while equipment is running, because shorting "hot" leads together can damage an amplifier. For that reason, it will be prudent to cut the receiver and/or all outboard amplifiers off during these manipulations unless double banana plugs are used (see page 81).

treble of the three front speakers is in phase (and will be irrelevant if a system-wide subwoofer is employed), it should work if the speakers are all made by the same company. If the center speaker is not the same brand as the mains, you may still have problems due to incompatible woofer-to-mid-to-tweeter phase reversals within the speaker systems themselves. This is perhaps the main reason to have your center speaker made by the same company that made the main speakers.

THX SURROUND

A number of companies (Shure, Yamaha, Bose, Proton, and Fosgate come to mind) have tried to improve on what Dolby Surround in general, and Pro Logic in particular, can offer in terms of home A/V performance—without violating the "letter of Dolby law." The refinements have included improved steering (both at the front and rear), special circuitry to enhance the sense of "spatiality" of the effects channel (in some cases making it sound like more than one signal), and special center speakers designed to improve dialogue clarity (short "line source" and woofer-tweeter concentric designs have both proved popular). In some cases, these attempts were short-lived; in others, they are still working quite well.

Nevertheless, the most influential developer of video-surround-sound "enhancements"—and the one that has become much more successful at promoting its system than any of the mainstream audio companies—is not a hardware manufacturer at all: it's LucasFilm, with their *THX* system. THX is not a brand name or even a radically new type of surround sound. Rather it is a set of strict, licensed "amendments" to the basic Dolby Stereo parameters that have been formulated by Tomlinson Holman and the research team he headed up when he was at LucasFilm.[11] These specifications were originally designed to improve the sound of Dolby Stereo systems in theaters. However, later versions of the original THX parameters have been formulated for home-theater use and are designed to improve upon or at least "certify" basic equipment specifications—particularly for Dolby Pro Logic and AC-3 surround, video monitors, and laserdisc players.

amendments

Just as the manufacturer of home A/V equipment who satisfies DPL requirements—and pays Dolby for the privilege—can put the Pro Logic logo on the front panel of a product, so the hardware manufacturer who satisfies the additional THX parameters can add the THX logo—provided the

[11]The term "THX" is a tongue-in-cheek invention by the people at LucasFilm and was derived from both *T*omlinson *H*olman e*X*perimental and also, of course, from the title of one of George Lucas's earlier films, *THX-1138*.

required fees are paid to LucasFilm. A THX logo on the front of a component will nearly always mean increased sales, at least if the component is oriented toward serious A/V junkies.

THX goals

The goal of the THX guidelines is twofold. First and foremost, they mandate that any *complete* THX-approved system for home-video use will be able to deliver sound to the listening room that closely parallels what was originally heard by the engineers at the dubbing facility of the film- or video-production company. (Nearly all major films are mixed in standardized THX facilities or unlicensed clones that nearly duplicate their characteristics.) A THX sound system might include the necessary interconnect wire, speaker cable, speakers, surround processor, amplifiers, and even laserdisc player that meets certain standards of video performance. Second, they guarantee that individual THX components installed in non-THX home audio or video systems will achieve certain minimum, but still exacting, performance standards.

It is not necessary to have a complete THX system to benefit from the LucasFilm research. THX-certified speakers, surround processors, amplifiers, receivers, disc players, and accessories are all compatible in every respect with all noncertified A/V products. Because some of those noncertified products are quite good (surpassing THX standards in many cases), it is possible to "mix-and-match" effectively and save a considerable amount of money. It is also possible to use the guidelines as educational tools to create a first-class reasonably priced system that has no THX-certified hardware in it at all.

The perceptive equipment shopper will discover that certain components that do not exactly emulate THX guidelines (such as a substantial number of top-grade loudspeaker systems and surround processors) may be more satisfactory than the certified gear under some conditions. In many cases, it may be best to employ only those THX-certified products (such as power amplifiers, surround processors, or receivers) that guarantee that the more important, verifiable, and traditional high-fidelity standards are achieved, while eschewing controversial items such as dipole-radiating (and usually quite expensive) surround speakers, line-source front speakers (the latter, particularly for center-channel use, may be hard to place properly), or usually overpriced wire and interconnects that work no better than the far lower-priced hardware-store or Radio Shack equivalents.

Initially, home THX hardware was *very* expensive. It is still not cheap in comparison to a lot of the more basic gear available. However, expensive or not, more and more THX-certified A/V equipment is showing up all the time, and some of it is within the reach of dedicated mid- and low-budget enthusiasts.

THX performance guidelines have even been applied to video software, and the THX logo now appears on a number of upscale laser-video and DVD presentations that satisfy THX's film-to-video transfer-quality standards (both in terms of picture and sound). While the THX parameters for software performance mainly have been applied to laser-read material, LucasFilm also has produced a THX-certified tape version of the *Star Wars Trilogy*, as well as a few other worthwhile films. Although not up to laser-image standards, these tapes are certainly examples of the best in standard VHS tape performance. The LV and DVD releases and rereleases are state-of-the-art; any enthusiast wanting the very best in software must place these THX versions at the top of his or her shopping list.

The THX Standards

The THX standards can help us understand what makes for good home video and audio. Unfortunately, because they are proprietary and known in detail only by LucasFilm and its THX licensees, it's hard to give precise information here (you'll find no data on the exact dispersion requirements for speakers, for instance). However, there is much we can learn from THX even in assembling a budget-oriented system. Let's examine the standards for various components of a home-theater system, while listing some concerns about them.

FRONT-SPEAKER SPECIFICATIONS AND PLACEMENT

All three front speakers must be identical in timbre and frequency response. In theory, this insures that panning effects are realistic and that musical playback is balanced across the soundstage. The vertical and horizontal radiation patterns must be shaped to reduce boundary-generated images and horizontal comb-filtering effects that would inhibit clarity. Vertical radiation in particular is limited in order to reduce floor and ceiling reflections.

polar response

Horizontal polar requirements stipulate the response smoothness of the mid- and high-frequency energy radiated off-axis and even behind a speaker. The response may not vary by more than 1 dB over a ±30-degree arc in front of the speaker over most of its operating range. This is a tight tolerance, and it supposedly keeps the three systems sounding similar from normal off-axis listening positions. Over the 30- to 45-degree horizontal angle, I would imagine that the response is still decently smooth and only slightly attenuated. However, beyond 60 degrees off-axis the response of most THX satellites will be substantially attenuated at some frequencies, affecting power response.

The THX parameters also stipulate the best front-speaker *placement* in relation to the TV screen, with the left and right main speakers somewhat closer together than what many consider optimum for music-only playback. In addition, the front speakers must be able to handle specific power inputs and together cleanly produce peak sound levels of 105 dB in the reverberant

Figure 7.2 THX Front Speaker

(Photo courtesy of Boston Acoustics)

This THX-certified speaker from Boston Acoustics is an excellent example of what is required to earn the coveted logo. Note the arrangement of the two midrange–upper-bass speakers at each end, with two tweeters in the space between them. The two tweeters have good horizontal dispersion, while their vertical spacing simulates what a single driver several inches long in the vertical dimension would deliver in terms of vertical dispersion.[1] This reduction in upward and downward radiation curtails floor and ceiling reflections and enhances clarity, provided that you position yourself so that your ears are close to the same height as the center axis of the speaker to eliminate driver-interference effects. The larger drivers (which handle mostly midrange, but also the upper bass) are similarly arranged, and although this spacing would also ordinarily produce serious interference effects in the vertical dimension, the frequency range covered insures that the upward and downward radiation, while somewhat irregular, is subdued enough not to cause problems. The reduced level of midrange reflections from the floor and ceiling, as with the higher frequencies handled by the more narrowly spaced tweeters, enhances clarity—again provided that you are seated near the vertical center-axis of the speaker. (See Chapter 2, "Hearing the Speakers," page 62, for a discussion of "comb-filtering" effects.)

This system has a 3,000-Hz crossover, which keeps the tweeters in the safe, small-excursion operating range. While a crossover point this high up might ordinarily cause some problems with the mid-to-tweeter transition (see Chapter 2, "Three-Way Flat On-Axis Curves," page 45, "Two-Way Flat Power Curves," page 46, and Figure 2.13, page 52), the small, 5¼-inch diameter of each of the larger drivers insures that they have decent horizontal dispersion right up to where the tweeters cut in. The system requires a separate subwoofer to deliver the bass below 80 Hz (part of the THX specifications), and by using three identical models (left, center, right), vertically oriented and spaced at equal heights, a Pro Logic or 5.1-channel frontal soundstage spread can be impressively uniform.

Interestingly, enthusiasts with limited resources can take advantage of the most important aspect of the THX main-channel front-speaker specification by simply purchasing four small (but good) two-way non-THX models and stacking them vertically, two to a side—with the lower units positioned with the tweeter ends up and upper units positioned with the tweeter ends down. Wire each pair either in series or in parallel, depending on their impedances, to keep the combined impedance of either channel in the 4- to 8-ohm range.

[1] The standard D'Appolito configuration uses only one tweeter and has more vertical scatter in the treble.

field—which is extremely loud—down to frequencies as low as 80 Hz, where the subwoofers take over. All THX-certified front speakers must be magnetically shielded to keep them from adversely affecting a TV picture tube.

Possible Concerns

Because speakers cannot be mounted behind a TV picture (unless you have a front-projection monitor with an expensive, acoustically transparent screen), the implementation of the THX (or anybody else's) audio standard for speaker use, particularly the center unit, can be awkward.

line-source radiators

Line-source radiators *must* be oriented vertically to work properly. However, a tall narrow speaker mounted above a large-screen TV set (which may itself be on a stand) may be odd looking, particularly if the left and right main speakers are sitting on short stands. Because the three front speakers should be at similar heights to achieve satisfactory left-to-right pans, you would have to place the television monitor either too close to the floor for comfortable viewing or position the three speakers so high above their optimum ear-level locations that they will have to be angled slightly downward.

A second possible solution would be to locate the TV set fairly high up. However, assuming that it was not placed higher than the middle of the wall—or could be elevated off the floor in the first place—all three speakers would then have to be located near the floor. Dropping them too far down could have negative effects on the sound, because satellite speakers that are close to the floor will trigger "suckout-induced" colorations in the middle bass or lower midrange. No matter what, low placement requires angling the three systems upward somewhat to maintain proper vertical focus.

horizontal placement

A third—and popular—option is to place the center speaker horizontally on top of or below the TV set.[12] This obviously makes it easier to position all three front speakers for reasonably flat left-to-right pans and also improves visual aesthetics. Unfortunately, it seriously undermines the timbral similarities between the center system and even identical main systems, because of substantial radiation-pattern differences between line-source systems that are vertically oriented and those that are horizontally oriented. Acoustic comb-filtering effects in the horizontal plane, interacting with the placement of our ears, can create uneven response and vague imaging. Some manufacturers—such as Polk Audio and Cambridge Soundworks—compensate for this effect by offering horizontal line-source designs with four small midbass drivers, instead of two large ones.[13] Even the THX people have certified a few "shelf-

[12]Most conventional "center-channel" speakers are configured for this kind of placement.

[13]The more widely spaced of these drivers are attenuated as the frequency climbs, somewhat compensating for the comb effects.

type" models, featuring centered tweeters with two small midrange drivers above and below them. Although larger than most shelf-type systems, they do image well. Midbass overload is prevented by having a pair of small woofers flank this short vertical array.

If, for physical or aesthetic reasons, you *must* orient a center system on its side, either on top of or below a TV monitor, you may get better results by using a good, conventional, single-woofer–single-tweeter mini-type system. Comb-filtering effects would then only be apparent around the crossover point instead of over most of the operating range of the system.

Users should have few complaints about the midrange horizontal dispersion characteristics of THX-certified front speakers. Because they are to be used with a subwoofer, the midrange drivers are smaller than those in most other two-way systems, aiding uniformity. Some may prefer speaker systems with still wider horizontal dispersion, at least for the left and right main speakers, because of the spacious sound such systems deliver when reproducing music (and even videos that have lush soundtracks). However, the combination of limited vertical dispersion and decent off-axis response allows a THX-certified speaker in the center-channel position to deliver exemplary definition, a very important feature.

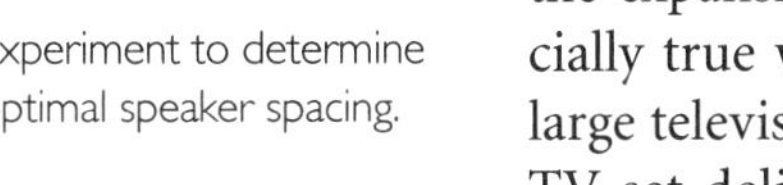

Experiment to determine optimal speaker spacing.

The spacing requirements of front-speaker systems is controversial. While the narrow sound stage that results from the close spacing of the main speakers may enhance the image formed near the TV screen (and insure that side-to-side pans stay closer to the confines of the TV set), the loss of spaciousness and broad-staged imaging will compromise listening to music and the expansive sound stage required for some motion pictures. This is especially true when you consider the subjectively small "theater" size of even a large television monitor. The wide spacing of the main speakers will help any TV set deliver a more theater-like impact. Remember, the center-channel speaker will see to it that the center image remains focused.

THE SUBWOOFER

In a THX system, bass between 20 and 80 Hz is handled by one or more subwoofers, making it possible to employ modestly sized satellite speakers while still achieving good deep bass. Like the three main satellite speakers, the subwoofer(s) will be able to generate peak levels of up to 105 dB down to at least 30 Hz. A THX processor will have an internal crossover network and subwoofer output to separate the sounds going to the subwoofer amplifier from those going to the main-speaker amplifiers. Typically, the crossover will use 24-dB-per-octave low- and high-pass filters, although some models use 12-dB-per-octave filters for the high-pass sections, combined with the natural 12-

dB-per-octave rolloff of the woofer system itself. Crossover curves this steep will minimize interference effects between the satellite and low-range systems.

The THX subwoofer parameters allow a sound system to achieve the kind of room-shaking bass performance that blockbuster films often demand. Remember that some manufacturers have developed high-quality subwoofers that exceed the THX performance parameters by a substantial margin (particularly below 25 Hz), although they are not THX certified because of minor compatibility problems. For example, some brands are designed to work primarily below 40 or 50 Hz and are not at home "reaching up" to the THX-mandated 80-Hz crossover point, and some models have built-in crossovers that do not dovetail well with the crossover in a THX processor. A number of full-range systems also equal or surpass THX bass-performance standards, making the use of an outboard-mounted subwoofer superfluous. To capitalize on this, THX processors have the option of bypassing the subwoofer function. Note that while the THX standards call for only one subwoofer (usually a big one), they do not preclude the use of two possibly smaller units, which some feel are more effective.

THX-certified subwoofers usually work best in corners.

SURROUND SPEAKERS

In a THX configuration, the two surround speakers are located on the walls to the sides of the main listening area and somewhat above ear level.[14] Each speaker has two tweeters, with one aimed down the side wall toward the front of the room and the other aimed down toward the rear. In most cases, a single midrange or semi-woofer driver in the same cabinet faces outward, toward the listening area, although some models have two, placed on the same panels as the tweeters. Each speaker's two tweeters are wired *out of phase*, creating a "dipole" effect that generates an acoustic "null" in the area between them—at least at upper-mid and high frequencies. The consequence is a diffuse, reflected, somewhat "phasy" ambience at the listening position, which limits directional cues that would call attention to the surround-channel-generated sound. Properly positioned, these speakers subjectively diffuse the sound around the room, supposedly delocalizing it more effectively than two conventional miniature speakers would.

dipole surround speakers

Possible Concerns

While the dipole systems work remarkably well when the listener is optimally seated, they do not work as well for those sitting somewhere else (although they work at least as satisfactorily as a pair of conventional speakers similarly

[14]This is in contrast to the more rearward placement suggested by a number of Japanese equipment manufacturers.

placed). Surprisingly, you can obtain very good results by using conventional speakers (even cheap mini-models) and mounting them so that they aim upward, bouncing the sound off the ceiling, or sideways, either towards the front or rear, "washing" their sound down the walls. This configuration should be less dependent on your listening-position than the THX standard, and considerably cheaper to buy.

A neater, if somewhat more expensive, trick might be to use more than two surround speakers: say two directly to either side or at the sides but more toward the front of the room and two more at the sides but toward the rear of the room. This is similar to what is done in theaters, including THX-certified ones. Configured this way, speakers with conventional radiation-patterns would probably diffuse the sound even better than the dipoles, especially for listening-viewing positions away from the null area.[15]

Even Radio Shack models may work OK.

If the dipole concept has caught your fancy and you have limited funds, you should know that it is possible to achieve a THX-like surround-speaker effect while operating on a shoestring budget. This can be done by placing a pair of reasonably cheap, small speakers on each side wall back to back, with one speaker facing down the wall toward the front and the other facing toward the rear. The two forward-facing units can be wired out of phase from the two rear-facing ones, allowing each pair to achieve a dipole effect. (If you are reasonably adept, the cabinets could be opened up and only the tweeters wired in reverse, keeping the woofer and/or mids in phase.)

One last note about budget surround sound: Cambridge Soundworks has long marketed some fairly low-priced, noncertified, dipole-surround speakers that may approximate THX standards. Because Cambridge is a mail-order outfit and LucasFilm requires products displaying the coveted label to be sold in retail stores, certification is not possible.

AMPLIFICATION

While most amplifiers, if operated within their power limits, work well for both audio and audio-video applications, there are some models that do not meet basic high-fidelity—let alone THX—standards. The THX parameters ensure that an amplifier or receiver will deliver all that most listeners require in terms of low distortion, overload recovery, input sensitivity, output impedance, noninversion of polarity, heat dissipation, mechanical noise, and power delivery to a variety of speaker loads.

[15]In doing their initial research, the people at LucasFilm discovered that in typical home-listening situations four surround speakers may display undesirable interference effects. Consequently, some experts (notably those at Yamaha) think that, when employed for home-theater surround use, four surrounds are only beneficial if each of them receives slightly different DSP-processed signals.

At times, the THX power requirements may seem odd. Initially, *each* channel of an approved six-channel system (left, center, right, right surround, left surround, subwoofer) had to produce at least 100 clean trouble-free watts, or 600 watts total. While this power level may be appropriate for the three front speakers and any subwoofer, it appears to be overkill for the surround hookups. After all, the surround-channel sound output level never equals that of any of the main channels. However, some surround speakers have less input sensitivity than the larger, more efficient main speakers and may demand a surprising amount of juice. The initial THX amplifier parameters also took into consideration future 5.1-channel hookups (see "Other Discrete-Channel Theater and Surround-Sound Systems," page 292, and "THX and 5.1-Channel Audio," page 296). The parameters also simplify hookups, since any channel of a THX amplifier will be able to drive any channel of a THX system. Later, the surround-channel minimum-power requirements were dropped from 100 to 50 watts per side—not a serious failing by any means.

Initially, all THX-certified power amplifiers were stand-alone units; the very high-power, multiple-channel requirements were thought to work against putting so much hardware in one box. However, a few THX-certified integrated amplifiers and receivers have appeared, although they are large, formidable—and expensive.

Possible Concerns

Before you race out to buy a super-powered amplifier for video use, remember that THX amplifiers and speakers are designed to produce high sound levels to insure delivery of the full dynamic range of wide-band formats like videodiscs—duplicating what the mixing engineers heard at the film-dubbing stage. However, many listeners will not feel comfortable playing their systems at such high levels, and apartment dwellers invite the wrath of their neighbors if they do so. I have played a lot of "dynamite-sound" A/V material on my big system—in a big room—and even when the subjective volume level is wall-rattlingly loud, the measured level rarely reaches even 100 dB. If you don't need super-high volume levels, this aspect of the THX amplifier standards may be overkill.

center channel power

Normally, the center channel requires as much, or even more, power as the main channels, but you can fudge a bit by running your Pro Logic processor in its "normal" mode, which routes the power-consuming center bass to the main speakers. A unit delivering 50 watts to three front speakers having typical input sensitivities will produce peaks in excess of 100 dB in a 2,500-cubic-foot living room—more than enough power to enjoy big-theater sound. Many modestly priced receivers can deliver 100 watts to each of the left and right mains, and 50 or more to the center—and supply 20 or more watts to each of the surrounds. A receiver with this kind of power can play pretty loudly even in a fairly big room.

SURROUND PROCESSOR

THX enhancements

While the Dolby Pro Logic standards for surround processors are specific, the THX enhancements are designed to make them still more effective in typical home-listening environments. Besides insuring that those basic Dolby parameters are themselves rigorously followed, three additional THX surround-processor enhancements are made: (1) a re-equalization of the main channels to compensate for the treble emphasis typical in film soundtracks, (2) timbre-matching circuitry to make sure that the tonality of the surround and front speakers is well matched, and (3) a "decorrelation" circuit (using time delays and pitch shifting) that helps to further delocalize the sound of the surround channel.

Figure 7.3 Onkyo THX Receiver

(Photo courtesy of Onkyo)

In its theater surround-sound mode, this remarkable THX-certified receiver by Onkyo delivers 110 watts to each of the three front channels, plus 100 watts (50 × 2) to the two surround speakers, enough for all but the most expansive listening-viewing room. Given "thumbs-up" reviews by several audio and video journals since its introduction, this unit has full audio-video switching capabilities and enough inputs (including the ability to handle an outboard 5.1-channel adaptor) to satisfy all but the most accessory-hungry enthusiast. In addition to the video surround modes, this receiver has a number of DSP surround functions for conventional audio recordings, including three that I deem essential: "Concert Hall," "Theater," and "Night Club." While not as high profile as some of the other makes from Japan, the Onkyo product line has proven to be their equal.

Possible Concerns

With THX-certified speakers, there is no doubt that re-equalization of the front channels is necessary with a lot of theater-equalized motion-picture program material. However, a number of competitive speakers already intentionally roll off or downward-tilt the highs (often by means of adjustable treble controls of some kind). Indeed, almost any extremely wide-dispersion speaker system that has a flat output in a reverberant environment will display a high-end room-response rolloff if listened to in a typical, well-damped living room, since the strong off-axis highs will be partially absorbed.

Perhaps it would be better if the speakers and the processor were *both* adjusted for flat-power output, with the system owner given the option of controlling the high frequencies to taste (or to conform to specific room acoustics) by means of simple, variable tone controls. After all, this same feature has been available on audio preamps, integrated amps, and receivers for years. While the frequency-contouring feature is by itself defeatable on some THX processors, on most of them it is not.

timbre matching

Timbre matching a pair of surround speakers to the main speakers initially sounds like a marvelous idea, because it takes into consideration the different frequency-response characteristics of the ear when receiving sound from different directions. However, what happens if—following Yamaha, Fosgate, and a few others—more than two surround speakers are used to further diffuse the sound? The THX designers discovered that it is nearly impossible to achieve proper timbre matching in this situation—no matter what kind of equalization is tried. This furthers their belief that two surround speakers—properly sited dipole models—work best.

One problem with timbre matching involves its nonadjustability. Obviously, room acoustics and listening distances will alter the perceived timbre of *all* the speakers in an unpredictable way; a fixed, built-in circuit to compensate may be likened to the fixed "loudness" control found on many preamps and receivers. The latter is designed to compensate for bass losses at low listening levels but, more often than not, it just screws up the sound. Of course, one major difference here is that the loudness control can be defeated with a switch, whereas with most processors the timbre-matching circuitry, just like the high-frequency roll-off feature, cannot—unless the entire THX option is switched off.

Adjustability aside, the main problem with timbre matching is that in the real world the ear does not receive direction-oriented re-equalizations, and it would seem that doing so for theater or home-system playback would detract from realism, not enhance it. As a counterargument, the THX researchers have shown that a Dolby-surround environment where *identical* sounds from the surround channel are coming from *two* directions, is *not* a real-world situation. Timbre matching therefore compensates for an unusual acoustic condition. However, we will soon see in our discussion of surround-channel decorrelation that identical-sounding surround signals cease to be a factor in determining the need for timbre matching.

decorrelation

A decorrelation circuit is a marvelous idea, and I believe it is more important than either re-equalization or timbre matching. Possibly it would work even better if four surround speakers were employed. Yamaha does this with some of its processors and enhances the result by mixing in additional reverberation to simulate the larger space of a movie theater.[16] Decorrelation should also enhance the performance of a pair of conventional speakers used for surround work. One of the most significant improvements that the manufacturers of standard Pro Logic receivers could do for home-theater enthusiasts would be to incorporate a similar feature into the surround channel.

In any case, decorrelation, because it modifies the monophonic signal to the two surround speakers so that they sound different from each other, would seem to eliminate the need for timbre matching. It is self-evident that if the surround speakers are made by the same, hopefully competent, company as the main speakers (particularly if they employ identical high-range drivers), they should be close enough in spectral balance to satisfy nearly anyone without the use of timbre-matching circuitry, provided the signals sent to the surrounds—be those surrounds two in number or more—are decorrelated.

SUMMARY

There is little doubt that the advent of THX for home video has done much to eliminate the mysticism that enshrouds the performance requirements of audio-video hardware and software. Individuals who are operating on a shoestring budget may not always be able to afford certified equipment, but a study of THX performance will aid shoppers in selecting the best and most cost-effective components for them.

Yet, the statement by certain THX gurus that film sound is "totally different" from live-music sound misses an important point: *both* are different from what is possible in a typical home environment. Given our present technology, there is no way that a classical concert-hall or jazz nightclub experience can be duplicated in a typical home-listening room by even the best audio system. There is also no way that a THX-certified A/V system operating in a similar small-room environment can subjectively duplicate a large-room, movie-theater experience. Nevertheless, it is quite possible for a well-

[16]This is a controversial feature in some circles, because it overrides the Dolby parameters, but it is really no more manipulative than the decorrelation and timbre-matching features of THX.

thought-out, home-based audio-video system to produce sound that is, in my opinion, *aesthetically superior* to what is encountered in a theater—even a good one.

The THX theater parameters (as well as those of a top-grade sound-mixing facility) are designed to mimic the sound of the Motion Picture Academy's "judgment theater," a place that may not be equipped as well as and be

Figure 7.4 Full THX Speaker-System Array

(Photo courtesy of Boston Acoustics)

If you opt for THX-certified main speakers and also want decent video surround sound, you will be shortchanging yourself if you do not use certified speakers throughout your system. Better yet, every one of them should be made by the same company for consistent sound (it looks more stylish, too). This ensemble by Boston Acoustics (the TV, a Panasonic, is separate) includes three vertically aligned short-line-source main speakers (see Figure 7.2), two dipolar surround systems, and two subwoofers. While one certified subwoofer should meet THX parameters (and will probably deliver smoother deep bass if located in a corner), the people at BA feel that going with two—at least in rooms enclosing more than 3,000 cubic feet—will notably improve bass impact and reduce distortion, as well as enhance placement flexibility and smoothness. This is among the most reasonably priced THX-certified systems available.

as acoustically refined as a lot of home theaters, particularly in terms of treble balance. In addition, there is a good chance that theatrical films might actually be subjectively improved by playback on home systems that are different from those certified by LucasFilm—especially if your home decor makes it impossible for THX speaker-placement requirements to be followed. I believe that a system using a good, high-quality (but non-THX) processor, in combination with high-quality amplifiers, four or more surround channels, and very high-quality (but also non-THX-configured) speakers—including a subwoofer with strong bass output to below 20 Hz—could give even the best THX configuration a run for the money.

AUDIO-ONLY SURROUND SYSTEMS

Ambience Extraction

While Dolby Surround—and its Pro Logic and THX derivatives—is the best-known matrix-dematrix technique available for home video, it is not by any means the only one. The Ambisonic process that was designed in England by Michael Gerzon and Peter Fellgett is another matrixing system that some listeners feel does a remarkable job of replicating a real-world hall (see "The Ambisonic System," page 278). On a much more basic level, and more popular with some domestic and Japanese equipment manufacturers, are general-purpose left-minus-right extraction circuits. Commonly, these circuits will be included as part of mid- or even reasonably low-priced Japanese-made receivers that also offer Pro Logic. Indeed, these circuits may be nothing more than an elaboration of the basic Dolby decoding processor.

This kind of audio-oriented *ambience-extraction* feature can work wonders with conventional stereo (but, alas, not mono[17]) software, including some older, non–Dolby stereo, movies. Mainstream stereophonic audio-only recordings may not have specifically "encoded" surround information, but they still have left- and right-channel sounds that are different enough to allow a substantial L – R signal to be extracted and sent to the effects speakers. In addition, they will almost certainly contain plenty of noncoherent, electronically synthesized, or naturally recorded hall or studio ambience that can also be extracted. Most good stereophonic recordings—from minimalist two-microphone field tapes to those made with a large number of recording tracks

[17]You cannot extract ambience from older film or TV shows that are not in stereo to begin with, because there is no L – R signal for the decoding circuitry to work with.

The Ambisonic System

The Ambisonic recording and playback process is a British system originally developed in the early 1970s by the English engineers Michael Gerzon and Peter Fellgett (the principles were also analyzed independently in the USA by Duane Cooper and a few others). Because it requires special recording and playback equipment, it has not caught on widely in the United States, although a few audio buffs believe it delivers superior concert-hall-type sound.

The microphone design most identified with Ambisonic recording is the Calrec Soundfield, which uses four very close together subcardioid pickups, each aimed to simulate the way they would face if individually mounted on a four-sided pyramid with equilateral-triangle sides. This microphone captures the four acoustical components that mattered to Gerzon: absolute sound pressure, left-right, front-back, and up-down. By combining and matrixing these data in a specific way, a many-channel signal could be produced that would allow a decoder and a properly positioned group of loudspeakers to accurately simulate a true concert-hall environment in a typical home listening room—even in the vertical dimension.

As further refined by the BBC, the "UHJ" version of this system allows it to be compatible with existing two-channel processes and reasonably compatible with a few other surround decoders. With UHJ encoding, the surround information is in the front channels during stereo-only playback but sounds somewhat distant.

Currently, the only recording company extensively using the Ambisonic process is Nimbus (although Hyperion, Unicorn-Kanchana, Ondine, and Collins have produced a few discs—and even Arista is said to have turned out an Alan Parsons Project release in UHJ), and the recordings are variable in clarity and tonal quality, mostly because of the stiff placement requirements and pickup limitations of the Calrec microphone. Because Ambisonic encoding apparently works best when no accent microphones are used to supplement the Calrec array, it can be difficult for a recording engineer to obtain a good tonal balance in some halls, particularly large ones with large orchestras. Nevertheless, for solo material as well as small-ensemble works and small-orchestra "classical" material recorded in the right hall, the Calrec-Ambisonic process can sometimes deliver sensational results, even with conventional two-channel stereo or a noncompatible four-channel matrix-decoding system.

The big problem for shoppers in the U.S.A. is not software (Nimbus, Collins, and Hyperion recordings are available in the U.S.A.—at least in the bigger record stores), but hardware. Only a few companies have included Ambisonic decoders in their equipment, the most notable being Meridian, Onkyo, and Hitachi—and that equipment was fairly expensive. In addition, the proper playback of a decoded Ambisonic matrix requires identical speakers on all channels. While many satellite-subwoofer systems would handle this requirement with aplomb (see Chapter 2, "Five-Piece System," page 32), the tonal-balance requirements make it difficult for individuals with larger main speakers to comfortably add the additional speakers required. Because frequency-response contouring is used to simulate ear pinna and head-shadow effects, the surrounds should also employ identical tweeters—and possibly midrange drivers, as well—as the main speakers, to insure compatibility above the bass range. Of course, this is not a bad idea with any surround-sound system.

Figure 7.5 Yamaha Surround Processor

(Photo courtesy of Yamaha Electronics Corp.)

This Yamaha "add-on" surround processor employs DSP circuitry for conventional two-channel recordings and standard and "enhanced" Pro Logic for decoding properly encoded audio recordings and films. It has built-in surround- and center-channel amps, and in terms of Dolby performance it will hold its own against even the best. Working with a good preamp and front-channel amplifier, this processor may be easier to operate than some more elaborate models and will be better in many ways than many midpriced A/V receivers—particularly when simulating concert halls or jazz-club environments with 2-channel recordings. While the small (25-watt) power output of the center channel could be a problem with a low-efficiency center speaker operating in a fairly large room, the dilemma will be minimized if the system is run in the "normal" Pro Logic mode, where the center-channel bass below about 100 Hz is sent to the main speakers or a subwoofer. This essentially "biamplifies" the center channel and relieves the little amp from having to produce large amounts of power for a woofer.

and then mixed down—have a broad, non-phase-coherent soundstage, and many of the better ones have well-defined images spread across that area. L – R extraction techniques can work to good effect with this material.

In a typical low- or midpriced surround-sound receiver and/or processor, there might be several nonvideo dematrixing or extraction-surround functions.[18] When engaged, a typical ambience-extraction circuit will operate similarly to the Dolby system, routing stereo-difference information that has been extracted from the main channels to the effects speakers—possibly running it through a delay circuit for added effect—while not affecting the performance of the front channels. This picked-off signal can add a pleasant sense of space and additional realism to conventional recordings, although

[18]The control settings for these might be labeled something like "Hall," "Surround," or "Matrix." Be warned that, in many brands, these settings could provide something different than what we are discussing here, because these are not hard-and-fast technical terms.

the sometimes-cheap delay circuits may produce a bit more background noise (hiss) than some would like.[19]

On the way to the effects speakers, this L – R signal (which, in spite of being extracted from a stereo input, is monophonic just like the surround channel in the Dolby system) may be altered—via comb-filtering, staggered-delay, or frequency-contouring circuitry—to give it a pseudo-stereo or diffuse characteristic. This additional manipulation prevents it from being localized either directly behind the listener or at the effects speakers themselves (just as in a THX system). With some two-channel recordings, a simple L – R extraction technique may deliver the best and most realistic sound possible in a home-listening situation.

Ambience Synthesis

Although ambience-extraction techniques represent a merger between the audio and video-and-film worlds, there is still another category of very effective surround-sound production that was developed independently of both film and video: *ambience synthesis.* An obvious way to overcome the limitations of two-channel sound—without the cost and complexity of discretely recorded additional channels or the sonic limitations of ambience-extraction techniques—involves the creation of *synthesized* ambient sound fields from existing full-bandwidth stereo (and even monophonic) sources. Ambience-synthesizing systems are based upon a simple fact: during a "live" musical program, it is the multiple delayed boundary-generated reflections within the listening environment that "tell" listeners whether they are in a large or small room. A large concert hall sounds large, a smaller one sounds small, and a typical living room sounds very small.

First, a little background review. A standard two-front-speaker playback system can locate in space only part of what is recorded by an orchestra, combo, or soloist. Two speakers can do a fine job in terms of tonal accuracy and articulation, of course, and they can even do a decent job of reproducing the frontal hall reflections. However, the reflections that come from directions other than from the stage area in any concert hall, club, or auditorium during a live performance are simply going to be improperly located during two-channel playback. They may be present on the disc or tape as a phasy-sounding ambience, but during playback they will come from up front instead from the sides, ceiling, and rear.

hall reflections

[19]Surround-channel noise is a common artifact created by the delay circuits in cheaper surround processors—even those operating in the Dolby Pro Logic mode—although the negative aspects of this are sometimes overstated by audio journalists.

Two-Channel Surround

Several companies—notably Carver, Lexicon, Polk, Spatializer, and SRS—have attempted to produce surround-sound effects from two front speakers. Although technically quite different, some of these processes aim to limit the inter-aural cross-talk that exists in a stereophonic playback system.

When a normal sound occurs in space, each ear hears it once, for a total of two events. However, in a two-speaker stereo system, a sound that comes from any direction other than dead center, far left, or far right is an unbalanced composite created by the unequal blending of the two channels. Two speakers cannot accurately recreate a point-source sound, because the ears hear the sound from each speaker twice, for a total of four events.

Cross-talk cancellation generates two additional out-of-phase signals (one from each channel) that null the two unwanted signals coming from the speakers. In a sense, these configurations try to simulate what headphones do but with the sound still coming from out front. Properly implemented (which usually means the listener must sit in the central sweet spot so that the sounds from each speaker system travel identical distances) and with the right recording, these techniques can deliver very wide-stage sound. Not only will images between the speakers often take on enhanced three-dimensional characteristics, but hall and ambient (and sometimes instrumental) sounds that originated from off to the sides at the time the recording was made will be correctly placed in the home listening room.

While there are limits to what these systems can do—the most obvious being the difficulty of having the process work to perfection with more than one listener at a time—they often perform marvelously with nonvideo material, particularly pop music. They *can* work well with simply miked jazz and classical material, but at times the stage spread they present may seem excessively wide. In most cases, they are at their best when combined with additional processing that routes dematrixed sound to additional speakers to the sides or rear (some of the Carver and Lexicon processors can do this).

Where stereo cross-talk cancellation *cannot* work well is in conjunction with a Pro Logic processor that is downstream from the cancellation processor, because the cancellation byproducts will negatively affect the front-channel imaging and the extraction process.[1] When used with two front speakers *only*, cross cancellation can enhance a Dolby matrixed soundtrack and make it sound better than it would sound if it were being played on a basic two-channel stereo system. However, it cannot equal what a good Pro Logic system and surround speakers can do, particularly if a steered center is involved, and is no match whatsoever for any discrete 5.1-channel system.

To experience two-speaker surround sound of this kind before springing for new hardware, there are select recordings available—notably by Sting, Madonna, and Julian Lennon—that employ the "Q-Sound" or Roland RSS process. Basically a complex audio filter, Q-Sound encodes cross-talk-cancellation

[1]Polk has no problem, because the "processor" is inherent in the design of the speakers themselves.

(continued)

Two-Channel Surround *(continued)*

signals into the program material—and also performs other, less-potent phase manipulation—making it possible for any stereo system to sound ultra-wide-staged.

The NuReality SRS (sound retrieval system) employed in some TV sets, desktop PCs, and stand-alone processors offers a different solution to this problem. Developed by Arnold Klayman, SRS adds matrixing and frequency-response manipulations to the signals delivered to the two front speakers, simulating a large-stage sound and "phantom" surround speakers without the need for extra channels. (The "Spatializer" system found in some VCRs and TV sets utilizes some aspects of this technology, and some Telarc CD recordings have had Spatializer contouring added during the manufacturing process, as have the motion pictures *The Lion King*, *Crimson Tide*, and *Broken Arrow*.) SRS is based upon head-shadowing effects and the frequency-response contouring of the outer ear itself as it receives sound from different directions. Unlike the Carver or Lexicon systems, there is no cross-talk cancellation and no need to be tied to the sweet spot. In some manifestations, this system also seems to boost both the gain and the bass slightly in comparison with the input signal, which might make it sound impressive even without the other manipulations. While often better than ordinary two-speaker stereo, and quite effective when incorporated into a TV monitor with closely spaced speakers, these systems are no substitute for properly set up Pro Logic and is obviously not in the same league as Dolby AC-3.

In a typical home listening room, the nearby room reflections that you hear (those that follow the sound after it has emanated from the two front speakers) "tell" you that you are hearing the sound of an ensemble recorded in a large room being played back in a small one. Unless the listening room is quite large and the recording was made in a small, dry-sounding studio—and features only a single instrument or at most a string quartet or small combo—the effect will not be a good representation of a live performance, no matter how otherwise excellent the playback system is. Although listeners concerned primarily with individual-instrument detail, precise imaging, ensemble tonality, and overall clarity may consider two-channel stereo adequate, those of us who want playback "realism" in our listening rooms want more.

The ambience-extraction technique discussed previously can go a long way toward negating this defect by putting the delayed L – R information from a stereo signal into the effects speakers, simulating a sense of space around the listener. However, in addition to not being able to work with monophonic source material, this technique has a serious problem: the tonality of the extracted signal is at the mercy of the recording itself. If the transcription is complementary to the playback room and speaker placement and has plenty of rich reverberation (especially in the bass), the result may be excellent. If it is not complementary, it may not sound so good. Indeed, quite often the L – R part of a recorded signal is limited in bass content (because the recorded bass is usually monophonic), giving the surround signal an emaciated sound. If the extracted reverberation is turned up loud enough to create a decent sense of hall space, it may negatively alter the frequency balance of the reverberant-sound field, making it more tinny sounding than it should be (unless the bass tone controls are carefully adjusted to compensate for this effect).

In contrast, in their most basic form, ambience synthesizers feed separate effects-channel amplifiers and multiple effects speakers with delayed, often redelayed, and sometimes frequency-contoured duplications of what was already on the recording, drawing from both the L – R *and* the L + R material, at least at low and midrange frequencies. Depending on the philosophy of the processor's designer, this signal is delayed an appropriate amount (depending on how large and lively the listener wants the processor-fabricated hall to sound and how subjectively distant the ensemble should appear to be), given a bit of additional, user-adjustable reverberation (also affecting how large the "hall" sounds), and frequency-contoured at the high-frequency end (affecting the sense of reflectivity or "liveliness" of the hall).

Early versions of this technique were pioneered by Sound Concepts and Audio Pulse—and further refined a bit later by Advent and ADS—with their delay-line and bucket-brigade circuits. All of these made use of analog circuitry and were plagued by noise problems, as well as delay and reverb-generated negative artifacts. However, with a good recording and the surround speakers placed properly, the synthesized reflections could do a surprisingly good job of simulating the acoustic space of a much larger room. Indeed, some demonstrations had dramatic impact, particularly when the surround circuitry was turned off and the sound field collapsed into the two front speakers.

real halls

In 1986, with its DSP-1 processor, Yamaha introduced a new generation of ambience-synthesizing hardware. This circuitry did not just synthesize hypothetical sound fields; instead, Yamaha engineers measured the reflective characteristics of a number of notable real-world halls, and the system was programmed to mimic those sound fields. It also used digital techniques, dramatically reducing the distortion, noise, and other negative artifacts that had plagued early ambience-synthesizing systems. The term *DSP* (digital signal processing) has since become a part of audio jargon. These days, DSP not only is used to produce surround effects but also is used in noise reduction, perceptual-code processing, equalizer circuits, and, of course, in video applications.

DSP problems

However, as with ambience extraction, DSP ambience-synthesizing techniques create potential problems of their own. Most of these are related to the material on the recordings themselves—but for reasons different from those which may negatively affect L – R extraction processes.

With a dry-sounding and nonreverberant recording, DSP-type ambience synthesis can add a sense of hall, nightclub, or studio space to the performance. In addition to spatial enhancements, it often highlights the "subjective" clarity of a recording by improving the sense of frontal depth and instrumental separation. (This has been the case with a fair number of small and medium-sized ensemble jazz recordings that I have heard.) Indeed, ambience simulation usually works better with a dry recording than ambience extraction, because the latter will not have enough noncoherent material to work with.

However, on a recording already containing a significant amount of recorded ambience, the additional reverb produced by ambience synthesis may lend a cavernous quality to the sound. A number of fine recordings already do a good job of reproducing the "signature" of the hall, studio, or club environment. Adding more reverb—particularly if it mimics a location that is unlike the one where the recording was made—can result in a net loss of realism. The ambience takes on a bloated, overly expansive quality, and the sound may get lost in the synthesized environment.

Another problem involves solo performers, particularly singers. A recording that almost *exactly* simulates how some vocalists sound in relation to an orchestra in a large hall would probably be objectionable to anyone who wanted to hear the singer clearly. The technically desirable "realism" of the blend would deeply submerge the voice into the mix. Soloists are nearly always miked separately to deliver improved clarity that, while not mimicking what would be heard at a live performance, may be what we want in a home-listening situation. The engineering goal is to "move the singer forward," improving clarity and articulation.[20]

solo performances

Ambience synthesis may undo all of the carefully balanced highlighting accomplished by the recording engineer and submerge the singer into the orchestra. The voice becomes too reverberant, and articulation suffers. With some recordings, if the singer has been electronically moved quite close and the orchestra pushed well into the background, ambience synthesis may create a monster, simulating a reverberant-sounding elephantine soloist working with a small, reverberant orchestra. While those who demand realism "all the way" might enjoy this additional reverb, others may feel that the featured vocalist (or instrumental soloist) has lost the very detail and clarity that the recording engi-

[20]Of course, many nightclubs and halls now use microphones and amplification themselves to improve on what would be the natural live sound.

neer extracted by careful microphone techniques. For some listeners, at least in some situations, a straight stereo recording may be more satisfying than one manipulated by DSP—or even better than a live performance.

For this type of recording, ambience extraction can have an edge over

Pro Logic for Audio-Only Recordings

While the Dolby Surround process was conceived for motion-picture-use, there are a few excellent audio-only recordings that take advantage of it. When reproduced on a system that combines Pro Logic steering and good ancillary hardware (particularly speakers), these recordings can be quite satisfying. Pro Logic works best if you sit well away from the centrally located, "best-listening-position," sweet spot. Under these conditions, Pro Logic can (1) keep the primary performers in the center of the stage where they should be, (2) make the left-right spread seem more uniform, and (3) restore the sense of front-to-back depth that is sometimes lost when the listener is sitting away from the central axis. Even if you sit at the sweet spot, steering enhancement can frequently deliver an improvement of sorts, particularly if the recording was made with wide-spaced microphones and contains a single centrally positioned soloist backed up by an instrumental group that is spread across the sound stage. The soloist may become more focused, as well as more realistically proportioned, and the ensemble is often only minimally affected by the steering process.

However, Pro Logic is not a cure-all. Often, the steering circuit produces a vague pinched-stage image that lacks depth compared with simple two-speaker playback, particularly if you sit near the central sweet spot. The sound stage may also vary in size as the steering tries to cope with material it was not designed to handle. The condition and layout of the playback system also can be critical, particularly if the center speaker is not compatible with the main ones or is positioned too high or too low in relation to them. With a poor center-speaker setup, the sound may be substandard no matter how much of a left-to-right blend or central-image improvement the steering delivers.

A steering-induced frontal squeeze often can be offset by turning down the center volume slightly and advancing the overall gain to compensate. (This trick can also be used to add lateral spread, and even more dynamic "punch," to regular Dolby videotape or laser videodisc motion-picture soundtracks.) The negative effects of steering-induced stereo-program squeezing will also be reduced if the main speakers are placed fairly wide apart in the traditional stereo-system manner, rather than with the closer spacing advocated by some videophiles, particularly those at LucasFilm/THX.

With Pro Logic processing, the surround channel can also influence the sound of standard stereo recordings, but—unless the extracted signal is run through some kind of additional manipulations (like THX decorrelation or Yamaha DSP-enhanced Dolby)—its contribution to the illusion of realism will vary from only moderate to nearly inconsequential. Indeed, under some conditions it might be best to shut off the rear speakers and set the processor in the "Dolby Three" mode, keeping all the sound up front.

ambience synthesis. Extraction techniques usually have little trouble with stereo recordings of solo performers of any kind—be they backed up by an instrumental ensemble or not—because engineers will place them at center stage where the L – R extraction process will have no effect on them. The sound of the orchestra and the noncoherent hall ambience on the recording will not only be in the left and right main speakers but will also be extracted from the stereophonic part of the program and sent to the effects speakers—leaving the phantom-center soloist as clear, as forward, and as detailed as ever.

For any solo-performer recording, the use of ambience extraction or synthesis, if combined with Pro Logic or another type of center-channel steering, may produce even better results. This will almost certainly be the case for listeners sitting some distance from the central axis between the main speakers, when not only the central focus but also—hard as it may be to believe—the left-to-right spread and the impression of stage depth may be improved. In addition, the judicial use of the center-channel volume control will allow you to control the loudness (and, consequently, the size) of a soloist in relation to the rest of the ensemble and can also allow you to subjectively control the "width" of the sound stage, whether a soloist is featured or not. The upshot here is that it is important to have a processor that has multiple options, allowing for a variety of playback situations. There is no substitute for flexibility in a surround processor.

TWO CURRENT STANDARD SETTERS, PLUS A LOOK AT THE FUTURE

Now that we have examined surround sound's capabilities, let's take a look at the theoretical and practical philosophies of two notable manufacturers involved in surround-sound audio. While there are other, often very large and financially influential, companies involved in DSP and surround-sound development, these two have important but divergent views concerning digital-ambience synthesis and video sound in home-theater playback systems, particularly regarding speaker arrangements and signal manipulations. Finally, we will take a quick look at Dolby AC-3.

Lexicon: A Major American Player

Lexicon, a prime producer of studio gear, is also responsible for some highly regarded consumer-level surround-sound processors. The earlier Lexicon surround processors were designed to be easily integrated into an existing stereo system that already had either a preamplifier and several power amplifiers or a receiver or integrated amplifier having external preamp and amp connections and additional power amps for the center and surround speakers. The first "consumer product" produced by the company, the CP-1 (hence the name; later updated to the CP-1 Plus), required a preamplifier or input switcher, as did the less-expensive subsequent model, the CP-2. However, the more advanced (and considerably more expensive) CP-3 Plus was able to operate as a full system control center, provided you were willing to give up a phono-preamp function, limited yourself to four or fewer inputs, and didn't need built-in tone controls. The latest-generation processors, the DC-1 series, carry forward the total-control aspect of the CP-3, with each of three variations offering DSP ambience synthesis, in addition to standard theater, THX theater, and THX/AC-3 modes, depending on cost.

Lexicon's extensive research into the human hearing mechanism has led them to develop strict ideas about speaker placement. For example, at its optimum, the Lexicon configuration has the primary surround speakers to the *sides* of the listener, with a second pair to the rear. These placement guidelines flow directly from studies[21] that demonstrate that the sense of space presented by real-world concert halls is the result of strong *lateral* sound clues and not simple, amorphous hall reverberations. The side-speaker requirement is not only a must for proper concert-hall sound synthesis but is also critical for proper playback of Dolby-encoded *video* material, particularly if supporting the THX parameters. The CP-3 Plus and DC-1/THX are not only among the best audio-program surround processors available but also perhaps the best of the home THX processors, with the AC-3 version (see Dolby AC-3, below) setting new standards for video-theater sound.

side surround speakers

Another feature of the Lexicon system is the stereo treatment of the surround-channel signal in the audio-only playback mode. Most other systems combine the derived signals for the surround channels to mono (L + R) and then run the result through ambience-synthesizer circuits before routing them to the surround amps and speakers. The result is synthesized-stereo sense of space from the surround speakers that is unrelated to the positioning of the front speakers. Lexicon processors do not combine the signals for the surround channels to mono and regenerate them as pseudo-stereo. Instead the processor derives the right-side reverb from the right-main channel and the left-side reverb from the left-main channel. This is said to achieve a better correlation of the main signals up front with the synthesized reverb.

Processors generally similar to and often cheaper than the Lexicon models are produced by a number of other companies, including Harmon Kardon,

[21]By Manfred Schroeder, Lexicon's David Griesinger, and others.

Fosgate, Proton, AudioSource, Pioneer, and, of course, Japan's main contender, Yamaha. However, it can be argued that none work as effectively, or are at least as flexible, as the all-digital Lexicons. Some companies, such as Kenwood, Carver, and Sony, have full-featured surround-sound preamps that rival any of the Lexicon models in decoding abilities, and Yamaha has a number of fully self-contained integrated amplifiers with DSP and theater-sound features that are also formidable. For a while, Onkyo even had a powerhouse of a receiver with DSP and seven channels! While none of these are in a class with the more expensive Lexicon designs, they often rival the cheaper models. It could be argued that, at least in some speaker-placement and room-configuration situations, their performance may be superior to some Lexicon models.

Yamaha: The Other Kid on the Block

As anyone who reads the popular audio and video magazines knows, Yamaha is a major participant in the audio-only DSP game—indeed, for a while it was *the* major participant. Yamaha started it all back in 1986 with its DSP-1, and with their current top-of-the-line models, they continue to lead the pack in terms of the variety of hall selections available from their processors and the number of enhancements for Dolby Surround.

During most of its existence as a maker of surround-sound hardware, two characteristics have differentiated Yamaha's products from most others. First, as we have noted previously, rather than fabricate a group of theoretically workable delay times, reverberation decay rates, and reflectivity characteristics for their digital synthesizing programs (the Lexicon approach), the Yamaha engineers actually recorded the three-dimensional "sound" of a group of real performing venues around the world. They incorporated that data in varying degrees into their processors. Yamaha believes that using real-world halls as models will at least result in a better synergism of hall ambience characteristics when playing back recordings than using theoretical archetypes.

Second, while most processors (including those adhering to THX video specs) supposedly work best if the primary surround speakers are placed to the sides of the main listening area (often with supplemental support from additional speakers to the rear), Yamaha continues to mandate that the principal surround speakers working with its DSP equipment be placed behind the listening-viewing position, near the room corners, with the option of having two additional "effects" speakers situated up front, somewhat above and to the sides of the left and right main speakers. Yamaha claims this is a mandatory positioning for their audio DSP programs, because the halls that were used for modeling them were recorded by means of a four-corners technique. They also claim that the positioning works best for their assorted movie-theater modes, because those enhancements to the basic Dolby parameters are themselves based on the audio DSP programs.

While it has been proven by Lexicon's Dave Griesinger (and others) that lateral reflections are what make a DSP-enhanced audio program correctly simulate the required hall spaciousness, the engineers at Yamaha apparently feel that four corner-located speakers will subjectively form "phantom"

Figure 7.6 Yamaha Processor/Integrated Amplifier

(Photo courtesy of Yamaha Electronics Corp.)

Top-grade integrated amplifier/processors can do just about anything a diehard A/V buff could ask for, and this particular Yamaha model even has Dolby AC-3. It also sports seven well-amplified channels: the usual three across the front and two surrounds to the sides or rear, with two additional surrounds (that are best placed at the sides also but more toward the front of the room).

The unit has a multitude of DSP functions for audio-only use, in addition a number of normal and "enhanced" Dolby Pro Logic and AC-3 possibilities, including combining DSP with standard Dolby decoding to simulate the larger space of a movie theater. While purists may decry this kind of manipulation, it is no more radical than what is mandated by THX and is in some ways superior, because of the additional surround speakers toward the front. All the surround parameters can be altered by the user to taste or to compensate for listening-room anomalies.

In its audio-only modes, while the left, right, and L + R components—in addition to being reproduced by the front-main speakers—are delayed, reverberated, and then sent to the surround speakers by the DSP audio processors, the 180-degree out-of-phase signals are not. This insures that L – R hall reverb already on a recording is not re-reverberated and sent to the surround channels to create too much of a good thing.

images to the sides that are more effective at mimicking real-world lateral reflections than a pair of fixed speakers placed straight to the left and right, even if they are dipole—THX or otherwise—radiators.[22] In addition, they apparently believe that four DSP-controlled surround channels, working with the three front speakers to simulate a movie-theater environment, do a better job of delocalizing video surround-sound effects than a pair of conventionally placed surround speakers—even THX models with a decorrelated input. This may especially be the case for those sitting forward of or to the rear of the cross-room sweet spot, because that area is not effectively covered by a single pair of aimed dipoles.

The Yamaha "35-mm enhanced" mode of Dolby Pro Logic used in many of its processors takes the extracted surround signal and runs it through a DSP hall-simulation program that supposedly duplicates the kind of large-scale space experienced at a movie house. All four surround speakers are involved in the simulation, in addition to the three front speakers. A direct comparison of this embellishment with standard DPL shows it to be much more effective at delocalizing the surround signal, while having no effect on the spread up front or the Pro Logic–steered dialogue. Unlike THX, however, the overall sound also displays a large-room ambience that will appeal to some home-video fans.

In addition to AC-3, upscale Yamaha processors also have a number of so-called "70-mm theater" modes for Pro Logic enhancement that supposedly go the 35-mm "enhanced" mode one better. The 70-mm embellishments apply DSP to the Dolby-extracted surround material, routing it strongly to the rear as with the 35-mm enhanced mode, and at a somewhat reduced level to the front surrounds. In addition, the front surrounds receive DSP-manipulated sound from the left-*plus*-right signals up front that is processed in a similar manner to audio-only programs. This additional potpourri of effects from the front surrounds adds a sense of big-room space to the action up front.[23] While this kind of heavy manipulation may seem sacrilegious to hard-core Dolby addicts, it does appear to be convincing with some older stereo movie soundtracks, Dolby encoded or not, and particularly with those deficient in lateral spread.

Possibly the strongest case for seriously considering the purchase of

[22]This side-phantom-image theory is disputed by some. A stable side image is said to be difficult to create because of the way the outer ear's pinna contours the sound coming from two different directions at the side, even if those sounds are identical. However, if the sounds are digitally reverberated somewhat and the rear surrounds are adjusted to play a bit louder than the fronts, or are closer to the listener, the image may tend to stabilize.

[23]The descriptions of the effect of the assorted 70-mm modes in the numerous Yamaha brochures and operational manuals is vague and at times almost cryptic.

Yamaha DSP gear is its operational simplicity and the self-contained nature of the processors. The latter are usually included in receivers or integrated amplifiers that have all the necessary amplification. (However, outboard amplification can be added if more power is required.) In particular, their top integrated amplifiers and receivers leave little to be desired, particularly when involved in typical living-room speaker-placement situations. Just add the seven speakers (plus a subwoofer/amp or two if you want) and the source material, and be on your way to near-top-drawer surround sound.

Perhaps the weakest aspect of the Yamaha philosophy, compared with user-information-oriented outfits like Lexicon, Shure (who were into surround sound for a while but have since bowed out), and Fosgate (the latter, like Lexicon, a Harmon International company), shows up in their operator manuals. While the Lexicon instruction booklets are paradigms of clarity and theoretical knowledge, the Yamaha handbooks describe the assorted audio and video-sound functions in terms that vary from moderately helpful to vacuous, misleading, and even inane.

Dolby Digital AC-3 Surround

Dolby Pro Logic's steering process is based upon identifying a single dominant channel at any one time (L, R, L + R, and L – R) and suppressing unwanted cross-talk in adjacent channels. Unfortunately, the dematrixing, even with advanced motion-picture theater systems, cannot produce wide-bandwidth steering with all four channels simultaneously, making it less than fully effective for some kinds of complex-signal situations, particularly program material that has not been specifically adjusted when it is recorded to allow the steering circuitry to operate effectively. Pro Logic cannot give us four fully discrete channels.

However, discrete-channel home A/V and theater sound *does* exist, and its implementation has resulted in a home-entertainment revolution. Commercially dominating this phenomenon, at least in the home-video realm, is Dolby AC-3, a digital perceptual-coding technique that delivers not just four but *six* discrete audio channels. If AC-3 (the AC stands for audio coding) is not the whole future, its primary competitors in the motion-picture world—the American DTS Coherent Acoustics, Japanese SDDS, and the Philips-developed MusiCam Surround systems—will all have to reckon with it, if only because it dominates in the home-theater realm.

As configured for theater and home-video use (including videodisc, digital videotape, digital cable, and PrimeStar satellite), the Dolby Digital AC-3 system was designed from scratch to be a multichannel system and is some-

Other Discrete-Channel Theater and Video Surround-Sound Systems

While the Dolby Digital Surround system has been accepted as the standard for HDTV and DVD, and the Philips MusiCam system is used for DSS and is also popular in Europe, there are a few other surround-sound theater systems, primarily found in movie theaters. Good as these systems are, they have not made serious inroads into home theater, mostly for financial and/or commercial reasons.

Digital Theater Systems offers the most noteworthy theater version. The DTS Digital Surround format employs five full-bandwidth channels, with a sixth usually reserved for deep bass. In its theater version, DTS uses much less data reduction than Dolby's AC-3 (about 3:1 instead of the 10:1 of the latest version of the Dolby system). However, believe it or not, for motion-picture work the audible effect of this format's 1,400-Kbps data rate is close to nil compared with the 384-Kbps version of AC-3 and is meaningless considering the technical requirements of home-video HDTV and DVD systems. Both DTS and Dolby Digital are found in thousands of theaters. Because it was developed by a subsidiary of Universal Studios, DTS has been used for scads of movies, including such sonic blockbusters as *Apollo 13*, *Backdraft*, *Gettysburg*, *Speed*, *Die Hard with a Vengence*, *Waterworld*, and *Jurassic Park*.

Theater DTS employs two digital surround systems. The more complex version—the one that is Dolby Digital's main competitor—can be reconfigured for eight channels, instead of six (see comments on the Sony system, below.) The second form is actually a two-channel matrixing system, which can feed directly into the optical preamplifier of a conventional 35-mm optical-stereo processor. The system is essentially a digital variant of standard Dolby Surround, and the two signals are decoded into the standard left, center, right, and surround channels. While not "discrete" surround sound, this format is easier and certainly cheaper to integrate into a conventional theater-surround system than the six-channel version, and yet it retains many of the advantages of digital audio: low noise, wide bandwidth, and tremendous dynamic range.

Another contender is the Sony SDDS (Sony Dynamic Digital Sound) system, which, like Dolby Digital and DTS, uses data reduction, in this case operating at about 5:1, the same as Sony's MiniDisc. *The Last Action Hero* was this format's opening shot, and *The Professional* also employed the system. However, SDDS not only has the de rigueur stereo surround channels but one-ups the competition by using five channels across the screen (left, center-left, center, center-right, and right), plus the subwoofer channel. That's *eight* discrete channels (OK, 7.1, which supposedly can be matched by DTS if the latter is reconfigured). While the Sony system is overkill for theaters with screens less than 40 feet wide, as well as for home video, it should be impressive in a larger theater. It can be reconfigured for 5.1 channels, if necessary. Like the other formats, SDDS films retain the matrixed Dolby Stereo tracks, allowing theaters that have not converted from standard dematrixing hardware to present these features with conventional

(continued)

Other Discrete-Channel Theater and Video Surround-Sound Systems *(continued)*

surround sound and dialogue steering. The Sony system, a late starter, has not caught on like DTS and Dolby Digital, although the AMC theater chain has used the system extensively.

There was another system, Cinema Digital Sound, developed jointly by Kodak and Optical Radiation Corporation, which was also first class. It was not commercially successful, because it required that separate prints be made for analog or digital playback. ORC's most notable achievement, and proof positive that this method could also deliver the goods in spite of an inherent lack of theater-to-theater flexibility, was *Terminator II*, a film nobody badmouths for bad sound.[1]

Good as DTS is in its theater version, the Dolby home-theater variant of 5.1 stereo became the system of choice for HDTV, the 12-inch LV[2], and DVD, because of company prestige, a marketing headstart, technical flexibility, and the fact that for all of these formats except the analog videodisc, DTS—for technical reasons involving digital-video requirements—would have to use the same average data rate as AC-3. Hamstrung by the latter requirement, DTS has no technical edge at all over AC-3 for most home-video uses.

DTS may still have a future as a home-audio-only system, because it lends itself well to ultra-high-quality surround-sound applications. Several software companies, including DMP and Telarc, have released audio-only discs for playback on systems with CD or videodisc players employing suitable digital outputs and processors incorporating DTS decoders, and a number of companies have produced the required home-playback hardware. This is high-end audio at its best.

[1]DTS, SDDS, and ORC film soundtracks can be converted to AC-3 for videodisc use.

[2]Although a few excellent specialty discs have been released with DTS tracks.

times called Dolby 5.1.[24] Initially, the system was the result of research by MIT graduate Dr. Mark Davis, who was also mostly responsible for the first "Stereo Everywhere" dbx Soundfield speakers, the MTS stereo television noise-reduction process (also under license to dbx), and the Allison Electronic Subwoofer (see Chapter 2, "Allison ESW," page 69). While DPL and ordinary stereo will remain the default choices for standard broadcast signals, VHS tapes, and some C- and KU-band satellite service, AC-3 will be the preferred system for individuals wanting the best in home-theater performance.

The system employs data-reduction and perceptual-coding (masking) techniques analogous to those of DCC and the MiniDisc formats discussed in

[24]The "point-one" nomenclature is a somewhat humorous reference to the deep-bass "effects" channel, which lacks the bandwidth of the other five.

Chapter 4. In essence, it is a very selective and powerful noise-reduction process. However, instead of cramming two channels of sound into a very small space as those systems do, the motion-picture-theater version of this digital technique allows six channels of sound to be reduced into a space that would ordinarily hold substantially less than a two-channel 16-bit CD. AC-3 also uses a "shared bitpool" process, whereby channels with greater frequency content can demand more data than sparsely occupied channels. The coder looks at what all the channels are doing together. The auditory masking model insures that a sufficient number of bits are used to describe the audio signal in each band.

Since its introduction as a theater playback system in June 1992, the design has been refined, and the latest theater, HDTV, and videodisc versions deliver 5.1 channels from 384 to 448 Kbps (the original theater version topped at 320 Kbps). This allows slightly better results with complex signals, smoothing out the occasional treble and transient harshness that some critics detected in the original version. However, even the latest upgrade exhibits a substantial 10:1 data reduction (the rate varies from 8.5 to 12:1), which is obviously much greater than DCC, MD, or, for that matter, the DTS and SDDS theater-sound systems (see "Other Discrete-Channel Theater and Surround-Sound Systems," page 292-93).

In either form, Dolby Digital displays five separate, "primary" channels: left, center, right, left surround, and right surround. The configuration is sometimes called a 3/2 system (three front, two rear), as opposed to Pro Logic Surround's less-than-perfect 3/1 system (three front, one rear). In addition, it utilizes an all-important sixth ("point one"), low-frequency effects "channel," covering the range between 20 and 120 Hz.

With standard 12-inch laser-video or DVD source material, the encoded signal is extracted by the AC-3-configured player and sent to a demodulator. The demodulator filters out the 2.88-mHz carrier frequency and converts the signal into a digital data stream that is then routed to a decoder. This is not as complex as it seems, and in most cases the demodulator and decoder are in the same box with a processor's or A/V receiver's other surround-sound circuitry, although stand-alone units are also available. The decoded data, now in analog form, is then sent to the required amplifiers and speakers. The required amplification (usually minus that needed for the subwoofer) will, of course, already be contained within an AC-3-ready A/V receiver or A/V integrated amplifier.

For many listeners, Dolby Digital's most significant contribution to the concept of true stereophonic sound reproduction is its ability to produce *three truly discrete channels of sound across the front* without the artifacts associated with steering systems like Pro Logic. The algorithm also gives us two independent surround channels (which many Dolby engineers and journal-

Dolby's AC Codes

Note that AC-3 did not spring full-grown like Athena from the head of Zeus. It is the result of an evolutionary process. AC-1, a refinement of systems developed by ADS and Audio Pulse for their early-generation time-delay devices, has been used for TV-sound distribution and was first employed in 1985 for DBS applications by the Australian Broadcasting Corporation. It has also been used for satellite communication networks and digital cable radio systems, and even now there are more than half a million AC-1 decoders in use. The data rate for this process varies from 220 to 325 Kbps, depending on the application. AC-2 (which has a reduction bit rate of either 128 or 192 Kbps per audio channel), is used for professional audio transmission and storage applications and was also used to link remotely located recording studios and/or film postproduction facilities for long-distance, real-time recording, mixing, and ADR sessions. It is also used in the Dolby DSTL system for linking broadcasters' studios and transmitters.

AC-3, the latest version, samples at 48 kHz (although rates of 44.1 and 32 kHz are also supported), is quantized with 18 bits (with 20 bits of dynamic range possible), and is then data-compressed with the AC-3 algorithm into the required six channels. AC-3 is specifically designed for multichannel use and effectively masks noise between channels as well as within them. All three of these systems are outgrowths of Dolby's experience with audio signal masking in the design of its analog noise-reduction systems.

ists consider to be its most important attribute), instead of the sometimes vague, extracted, monophonic signal of standard Dolby Surround or theater Dolby Stereo. With this capability, the two surround speakers so long utilized in typical home surround-sound installations are now able to actually deliver the stereo information that their physical placement demands. In addition, the AC-3 surround channels have a lower background-noise level than is possible with earlier surround systems, allowing for a considerable increase in the dynamic range of certain special effects.

With appropriate decoding gear and the speakers set up properly, Dolby Digital material can improve upon the sense of envelopment presented by the old 4:2:4 matrix system, increase the dynamic impact of off-screen artifacts, and also distribute precise localization clues that make "flyovers," "incoming" artillery fire, smooth left-to-right pans, and other subjectively near and more-distant screen sonic embellishments much more effective. I should note here that material previously mastered for conventional Dolby Surround, or even regular stereo, *can* be remastered and encoded to AC-3.

While digital-video formats have Dolby Digital as their main source of audio, all laservideo software that is AC-3 encoded also contains separate

THX and 5.1-Channel Audio

Extending their work in improving home-theater sound and image quality, LucasFilm has established hardware and software certification guidelines for AC-3 systems. They may be more beneficial to serious home-theater enthusiasts than the Pro Logic specifications.

1. "Dynamic" decorrelation is included to remove in-the-head and nearest-speaker-location artifacts from the *monophonic* part of AC-3 surround signals. This is because not every AC-3 surround program (particularly older reissues, even if they are DVD items) will have a "stereo" surround track, and even up-to-date films with stereo surround have a monophonic complement to that signal at times.

2. Stereo timbre matching between the main and surround channels is required.

3. A "bass-management" parameter includes 5-channel crossover functions and low-frequency effects-channel summation, along with selectable bass-output alignments to accommodate a variety of installations. A subwoofer peak-level-control network is also required to prevent overload of subwoofers.

4. A position/time alignment circuit compensates for nonoptimum speaker placement (an excellent idea).

5. Re-equalization is employed to compensate for occasional overly bright front channels.

6 Laser-video players must have an undemodulated output for AC-3.

Finally, while some individuals initially felt that the "dipole" surround speakers mandated for earlier home-THX systems would be at a disadvantage with AC-3 surround signals, it appears that the dipole-surround concept dovetails as well with 5.1-channel audio as it did with Pro Logic.

Dolby 4:2:4 matrixed sound on the two usual PCM-digital tracks. This is vitally important for individuals who have older surround processors and/or LV players that do not have AC-3 capability. These discs, when played on a conventional Pro Logic system, will sound just as good as ever but will not have 5.1 discrete channels. In addition, Dolby has mandated that AC-3 decoder chips also decode DPL conventional soundtracks, allowing for very high-quality surround sound from standard Dolby Surround sources. This insures that most analog videodiscs in your collection will be compatible with the new format.[25] The new digital videodisc system (DVD) dovetails

[25]Note that I said "most" and not all. Older laservideos that don't contain PCM soundtracks (that is, those having FM-analog audio only, with or without CX encoding) will *not* deliver proper sound on AC-3-equipped players. This is because on analog videodiscs with AC-3, the track that previously carried the right-channel FM-analog information carries the AC-3 data. On these discs, the left channel can still be used to carry a mono version of the soundtrack or can present a voice-over commentary on the film by the director, special-effects people, etc. Note too that because AC-3 discs contain no analog sound at all, older laser-video players that have no PCM-audio capabilities will reproduce none of the audio on the new discs—except the mono FM track.

with AC-3 even better than the older analog videodiscs, and with proper playback decoding, the data stream will allow either 5.1-channel, Pro Logic, stereo, or even mono playback. Dolby Digital's cross-format flexibility will also allow a 5.1-channel video program source, once decoded, to be matrixed to 2 channels and be compatible with a standard DPL dematrixing processor or be copied to a VHS tape as Pro Logic–compatible material.

effects channel

While a textbook-configured home AC-3 system will employ a big subwoofer to handle the "effects" channel's bass—with all the other speakers being freed from the job of dealing with really powerful low bass—all processors have the ability to shunt those signals to any combination of the other speaker systems if no subwoofer is available. If the mains are potent full-range models, the "point-one"-channel bass can be routed to them. If the surround and center speakers are too small to handle the load, their bass can also be routed to the main speakers, instead of to a subwoofer. For the adventurous, it is conceivable that an AC-3 configuration could have as many as five, or even six, subwoofers—with even surround and center channels getting loads of deep bass when the soundtrack calls for it. AC-3 processors are very flexible at sending the deep bass where it works best with your speaker setup.

While Dolby Digital has mainly had impact on film and video programming, its potential for audio-only use, particularly since DVD players will also play CDs, should not be discounted. For while it is not overwhelmingly better at simulating hall ambience than top-grade ambience-synthesizing hardware working in combination with good standard 2-channel source material (let's face it, both configurations must confront the acoustic realities of small-room playback), the ability to provide three discreet channels across the front *is* very significant—especially for those sitting anywhere else but the central-axis sweet spot.

For some time, assorted matrix-surround and ambience-synthesis systems have been the backbone of surround-sound audio in the home. The advent of Dolby Digital and other 5.1-channel formats may be a home-entertainment milestone as significant as the CD and the high-fidelity video recorder—and may even rival high-definition television in impact. In conjunction with advanced-digital video, this new "high-definition sound" will make future A/V formats *HD-everything.*

IS CHEAP HOME THEATER POSSIBLE?

Once you have a fairly large or, hopefully, wide-screen TV monitor and appropriate source-material playback hardware, the key to good home theater is sound. While a giant-screen TV set working by itself may be quite impressive, it will not be nearly as awe inspiring as a somewhat smaller set operating in conjunction with an appropriate surround-sound system—especially when the program material is blockbuster-grade motion pictures and the system includes a decent subwoofer.

See Figures 5.3 (page 161) and 6.4 (page 203). Just add the center and surround speakers!

Cheap home theater *is possible* because it is not all that expensive to put together a decent surround-sound video system. Combine an existing pair of good speakers with a moderately priced new or used Pro Logic receiver, a good (magnetically shielded) center speaker, and a pair of mini speakers for the surround channels (even cheap Radio Shack models should be surprisingly effective), and you have the necessary audio components. Correctly stationing this combination in a decent room and then adding an existing 27-inch, or somewhat larger, TV set (particularly if it is a monitor/receiver) and a hi-fi VCR should yield the theater experience in spades. True, it will not be a 56-inch-wide screen, Lexicon-processor-equipped, DVD picture/sound, Velodyne-subwoofer experience, but it will positively transcend what a typical stand-alone "TV set" can deliver—even a large, console model. If the speakers are good ones and they are well-matched to the room, the subjective impact of this system may equal or even surpass that of many big-buck, decor-oriented high-end systems—the kind featured in those glitzy home-theater magazines. Also, if a small-dish satellite receiver and processor are installed, the picture and sound quality of your system will be improved substantially at a reasonable cost.

A pleasant byproduct of this kind of configuration will be the system's performance with audio-only recordings. In most cases, this so-called "mid-fi" system will be superior in terms of subjective "realism" to even the most esoteric, outrageously expensive, high-end, two-front-speakers-only stereo systems—particularly if the speakers are good ones, the surround processor includes high-grade DSP circuitry, and a good subwoofer is thrown in. As we said at the beginning of this chapter, there is no high-fidelity in home-listening situations without surround sound!

Few low- or medium-priced receivers (and outboard processors) have the kind of digital-signal processing required to simulate the enveloping concert-hall or nightclub experience with audio-only program sources. However, a lot of them—including a surprising number of reasonably current ones that may be available as used items down at your local hi-fi shop—at least do a decent job. What's more, those with Pro Logic or some other kind of proprietary steering may be able to greatly enhance the sense of frontal-space realism and imaging found on some nonencoded, but still stereo, audio recordings—particularly if you are sitting some distance from the ideal central-axis listening position and the main speakers are far enough apart to

Surround-Room Diagram

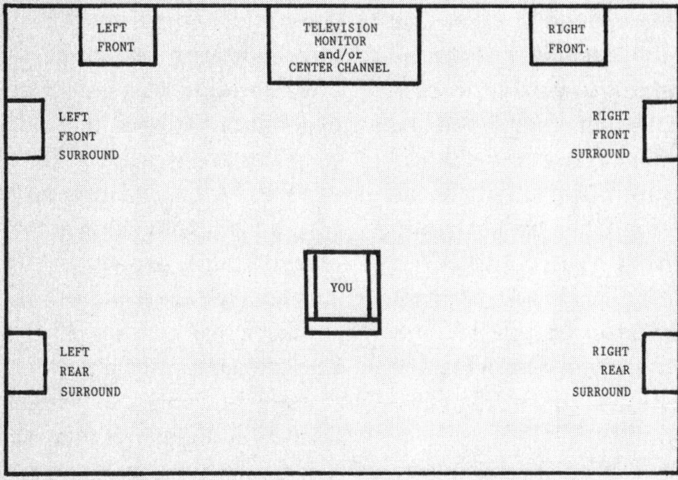

Some experts prefer that a TV set be located along the narrow wall of a room (with the main speakers closely flanking it), with the center speaker centered, and the surrounds placed to the sides and/or rear. I favor putting the monitor on the *long* wall, with the main speakers fairly wide apart (after all, the center channel will take care of the central focus, and widely spaced main speakers add an important degree of "bigness" to an often-small TV picture) and with the surrounds placed so that they are fairly far away, to the sides or slightly to the rear. This allows the entire ensemble to do a better job of diffusing ambience, allows more effective use of a pair of TV-monitor-bracketing center speakers, and also smoothes the reverberant-field frequency response of the main speakers up front, particularly in the bass.

When the main speakers are widely spaced, they may be substantially farther from the listener than the center speaker. When performing a Pro Logic balancing test, it may be wise to slightly reduce the gain of the center channel relative to that of the mains, in order to offset the impact of the nearer speaker's precedence effect.

THX standards notwithstanding, I also favor using four surround speakers, rather than two, to diffuse the ambience even better. After all, theaters use more than two speakers, and there is no doubt that lining the walls with well-positioned effects speakers will make it difficult to localize the surround images, even if those speakers are facing into the room and not spreading their sound along the walls THX-style or bouncing it off the ceiling.

Yamaha uses four surrounds in its upscale processors and minimizes the timbre-matching and side-image-shifting problems that vexed the THX engineers by electrically decorrelating the four surround signals via one of its "enhanced" or "70-mm" Dolby modes. While companies like Lexicon require

(continued)

Surround-Room Diagram *(continued)*

that the forwardmost surround speakers be to the sides of the prime listening area, with a second pair of surrounds towards the rear for delayed digital-processed reverb, Yamaha specifies (at least in rectangular-shaped rooms) that the forwardmost surrounds be in—or at least near—the front corners, with the other surrounds in the rear corners. With the Yamaha configuration, each pair of front and rear surrounds is said to be able to form phantom images to the sides, achieving the lateral spatial impression that the Lexicon people say is so important.

The debate over surround-speaker placement with basic two-surround-speaker Pro Logic goes on and on, with Japanese manufacturers favoring rear placement and American ones tending to favor side placement. My experience has been that side placement tends to better propagate a sense of ambience, while a more rearward placement tends to do a better job of delivering occasional "flyover" effects, at the expense of a more enveloping ambient field. The use of four properly placed speakers may successfully split the difference. Unfortunately, with either matrixing or synthesizing-type surround systems (as opposed to discrete-channel systems like AC-3), no placement scheme can work to perfection with every kind of source material.

minimize the tendency of steering systems to somewhat collapse the sound field toward the center. In addition, used carefully, even the most rudimentary ambience-extraction circuitry, working with a pair of cheap surround speakers, will improve the sense of subjective space above and beyond what would be possible with only two front speakers.

Finally, once a basic surround-sound/video package is in place, it is easy to upgrade a bit at a time (disc player, better main speakers, bigger TV monitor, advanced processor, subwoofer, etc.) and gradually bootstrap yourself into the big leagues. The thrilling part of this is that every upgrade—short of 5.1 discrete digital channels—results in an improvement to your existing collection of videotape, videodisc, and audio recordings—not to mention everything you rent down at your local video store or receive over your satellite dish. (An AC-3 hardware upgrade can improve the sound of any AC-3-encoded discs you might already have.) Old recordings or tired hi-fi videotapes will appear new when experienced with your new, enhanced setup. Once you have become hooked on home theater and surround-enhanced audio playback, you will never want to go back to regular TV and plain old stereo sound.

8 Embellishments, Modifications, and Manias

NOW THAT YOU'VE COMPLETELY assembled your home-theater system, you can sit back and enjoy your favorite blockbuster movie or bass-thundering symphony. However, once bitten by the advanced A/V bug, you may begin to lust after more esoteric additions to your basic system. For more than a few enthusiasts, particularly those preoccupied with high-quality audio, these accessories can be as important as speakers, CD players, amplifiers, VCRs, and television monitors for the full enjoyment of a home-theater system. In previous chapters, we discussed some of the accessories that are available to A/V enthusiasts, including a sprinkling of information about equipment modifications and A/V manias. This chapter attempts to tie up some loose ends by reviewing additional accessories, viewpoints, and theories—from the rational to the harebrained—that are currently prevalent in the A/V market.

While there are a fair number of good accessories available for the serious *video* enthusiast, many of the *audio* accessories currently touted by independent manufacturers are less useful. Sound reproduction is more susceptible to subjective interpretation than video and easily lends itself to assorted mystery panaceas and placebos. As with hardware, philosophical ideas, concepts, and manias dealing with audio are more likely to be offbeat, anecdotal, and unsubstantiated than those dealing with video. Video enthusiasts often tend to be more brass-tacks oriented and even better trained than audio enthusiasts and are therefore less susceptible to the influence of nostrums and mythology.

There is no doubt that some accessories and embellishments can be beneficial. However, many do no good at all, and a fair number are hazardous to hardware and software longevity, accurate sound reproduction, and/or decent picture quality. Because items of this kind are numerous and new ones are continually surfacing, we cannot hope to analyze them all. Some, because of their complexity and impact, are discussed in depth, while others are discussed only briefly—because their positive attributes (or potentially negative capabilities) are self-evident. To keep things simple, all of the hardware, principles, and speculations are categorized together in groups according to their effectiveness or perniciousness.

EXPENSIVE GOOD THINGS

There are many accessories available that are functional, educational, cost effective (in spite of their often-high price tags), and fun to use. A few of them are considered essential by some enthusiasts, but most of us will survive without them and still enjoy decent sound and good home theater.

Subharmonic Synthesizers

Many recordings, be they audio-only or movie soundtracks, lack the really deep bass necessary to add a sense of depth and realism. Imperfections of this kind commonly materialize in CD reissues of classic early recordings, as well as in videotape and disc releases of classic films, particularly those mastered from 35-mm optical tracks. In addition, even some up-to-date productions lack the kind of deep bass that makes the purchase of a subwoofer worthwhile.

While careful use of a bass tone control or multiband equalizer can often partially restore the bass line of these recordings, there are times when there is simply no deep bass to boost. This is where subharmonic synthesizers can come in handy. These fabricate *new* low frequencies at one-half the frequency of the originals already on a recording. For example, a 90-Hz tone will prompt the synthesizer to create a complementary 45-Hz tone to go with it, a 60-Hz tone will beget a parallel 30-Hz companion, and so forth. The output of both the original and synthesized signals can be varied, giving the operator somc control of the process.[1]

Most subharmonic synthesizers are configured to work with original signals spanning the bass range between 50 and 100 Hz. Consequently, the sup-

[1]Interestingly, theatrical releases of some bass-deficient films (at least optical prints, which often have less bass than 70-mm magnetic master prints) are often given a low-frequency boost by a theater-type subharmonic-synthesizing system called the optical bass extension (OBE) module.

plementary frequencies cover the next-lower octave, between 25 and 50 Hz. These processors cannot restore lost bass, as some of their ads have claimed. However, they can creatively "fake" things well enough to make some prerecorded material sound richer and more up to date, particularly reissued analog-master pop-music CDs, as well as pre-digital-sound films.

Subharmonic synthesizers can also add a fair degree of subjective dynamic-range enhancement to videos that feature a lot of artillery-type pyrotechnics. The additional deep bass, in addition to making the low-bass dynamics seem fuller, will increase the average bass loudness, making motion-picture explosions seem more potent than ever. A less expensive solution is the judicious use of a good bass tone control or the low-bass slider on an equalizer; either can soup up the bass dynamics of a slightly anemic action-adventure soundtrack, although when overdone they may also add too much bass richness to male voices.

Subharmonic synthesizers are not cure-alls. For example, they will not work well at all with classical music of any era, and I have never had much luck using one with jazz. When attempting to enhance the low-frequency acoustic instruments featured in this music, boosting the subharmonic frequencies may create a sort of thickened sound quality that seems artificial. Low-bass synthesizers will also not function satisfactorily with film or video sources having genuinely anemic soundtracks (with a depleted output over the vital 50- to 100-Hz range). You cannot make the explosions in the 1945 film *They Were Expendable* convincingly shake the room without advancing the deep-bass synthesizing control so much that John Wayne sounds like Darth Vader.

Only two companies that I know of have produced subharmonic synthesizers for consumer use: dbx and AudioControl. The discontinued dbx 120 worked quite well. AudioControl's cryptically named "Phase-Coupled Activator" has received excellent reviews, with some critics claiming it is superior to the dbx model. It has gone through several upgrades over the years, and the latest version is said to be very effective. Used samples of any of these models would be good buys. There is no reason that a subharmonic synthesizer should be prone to serious wear and tear, so if you find a decent used one for sale cheap and you find the concept attractive, snap it up. Most sell brand new for about \$250–\$300, and a used one should not cost more than \$100–\$150.

Dynamic-Range Processors

Program sources that use Dolby noise reduction, whether they are tapes made at home or commercially produced, have their dynamic range compressed at higher frequencies during the recording process. During playback, the decoding circuitry expands the signal in a complementary manner over the same bandwidth. The result is attenuated background noise. Variations of this type of "double-ended" operation are also found in the dbx noise-reduction process that is part of the MTS system for conventional stereophonic video sound transmission, VHS hi-fi, and the analog-audio CX noise-reduction circuitry found in most non-AC-3 laser-video players.

In addition to the usual noise-reduction compression that requires complementary expansion during playback, many audio engineers apply additional compression during the recording process to solve a variety of problems. Much of the pre-digital media—including LPs, motion-picture soundtracks, analog recording tape, and radio transmissions—featured (or still feature) background noise that could be reduced through some type of "single-ended" compression. Moreover, the environment in which the media were played (or are still played) offered sound problems—automotive road noise, theater noise, crowd noise at discos, or appliance noise in home environments—that also could be tamed through this technique. A dynamically compressed program source (be it CD, LP, audiocassette, videocassette, radio program, or TV program) will have its louder passages made softer and the softer passages made louder to help mask this background noise.

Unfortunately, certain kinds of music—classical and acoustic jazz, for example—suffer a loss in playback realism when compression is applied. What's more, films with a lot of rip-roaring action and explosive pyrotechnics lose substantial impact when the roars do not really rip and explosions do not seem to *explode*. That is where electronic dynamic-range expansion may come in handy.[2] A dynamic-range expander may solve program-compression problems by making the loud sounds louder and the quiet sounds quieter, although it usually requires intelligence to properly adjust even the best ones.

However, used to excess—particularly on programs that are severely compressed to begin with—an expander, just like a subharmonic synthesizer, can wreak havoc. Heavy-handed use will always result in audible "pumping-like" artifacts. Some action-adventure films that feature loud dialogue along with thunderous explosions will be almost unwatchable when sonically expanded, because the dialogue may become nearly as ear-shatter-

[2]More often than not, dynamic-range signal processors are stand-alone units that function strictly as range *expanders*. However, a few operate strictly as *compressors*, and a few more have been built to perform both functions.

ing as the effects. Also, most "single-band" expanders work mainly with the midrange frequencies; while explosions may gain extra punch, they may lose some of the deep-bass rumble that makes them sound like explosions. For this reason, expanders often are at their best when operated in tandem with a subharmonic synthesizer. If the latter is not available, a slight boost of the bass tone control will often restore body to the dynamically expanded pyrotechnics.

Because of their inherently good signal-to-noise ratios and wide dynamic range, the CD, hi-fi videotape, laser videodisc, and, of course, the Digital Video Disc have mostly eliminated the need for a dynamic-range expander. However, a lot of source material—broadcast audio or video especially—still contains a substantial amount of dynamic-range compression. Compression is found in many CD recordings (partially reissues of older, LP recordings). A lot of video material is also compressed during production, particularly film-to-video transfers of older motion pictures, because there would be substantial soundtrack noise if this were not done. Most satellite broadcasts are not compressed appreciably, but some are—mostly for aesthetic rather than technical reasons.

Some dynamic-range processors, in addition to being able to expand the range of an input signal, may also feature compression circuitry to *reduce* the dynamic range of a recording. Dynamic compression can be useful if you want to listen to (or in some cases, record) material for background-music or automotive playback, where an inherently wide dynamic range would be a problem. A few receivers and some CD players feature such circuitry.

Dynamic expanders have gone into and out of vogue, and sometimes new ones may be hard to locate—although they will certainly be available used from a variety of sources. Fortunately, used items should be safe buys, because they are mostly electrical and not inclined to deteriorate. Pioneer, Phase Liner, RG Systems, and dbx all have produced "single-ended" dynamic expanders, with the dbx 3BX model being particularly effective.

Equalizers

Equalizers are used to in some way modify the sound of an A/V system. They come in a variety of styles. The most common are multiband *graphic* models, with one or two sets of slider controls for shaping the frequency response of a sound system. They are called graphic because the position of the sliders simulates a graphic display of the response adjustments selected. Single-set types influence both channels identically and are often built into even inexpensive receivers. Double-set versions, which have greater flexibility because of their

graphic

ability to independently adjust each channel, are usually available only as stand-alone separates.

While some professional-caliber graphic equalizers have a slider centered at one-third-octave points for each channel, most consumer-grade models

Figure 8.1 RTA/Equalizer

(Photo courtesy of AudioControl)

A one-octave equalizer, like this AudioControl model (which also incorporates an RTA), can work near-wonders at solving placement problems if your speaker systems are decent to begin with. While this model is not something that an average consumer would buy to do a one-time measurement of their sound system, a lot of dealers will lend out similar gear to assist customers in positioning their speakers for best results.

This model measures and equalizes at discrete one-octave intervals centered on 32, 60, 120, 250, and 500 Hz, as well as 1, 2, 4, 8, and 16 kHz. An LED panel array presents the readout, with the source signal coming from an inboard "pink-noise" sound generator feeding the sound system's amplifier or receiver—usually via a tape loop or external processor loop (EPL). While the unit can read electrical inputs directly, its main use is as a speaker measurement and equalizing tool, with the speaker output measured by a microphone on a long cable. An analyzer of this calibre will give only a rough approximation of room response; although I would never recommend something this basic as a testing tool for actually rating or comparing speakers, it can be helpful when setting up systems of known high quality and in determining or fine-tuning a good listening position.

This tool will also be valuable if you're trying to accurately level-adjust the multiple channels of a surround-sound installation and equalize the main channels. AudioControl actually has a seven-channel equalizer for really serious surround-system adjustments, and for its proper implementation it is hard to beat their model R-130 one-third-octave real-time analyzer. While channel-level balancing can be done by ear, there is no substitute for a job done with the aid of a good metering system.

Note that you do not really have to use an elaborate RTA/equalizer to do basic surround-system channel-level adjustments. A cheap but decent sound-level meter (see "Sound-Level Meters," page 309) or a simple hand-held RTA (see Figure 2.16) will do a more than adequate job of measuring the built-in pink-noise channel-balance signals produced by any Pro Logic A/V receiver or surround processor (see Chapter 7), as generated by a good surround-sound test disc.

have 5, 7, 10, or 12 bands per side, with one-octave spacing being the most common. This allows them to apply a variety of "lumps" or "sags" to the generally flat frequency response of the electronic components. A few models come with a built-in real-time analyzer, pink-noise generator, and remote microphone to allow one to visually observe the results of the equalizing process.

parametric

Another type of equalizer, which ordinarily has only a couple of adjustment points, is called *parametric*. However, this can be maneuvered to different parts of the spectrum and can usually be adjusted for width (the "Q" of the curve) as well as amplitude. Because most frequency-response problems are complex in shape but few in number, a parametric unit may be able to correct a good sound system's performance better than a multiband, graphic type, although it may not be as effective at fixing the really ragged response of a pair of poor speaker systems.

EPL

Most equalizers are designed to be installed either in a system's tape-monitor loop or in an EPL (the *e*xternal-*p*rocessor *l*oop is basically a relabeled tape loop and will mainly be found on feature-laden preamplifiers). The signal is looped through the equalizer and is returned to the preamplifier tape or EPL inputs via the usual connections. Equalizers also can be installed between a preamplifier and a power amplifier, or between the "main-out" and "main-in" connections on an upscale receiver or integrated amplifier (see Chapter 3). However, the "to-main-amp" outputs of some preamps may occasionally overload the inputs of an equalizer, whereas the output of a tape loop or EPL rarely will.

If an equalizer is installed in a loop, it can be shut out completely from the circuit through a front-panel switch on the preamplifier. This is important, because there are times when an absolutely phase-shift-free output is required (such as when a Dolby-encoded signal is involved that is to be subsequently routed to a Dolby Surround decoder) or when playing pristine, ultra-hi-fi programs that might not require equalization. If the equalizer is connected between the preamp and power amp, its own "bypass" function will have to be used; however, with some models, that may not completely remove all the electrically active circuitry from the signal path.

Equalizers do not suffer from any inherent design problems (at least with the better-built models), but many listeners do not know how to use them properly. Equalizers serve three primary functions: to correct for *room*, *speaker*, and/or *recording* deficiencies.

An equalizer will be a far more flexible tool if it is intelligently used in conjunction with a decent real-time analyzer. Some equalizers come with a built-in RTA and even have a remote microphone to simplify speaker positioning and adjusting (see Figures 2.16 and 8.1 for an overview of RTA behavior).

Equalizer's Three Functions: Pluses, Minuses, and Alternative Solutions

Deficiency	Plus	Minus	Preferred Solution
Room	Correct for low-, mid-, or high-frequency standing-wave/boundary effects	(i) Sound may be improved for one listening position, but hurt for others; (ii) when working to flatten bass nulls caused by boundary-induced suckout or standing waves, immoderate use of an equalizer may overload a speaker driver or amplifier	Try to position speakers for the most-balanced sound and possibly make some subtle decor changes to the room itself; then, adjust an amplifier's ordinary bass and treble tone controls
Speaker	Correct frequency-response errors	(i) Correcting for an irregular, close-up response will almost always result in maladjusted off-axis radiation; (ii) correcting for an erratic reverberant-field response will be more effective, but may still involve enough room factors to make the sound acceptable at only one or two locations	Obtain better speakers; correcting a poor speaker's performance by means of an equalizer is the classic band-aid approach
Recording	Correct for poor recording quality	(i) It's difficult to sort out the source of a defect (Is it the recording that's at fault, or your room, or the speakers, or some other variable?); (ii) results in a tedious listening session, with the corrections often being completed just as the recording comes to an end	Make broad-band corrections in the bass, midrange, and/or treble by means of the basic tone controls that come as standard equipment on most preamps and receivers

CHEAP GOOD THINGS

Computer Programs

Roy Allison has produced a computer program, *Bestplace*, which makes it possible for anyone to intelligently place speaker systems in fairly normal rooms for minimal boundary interference in the bass range. The program,

Sound-Level Meters

A simple analog-dial sound-level meter (SLM), not much bigger than a telephone handset and available from places like Radio Shack for about $40 (with a digital-readout version costing somewhat more), can be an educational tool as well as a means to obtain better audio performance from your system. As a lark, you can take the device around and measure the loudness levels of a whole variety of items, from automobile and motorcycle engines and the drone of your computer at work to aircraft flying overhead and concert performances. However, for the more seriously minded, an SLM can greatly simplify the level balancing that is necessary to set up any surround-sound installation, particularly one employing Dolby Pro Logic or Dolby Digital AC-3. While all good surround processors have a switch-operated circuit to supply a "random noise" signal for the channel-balancing process, performing the procedure by ear—particularly if the speakers are not well-matched themselves—will probably compromise surround-sound performance. Employing a sound-level meter to monitor the balancing job will insure much better results. Believe me, this can mean a lot.

which is Windows or Macintosh compatible and is available from Roy Allison Labs free of charge from his bulletin board (see page 424). If your PC does not have Windows capability, the PC-disc version had an alternate file, called *Boundary*, which works with Lotus 1-2-3. Although it does not deal with standing waves, once you work with this program, you will never again have to guess if the speaker positions you have selected will produce uniform bass input to a listening room. Programs by a few other companies work similarly, although most are not all that useful, because they deal only with axial room modes and not the tangential and oblique series.

Some of the consumer-oriented programs that I have seen are either needlessly complex (because they try to solve mid- and high-frequency placement problems as well as those related to bass suckout) or inaccurate. There are also a few programs for speaker designers—such as LMS from Audio Teknology and Low-Frequency Designer from Speakeasy—that will prove to be educational even for those who do not plan to build their own speakers. These are sometimes fairly expensive, however.

Audio Imaging Enhancements

Two popular systems designed to improve audio imaging are Dolby Pro Logic, which involves front-channel steering as well as surround ambience, and Dolby Digital, which has three discrete channels up front, plus stereo surround channels (see Chapter 7, particularly "The Ambisonic System," page 278, "Two-Channel Surround," page 281, and "Other Discrete-Channel Theater and Surround-Sound Systems," page 292–93, for other designs that work to improve imaging). In addition, there are other, more economical ways to enhance audio-system imaging.

One cost-free and fast way to enhance sound-stage imaging may be to simply "toe in" the main speakers. Depending on the radiation patterns of the two speakers involved, you may find that this will solidify the frontal spread when you are listening from off the central-axis sweet spot. Speaker tester David Moran once said that sound-stage imaging and spectral balance might both be enhanced if the main speakers were each toed in a full 90 degrees—to actually be *facing* each other. While this would be visually awkward in some situations, it would certainly be aesthetically workable in a long, narrow room, if the listener sat facing one end and the speakers were pulled out from the front wall a few feet and backed up to the opposite side walls. Note that when listened to from up close, much of the important direct sound from speakers so positioned would be coming from substantially away from each system's central axis. Consequently, designs with erratic dispersion might not work to advantage when facing each other.

Floor Coverings

While it seems like anything but an audio accessory, the floor of your listening-viewing room may have as much to do with the way a system sounds as the type and quality of loudspeakers used. Indeed, the addition of good carpeting over a bare floor will have immeasurably more impact on sound quality than "upgrading" from a budget-grade CD player or receiver to much more expensive versions.

slap echo

Floor reflections and the back-and-forth slap between a hard floor and plaster ceiling or undraped walls facing each other can add an unpleasant, short-duration—but still somewhat echo-like—sound to any room. This "slap echo" will not only annihilate the positive subjective playback qualities of the audio system but may even go so far as to make it difficult to hold a meaningful conversation unless you're close enough to each other to be holding hands. Adding a bit of carpeting—particularly if it is a deep-pile wall-to-wall type with a thick foam pad underneath—will dampen reflections to a proper degree and improve both a system's imaging and its clarity. If thick expansive carpeting offends your back-to-country sense of fundamentalist-decor basics or Danish-modern sensibilities, at least try to dampen the more flagrant reflections by placing a large area rug, or rugs, in front of your speakers.

Exotic and Mundane Wall Coverings

If you've gone so far as to pad the floor, you may be ready to continue spreading surface-damping or sound-scattering materials right up the walls. Let's face the truth: an acoustically friendly wall treatment may look pretty repulsive—while also being outrageously expensive. However, it is possible to doctor a wall fairly cheaply with everyday objects that do not look so bad at all.

Sample Listening-Viewing Room

The most elaborate of my three listening-viewing rooms is 18.2 × 22.5 × 8.5 feet. It has a heavily carpeted concrete slab floor and very rigid walls (6-inch studs on 16-inch centers, covered with a laminated sandwich of thick paneling over drywall), three of which are lined with twelve 3-, 4-, and 5-foot-high book, compact-disc, laser-disc, and equipment cases along the sides and rear. These cases, which are solidly filled with irregular-sized books, CDs, videodiscs, and of course equipment, do much to break up both slap echo and standing waves and can be considered as an excellent wall treatment all by themselves.

When you face the main speakers, there is on the right wall, in addition to a bookcase, a French door that has a large and thick, but moveable, drape covering it. To balance this large-scale sound absorber, there is a 2 × 6 foot section of thick hemmed carpeting on the opposite wall above two 3-foot-high bookcases. The wall covering is positioned to absorb reflections from the same angle—but from the opposite side of the room, of course—as the drape. (Actually, the carpet section is not right up against the wall, but is held away from the wall about 2 inches by means of a molded frame across the top, further adding to its sound-trapping capabilities.)

This rectangular carpet section does not stand out from its surroundings (certainly not as much as commercial acoustic foam would do), because it is the same color as the floor carpeting and occupies an area near some pictures and also near the equipment cabinet and, of course, is right above the bookcases. It almost resembles art work. In addition, the exposed sides of the bookcases and equipment cabinet (there are about 6 inches of space between each) that face towards the front of the room are also covered by carpeting. This material is not easily noticed, because of its relation to the listening-viewing position. The idea here is to have a mixture of sound-scattering and sound-absorbing materials. You could possibly have just as easily installed a second, identically shaped set of drapes on the opposite, nonwindowed wall—provided your spouse and family can live with this decor.

There is also a window behind my front projector's pull-down screen at the front of the room, and it is covered by an equally heavy and moveable floor-to-ceiling drape. This helps to control front-to-back sound reflections. (When closed, the two heavy drape sets are also able to block out daylight to improve video picture contrast.) With many top-flight speaker systems—particularly those with extremely wide, uniform dispersion—this kind of decor-oriented reflection control will prove to be optimum.

Indeed, all the job may require is the repositioning of materials that are already in the room.

Heavy drapery (not lightweight curtains) and other damping and/or scattering materials always produce better sound—provided that they are located in the right places. While wrapping floor-to-ceiling drapes or specialized acoustic padding completely around the front and sides of the room may be visual overkill—as well as aural overkill, at least with certain kinds of speaker systems—fairly dense window coverings and/or wall-damping or sound-scattering materials, judiciously applied, will reduce the annoying slap of front-to-back and lateral reflections, particularly if large window areas or other large, hard, smooth surfaces are involved (review Chapter 2, "LEDE-Type Room," page 47, Figure 2.11, and "'Normal' Room and Speakers," page 50 to brush up on room-treatment principles).

A popular rule of thumb for controlling horizontal slap echo is to have drapery or padding covering a substantial part of one wall, with the opposite surface left reflective. This works best in dealing with front and rear walls (the padding should be up front) and is less effective with side walls. One example would be to cover part of the front wall with a reasonable amount of drapery while leaving the rear wall with mostly hard surfaces (as would be the case with cabinets full of irregularly placed books, scads of hard knick-knacks, and haphazardly aimed, different-sized picture frames, so that those hard surfaces scatter the rear reflections in a multitude of directions). The abbreviation of reflections from the front wall will almost always improve imaging, although the midrange and treble of speakers designed for very wide or omnidirectional dispersion may be both roughened up and attenuated too much if the front wall is heavily padded (see Chapter 2, "Anechoic vs. Reverberant," page 38, "How Off-Axis Behavior Affects Speaker Sound," pages 39–40, "Near/Far, Direct/Reverb Explanation," page 42, and "Hearing the Speakers," page 62).

Lateral reflections can be more difficult to work with. If you have a large window on one side of a room (particularly if it is located somewhat toward the front), covering it with floor-to-ceiling drapes while leaving the opposite wall untreated will mitigate slap echo to an extent but may also cause lateral imaging problems and a sound-stage shift toward the more reflective wall. My solution for nonsymmetrical-reflection situations like this is to provide some kind of additional absorbing and scattering material on the wall opposite the drapes to restore lateral-reflection uniformity. Those with money to burn may opt for some kind of professional-grade wall-treatment padding to do the job; more economically minded folks may opt to try something a bit cheaper.

sound-absorbing

Sonex is one of the more famous sound-*absorbing* materials. This foam-like substance is commonly dark gray, and can be either attached to the walls or mounted on free-standing frames adjacent to your speakers. With its padded large-egg-carton look, Sonex works effectively at absorbing high- and mid-frequency sounds. However, using anything more than a few small sections of it will make your hitherto stylish listening area resemble a studio control room or the inside of a beehive. Those planning extensive room treatments should also be aware that Sonex is quite expensive. While not quite so effective at absorbing reflections, drapes may actually do a better job of absorbing the most troubling sounds, and they will certainly look better.

sound reflecting

Two of the better-known sound-*reflecting* materials are made by Systems Development Group (SDG) and RPG Diffusor Systems (the "o" in "Diffusor" is not a typo). While these specially designed and sculptured sound-reflecting panels can do an outstanding job of breaking up reflective hot spots, they are also quite expensive and are not overwhelmingly more effective in home-listening situations than properly positioned book and knickknack cases. However, if your budget can stand it and your decor includes a lot of modern art, it is possible that the "collage" look of specially built sound diffusers will appeal to your sense of visual aesthetics. The same certainly cannot be said for the Sonex panels!

CD Storage Racks

One important way to insure that your CDs remain undamaged—as well as easy to find—is to install them in a good storage container when they are not being used. While a good bookcase with bookends works well enough, a specialized case of some kind may save space and will also look more "high tech." Assorted types of CD racks are available, with prices ranging from fifteen bucks for a good crate-like container that can hold about 90 discs up to expensive glass-doored polyurethane-finished storage cases that can hold over 300.

Vertical and horizontal storage racks with disc-holding grooves suffer from an inherent design problem. The many small slots necessary to stabilize the discs eat up space and are usually sized to hold a single disc, which makes it a problem to install double- or triple-disc jewel boxes. Some racks have oversized slots at intervals, but they may not dovetail well with the alphabetical or chronological system you have for arranging your recordings.

Programmable Remotes

I have used a programmable remote for years, and I readily admit that these devices can be a handy if you have several dissimilar remote-operated com-

Figure 8.2 CD Storage Rack

There is no need to purchase expensive racks to hold your disc collection. I hammered together this semi-portable storage case, as well as two others just like it, from some old 1-by-6-inch pine boards (left over from another construction job) and wall-paneling remnants. I did have to purchase the finishing nails and the stain. A well-equipped do-it-yourselfer could do a tongue-and-groove job that looked a bit better; if you had to buy the material from scratch, the result might cost $20.

This particular bin holds about 130 disc singles and uses good old-fashioned bookends to keep them from tipping over. Actually, if you do not live in an earthquake zone, you may not need the bookends at all, since CDs stand up very well all by themselves. Each of the three cases fits comfortably into a large étagère at the back of my main listening room. Containers similar to this one are often available ready made—and fairly inexpensively—from record stores. When my homemade versions filled up several years back, I bought several to hold the overflow, stacking them on the bookcases lining my main A/V room's walls.

If you watch TV in a darkened room, get an illuminated remote.

ponents within your A/V system. Some manufacturers offer remotes that can operate all of their equipment, so if you buy different components from them you can control them all using just one remote. Other remotes are designed to be 100-percent programmable in order to run hardware from any number of manufacturers; these are usually quite economically priced and are available from many small outfits that specialize in electronic accessories. Perhaps the best designs split the difference, being capable of operating a certain amount of dedicated gear, with the option of being partially programmable for hardware made by other companies.

Most programmable remotes are designed to "learn" their commands from the dedicated remotes that come with CD players, VCRs, and TV sets.

To "teach" a learning remote the required commands, face the two remotes toward each other. Then, while pressing the appropriate button on the dedicated unit, you press the corresponding button of the programmable unit while it is in the "learning" mode. It will usually signal when it has "learned" the new command. Each design will have its own procedure, so review the instruction manual prior to making a purchase in order to see if it is easy to set up and operate. You should also study the keyboard carefully to see if the positioning, size, shape, and quantity of buttons make sense. The only problems I have had with some models is that their often-bountiful network of small, same-shaped, same-color buttons are confusing to operate in the dark; some are also hard to fathom in the light (although some dedicated remotes are not much better).

Test Discs

I have always loved test discs, and I believe that anybody who wants to know what one's system is capable of doing should obtain a few. While you can get a decent idea of performance by listening to or viewing standard entertainment software, a well-designed test disc will sometimes help to pinpoint the often-maddening little problems that would ordinarily have to be referred to a repair facility. Some discs are specifically designed for optimizing standard stereophonic systems, while many also have tests for setting up and calibrating surround-sound hardware. There is a good list of these discs, as well as a list of "dynamite" discs to show off your system, in Appendix A, along with a brief description of their strong and weak points.

Ear Care

There is a lot to say in favor of the lowly Q-Tip as a major audio accessory—beyond what it can do as a cleaner of local-area "grit" on CDs. Personalized ear care may be the least expensive way to improve home audio!

While the human ear can normally do a decent job of keeping itself clean, there are times when a minor amount of wax buildup may hinder your ability to hear well. Handle a Q-tip carefully for wax removal, because rough or indiscriminate use may result in packing excess wax against the eardrum instead of removing it, defeating the whole purpose. Also, overly vigorous use may injure the drum itself.

Water buildup in the ear (after a swim or even a shower) can be a problem for some people. One trick to prevent moisture-induced ear problems is to twist a tissue (toilet paper or Kleenex) into a long skinny wick and use it as a small narrow sponge to suck up any water from the inside of the ear canal. Insert and twist the tissue carefully while just touching the drum area lightly. This will slurp up excessive moisture but won't pack the wax the way a Q-Tip could. It should also not irritate the drum. Tissue "wicking" will efficiently help to dry the ear before moisture-induced problems can get started.

Another related accessory is an ear-cleaning kit. Produced by companies like Murine, these usually include a glycerine-peroxide solution and a syringe. While these kits can work wonders if your ear(s) has a heavy wax buildup, they must be used with care and should not need to be used often. If you are continually plagued by an ear-wax buildup or hearing problems, see a doctor.

SILLY THINGS

There are scads of subjectively benign, money- and/or time-wasting accessories, tweaking procedures, and equipment-oriented hysterias that have captured the imagination of A/V enthusiasts. Most of these are harmless, but none are worth wasting your money on, unless you have a need to engage in voodoo electronics . . . in which case you probably wouldn't have read so far in this book.

Compact-Disc Placebos

When the CD first appeared, the days of the audio "diddler" seemed to be numbered, and the era of the "happy music lover on a shoestring budget" appeared to be dawning. During audio's "bronze age" (the LP-record era), serious audio enthusiasts might spend hours fiddling with their *sound systems*, rather than employing the hardware as a means to an end: as a way to enjoy *music*. As a part of this activity, many would add aftermarket upgrades, particularly to enhance the performance of a turntable—a device that readily lends itself to fine calibrations, peculiar behavior, and arcane accessories.

The CD, because it was so straightforward, seemed immune to such manipulations. Unfortunately, aftermarket tweaking products and procedures for the so-called improvement of CD performance were quickly developed to fill the need for audiophiles to try to wring every last decibel out of their systems. Oddly enough, there seem to be more snake-oil products for CD and CD-player enhancement than there ever were for the LP record.

Here are a few of the harmless but unnecessary products that have been developed.

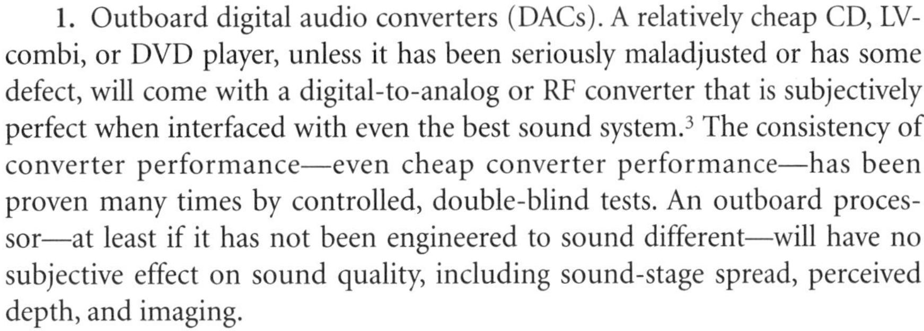

1. Outboard digital audio converters (DACs). A relatively cheap CD, LV-combi, or DVD player, unless it has been seriously maladjusted or has some defect, will come with a digital-to-analog or RF converter that is subjectively perfect when interfaced with even the best sound system.[3] The consistency of converter performance—even cheap converter performance—has been proven many times by controlled, double-blind tests. An outboard processor—at least if it has not been engineered to sound different—will have no subjective effect on sound quality, including sound-stage spread, perceived depth, and imaging.

A proper double-blind test requires very close level matchings.

2. Magic-Marker edge treatment. Compact-disc edge coating is said to reduce internal reflections within the plastic and aluminum substrata of the disc—thereby "cleaning up" the sound. While permanent-ink felt-tipped markers normally cost less than two bucks, at least one scam operation has sold a relabeled "audiophile grade" maker for over $15 a pop. Edge marking has no audible effect. Double-blind ABX tests have conclusively proven this, and engineer Fred Davis also has done extensive electrical testing of assorted disc and disc-compartment coating techniques without discovering any measurable benefits. Fortunately, edge coating should at least cause no problems unless excessive amounts are used. If you have a pathological urge to be artistic with a magic marker, try it out on a restroom wall rather than your CDs.

3. Disc magnetizing. There is a fixation among some audio gurus who believe CDs should be "demagnetized" to improve their sound. Most CDs are made of aluminum, plastic, and paint, and there is no way that magnetism will have any effect on these substances.[4] Paying somebody to demagnetize your disc or buying a "disc demagnetizer" will not improve its sound at all.

4. Disc spinners. At least one company makes a device that allows you to "prespin" a CD before putting it into your player. Apparently, the player itself does not spin the disc in the right direction or at the right speed or something.

[3]Some test discs will emphasize nonlinearities at very low playback-output levels; some nontest recordings, including those with musical fades to silence or near silence, may also show up certain low-level nonlinearities. However, these recordings will readily display these artifacts only if the amplifier gain is advanced to an abnormally high level during the low-level passages or musical fades. Obviously, this would result in material recorded at normal levels being much too loud for ordinary listening. Also note that headphones, as opposed to loudspeakers, are very revealing transducers and may show up certain kinds of low-level anomalies even when the amplifier gain is not advanced appreciably. However, most music will not show these artifacts even with headphone listening.

[4]Some disc manufacturers substitute gold—and at one time silver, which as we pointed out in Chapter 4 sometimes led to catastrophic cases of CD rot—for the aluminum, but those substances are just as immune to magnetism as the cheaper metal.

One alternative journal ascribed a prespinner's supposed ability to reduce "digital glare and harshness" to the removal of electrostatic charges in the disc that make it hard for the laser to see the pits—although the reviewer admitted that this was strictly his own speculation, because the manufacturer would not describe exactly how the spinner worked its magic. According to the reviewer, the spin effect was not permanent, and a prespin was necessary every time a disc was played. After reading his review, I did not know whether to call the Better Business Bureau, the FBI, or the American Psychiatric Association.

5. Gold CD recordings. Some CDs use a gold substratum instead of aluminum. While this expensive production technique can certainly improve the long-term durability of a recording (we're talking about hundreds of years here), it will have no effect on short-term durability (several dozen years, by which time the disc will be technologically obsolete anyway) in any normal, clean, reasonably dry home environment. Gold discs are also said by some to sound better than aluminum discs of any age, but this has not been proven to be true. Indeed, tests by Fred Davis have shown that some gold discs have substantially greater error rates than aluminum discs. While most players can correct for these defects, some may actually have problems cueing up these super-grade discs. These anomalies are the result of manufacturing difficulties and not defects inherent in gold, please note. In any case, there is little doubt that gold discs are not intrinsically better-sounding than aluminum ones. They sure look pretty, though.

Enhanced Power Cords, RCA Lines, and Speaker Wire

POWER CORDS

The electricity that powers your system moves though miles of heavy wire and transformer voltage boosters on the way to your house, then travels through assorted circuit breakers and dozens of feet of 12- or 14-gauge solid-core wire to reach your electrical outlets, and finally proceeds though a wall plug and three or four feet of usually 16-gauge lamp-cord wire to your equipment. For a few aftermarket manufacturers, this is not good enough. They offer substitute power cords and plugs—often costing hundreds of dollars—for the outlet-to-hardware connection—which supposedly improve the sound in many ways, including better sound-stage depth, better articulation, etc., etc. Substituting an expensive and exotic replacement for a component's existing power cord makes about as much sense as putting a rope-sized power cord on a table lamp, iron, or vacuum cleaner and expecting it to work better. It will not.

ELECTRICAL CONTACT ENHANCEMENTS

Advertising hype notwithstanding, liquefied "contact enhancement" applications do nothing worthwhile at all. The small degree of resistance at a proper electrical connection is inconsequential, and contact enhancers have no effect on either audio or video performance. Some are said to reduce contact oxidation, but in all the years I have fooled with audio and video, I have never had an oxidation problem with electrical connections; I doubt you will, unless your A/V room has high levels of humidity or you use your equipment outdoors. If you plug and unplug your connectors once in a while (like once a year), oxidation problems won't occur in any case. One well-known brand of contact enhancer, "Tweek,"[5] costs about the same per container as a good compact disc. Get a disc instead; audio is about music, not gimmicks.

RCA LINES

Some lunatic-fringe audio journalists and their followers feel that the RCA-plug-fitted cables that come with CD players to transmit the signal from them to an amplifier detract, not just slightly, but markedly from the sound they produce. Consequently, a number of profit-oriented accessory companies now turn out military-grade-looking interconnects for CD players (as well as LV combi players, DVD players, VCRs, preamps, and amplifiers) that cost tens (and sometimes hundreds) of dollars per foot. These cables are often built very well, and some are monumentally impressive looking. A few even come marked with arrows for directional orientation, as if alternating current somehow needed to know which way to flow to work at its best.

However, there are *no* scientific data whatsoever to support tweak audio fanatics' claims concerning interconnect cable performance. In spite of the anecdotal rave reviews of assorted alternative-press equipment testers, no controlled tests of any kind have ever proven that expensive cables are audibly better than standard, reasonably well-built interconnects. If you must have impressive-*looking* cables to excite your friends, go down to Radio Shack and buy their one-meter "heavy-duty" gold-tipped interconnect sets. They sell for about $10 to $15 per one-meter set and look sturdy enough to use as lynching rope.

Some manufacturers supply their CD players with fiber-optic connections in addition to the standard ones. They should work audibly as well as—but no better than—the RCA-plug-equipped shielded-copper cables that attach to the standard electrical fittings. Apparently, some aftermarket outfits

[5]A diluted form of a product called Stabilant 22, produced by D.W. Electrochemicals, Canada.

are also producing deluxe fiber-optic cables and making the usual wildly extravagant claims for them. Again, until somebody effectively proves that fiber cables—cheap or expensive—work better than cheap copper ones, you will do well to save your surplus cash for purchasing additional CDs, videotape, or videodiscs.

SUPER-DUPER SPEAKER WIRE

Most of the scientific studies concerning speaker-cable performance have come to the same conclusion: on runs of up to about 20 feet, the differences between even the "best" speaker wire and good 18- to 16-gauge stranded wire (lamp cord) is probably not audible—at least with musical program material. Even when audible, the differences will only manifest themselves as a minuscule loss in overall gain with the ordinary wire and possibly a slight softening of the highs. (The latter will depend on the acuity of your hearing.) If you must be absolutely sure about your system's performance, the experts would probably suggest going with 14- to 12-gauge stranded wire, particularly if the runs were going to exceed 20 feet.

Low-voltage outdoor lamp wire is also a good buy.

So, how much to spend? Well, good 16-gauge lamp cord (the smallest size I personally feel comfortable with) has been available for years from your local hardware store for under a quarter a foot. Those with a need for heavier wire might try the mail-order outfits. I have seen 14-gauge automotive audio speaker wire (fine for home use, too) in some catalogs for about 40 cents a foot, and 12 gauge going for about 60 cents. Radio Shack has for some time had a good "Monster Cable" look-alike wire for about a buck a foot. (If appearances are important to you, one mail-order catalog offers wire that glows in the dark; no claims have been made—yet—for improved audio performance!)

Of course, it certainly pays to shop around. Once, when strolling through one of those immense hardware-store emporiums that dot suburbia, I happened upon some 12-gauge stranded "speaker" wire selling for 18 cents a foot! Being in the process of rearranging (and by necessity rewiring) my two larger A/V systems and also installing some fairly distant remote speakers, I purchased an entire 250-foot roll for $45. If I had gone the usual premium-wire route, it would have cost me at least $250, while some of the more upscale cable available in high-end shops would have set me back over $1,000. At least one esoteric brand would have topped out at over $30,000! Once I got the wire in place, I felt pretty shrewd.

Of course, speaker wiring all by itself is not the only thing that attracts the attention of high-end audio loonies. The list on pages 322–23 details five more genuinely oddball speaker-wire subcategories that have caught the attention of wayward consumers, sales personnel, and audio journalists.

Speaker Cable Impedance

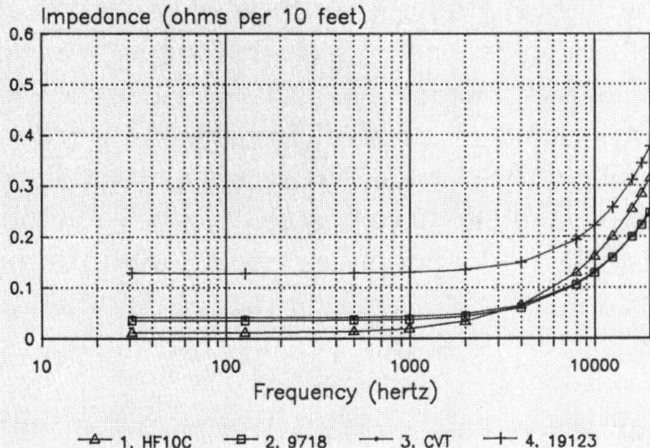

Here are the results of research by engineer Fred Davis that pretty much substantiates that 16-gauge lamp cord is perfectly acceptable for use as speaker wire for runs up to 20 feet. Four wires are being compared: *HF10C* (an expensive Levinson "audiophile" cable of 3-gauge size); *9718* (Belden 12-gauge standard, heavy wire); *CVT* (Music Interface Technologies' three-quarter-inch-diameter super wire, costing about $130 per foot); and *19123* (Belden 18-gauge "lamp cord," which is two-thirds the size of 16-gauge wire).

The three heavier wires are more or less equal in performance over the audio bandwidth. However, note that even though the small 18-gauge wire does cause signal losses, those losses pretty much *parallel* those of the larger-size wire. The result would be that the 18-gauge would cut the power delivered to the speakers somewhat but would still sound similar to the other wire, provided (1) that the gain of the amplifier was cranked up a bit to compensate for the resistance losses, (2) that the amplifier itself had an adequately low output impedance (common behavior with most transistor amps, as I have noted previously), and (3) that the amplifier was not being driven into overload while attempting to play loud enough. Note that the total mid- to high-frequency losses for all the wires amounts to less than three-tenths of a dB. This would be difficult for anyone to hear unless special test tones were employed and the listener had extra-sensitive ears.

Davis, who has done extensive research on speaker cables and printed his findings in both popular and technical journals, has indicated to me that most of the "features" claimed for exotic cables will not significantly affect the sound—in spite of advertising claims. Examples of cable design overkill include exotic materials (such as "oxygen-free copper," "linear-crystal-oxygen-free copper," "silver-plated OFC," and "platinum"), as well as special connectors and wire-stranding techniques to create separate sound paths, reduce spurious signals and "dispersion," and eliminate vibrations. These features have no effect on signals within the audio bandwidth.

Certain factors *are* important to more astute listeners. These include wire length (note that the graph is for cables of 10 feet) and thickness. The latter should be kept between 18 and 10 gauge. Cables of very large size may do more harm than good (note the mid- to high-frequency relationship between the 3-gauge HF10C and the 12-gauge Belden). For runs of extreme length (beyond 50 feet), the concerned buff will opt for fairly heavy cables. Anything from 14 to 10 gauge should be fine.[1]

[1] A note for those of you who are not familiar with wire sizes. The smaller the number, the larger the wire. Ten gauge is twice the size of 13 gauge (an odd size), and the latter is twice as large as 16 gauge. Seven gauge is about the size of auto battery cable. Common lamp cord is 18 gauge, although some is 16 gauge. Most wire will have the size written on it (such as "16 AWG," which is 16 gauge), although the printing may sometimes be hard to read. Always use a multistrand wire, not the solid-core stuff that is used for house wiring.

1. Cable support stands. Enter any number of high-end audio-video shops in America, and you will not only see hose-sized wire running from the big amplifiers to the speakers but may also occasionally observe that the wire is held off the floor by a row of miniature stands. These are said to reduce interactions between the wire and the ground—or even a wooden floor. These devices, their use based strictly on the anecdotal observations of recondite audio writers and profit-hungry sales personnel, are so absurd that no self-respecting researcher has bothered to do hard-and-fast tests. It would be akin to testing for the effects of motor-vehicle traffic on the earth's rotation.

2. Cable wrapping. This inane placebo involves wrapping speaker wire (and shielded interconnect cables, too, I assume) in *leather* soaked in some kind of damping fluid to reduce speaker-induced vibration problems. I have also heard of cables that were wrapped with pure silk cloth, run through exotic tube-like conduits, or enclosed with a water jacket! I am not too sure just how these fixups are supposed to work, but it's certain that some kind of rationalization for it has been printed out in detail in one of the alternative-press publications, somewhere.

3. The skin effect. This relates to the tendency of higher-frequency signals to travel down the outside surface of a conductor, rather than through the interior. When it manifests itself with conventional speaker wire, this phenomenon (which deals with radio frequencies rather than audio) supposedly colors the treble response of a system and is the reason that some high-end accessory companies use multi-sized strands in their wire. Research by Fred Davis has shown that the change in wire impedance with a 12-gauge stranded cable 100 feet long, working with an 8-ohm speaker load, would result in a modest (I mean really modest) treble rolloff beginning at about 4 kHz. This attenuation would reach minus 0.05 dB at about 8 kHz, minus 0.1 (that's point one) dB at about 12 kHz, and a maximum of minus 0.18 dB at 20 kHz. This is almost certainly inaudible, and at normal wire runs in the 10- to 30-foot range, the skin effect will be functionally nonexistent.

4. Breaking wire in. Wire is wire, unless something comes along to melt, shred, or break it. However, some buffs think that speaker wire (and I guess shielded interconnects, too) works better if it is "broken in" for some time. When this audio pastime first appeared, it was supposedly a pleasurable thing. One would listen to a variety of musical programs, and over a period of weeks the wire would get better, and the sound would get better, and life would get better. Now, the aftermarket people are in on the game, and at least one company has a device that generates a broad-band signal that is designed to break in wire "correctly." In addition, at least one test disc supposedly has special signals on it to break in wire—as well as speakers. Once wire is broken in, you're urged to make sure that it is always hooked up the same way to the amplifier and speakers. If it is removed and reinstalled backwards, the new orientation will screw up the sound. These people treat speaker wire the same way smart automotive buffs treat radial tires: once the latter are broken in, they must always rotate in the same direction. However, while many mechanical devices (tires, for instance, or gasoline engines) really do "break in," electrical conductors do not. Wire is wire.

5. Bi-wiring. Some enthusiasts, even a few outwardly level-headed ones, feel that there are serious problems with high- and low-frequency electron interference when a speaker is connected to an amplifier with only one pair of wires. Bi-wiring supposedly corrects this by running two wire pairs from each amplifier channel to each speaker instead of the usual single pair. If a system has separate inputs for the bass- and higher-range drivers, one of each pair is hooked to each. The divided hookup is said to reduce electron interference. Fortunately, electron flow does not work this way, and if bi-wiring does improve things, it does so by affecting total series resistance. A bi-wire hookup will not reduce the series resistance for single frequencies except at the crossover point (where both pairs are reproducing identical signals), but it can reduce the overall resistance when reproducing complex signals. Obviously, then, the basic advantage of bi-wiring would be merely one of reducing overall resistance with long speaker-wire runs. Consequently, less complex, less unsightly, and equally effective results will be obtained by opting for a single larger-sized wire.

Spikes, Stands, and Springs: Vibration Control

For years, sundry audio purists have promoted the idea of solidly coupling speaker systems to the floor or front wall for the sake of better sound. Some speaker companies include sharp-pointed "spikes" or spike-attachment fittings on the bottoms of their cabinets as a matter of course to insure a solid contact with the floor (the spikes punch right through typical carpet material). One reputable manufacturer told me that he had to install spike-attachment fittings to his English-export models to humor assorted audio-mystic magazine reviewers across the ocean, even though the spikes would have no effect on the sound. A number of accessory companies also market metal cone-shaped spikes for speaker systems not factory equipped. At least one big-time speaker company has on occasion offered special braces to affix its floor-standing speaker cabinets solidly to the wall behind them. However, research by techies like Tom Nousaine (see Chapter 2, "Speakers and Stands," page 58) has proven that speakers will not have their sound improved by coupling them to the floor via special feet or to a wall with a special brace.

This idea also has been applied to electronic components, such as amplifiers and CD players. Two categories of under-chassis supports have been developed: soft-and-spongy and hard-and-pointed. Soft supports are usually special puck-like feet that absorb feedback vibrations from speakers or more straightforward oscillations from footsteps. Often the spongy feet are included as part of massive sub-bases upon which you are supposed to place components needing protection. Rather than isolate components from outside vibrations, hard supports are said to reduce self-induced "electronic" vibrations that muddy up the sound. Some otherwise reputable companies (both American and Japanese) have joined in this "battle-against-vibration." We have seen electronic components with designed-in ballast, massive sound-deadening bottom plates and covers, and special platters for CD players to solidly clamp a disc and keep it from oscillating like a tuning fork during playback.

Now, unless you place a CD player on top of a speaker system and crank the volume up to ten, there is little chance that external, vibration-induced feedback will be a problem. Error-correction circuitry is designed to handle this problem. If a vibration problem *does* rear its ugly head, it will not do so in a subtle manner but will trigger a rather catastrophic response such as skipping. When placed at any reasonable distance from the speakers (say beyond 4 or 5 feet), CD-player feedback should never be a problem at all.

As for feedback with amplifiers and other solid-state components (externally generated or self-induced), unless there is a wire loose somewhere or some kind of truly bizarre malfunction, vibration simply cannot produce audible effects. No serious research has shown that vibration-isolation devices help solid-state components to perform better. The feet that are permanently attached to your amps, preamps, surround-processors, or CD players are more than adequate to keep those items out of vibration trouble—and are also usually essential to keep cooling air flowing in from underneath.[6]

A few years back, a few self-deluded tweakers (abetted by assorted alternative-press journalists and profit-motivated salespeople) discovered that putting heavy weights on amplifiers appeared to make them sound better. (All comparison "tests" were anecdotal, needless to say.) Initially, books were employed as ballast, but a few daring enthusiasts preferred bricks. In time a few accessory companies produced special "amplifier weights" to do the job. One outfit has even designed a shelf rack with special clamps to lock assorted

[6]Vibration *can* produce microphonic feedback problems with tube equipment, so if you own that kind of esoteric gear you may need vibration isolators; see Chapter 3, "Tube Amplifiers," page 111. Needless to say, you may also need them to tame LP-turntable feedback.

electronic components into position to supposedly channel both self-contained and room-induced resonances into the shelf structure so that the sound is actually enhanced beyond what it would be if no vibration existed at all! People who believe stuff like this probably still set traps for the tooth fairy.

The Preposterous

And now for a night gallery of electronic gizmos that have appeared supposedly to improve audio-video performance.

1. Audio system demagnetization. Another absolutely outrageous scam involves using a specially treated compact disc to "demagnetize" your entire audio system. The disc is supposedly encoded with a "special" signal to remove harmful magnetism from all the audio hardware you own—including, possibly, the speakers. From what I can gather, the company that hawks this device claims that, like a tape recorder, an entire audio system can pick up magnetism that will color the sound. While I hate to burst a marketing balloon, simple science reveals that demagnetizing CDs couldn't possibly work. Tape recorders pick up magnetism because the tape passing through them is magnetized. Because the magnetism can affect the magnetic properties of the tape itself, it is a good idea to demagnetize a deck once in a while (see Chapter 4). However, the rest of an audio system is unaffected by such small amounts of flux and does not need be to demagnetized. Even if it did, a signal from a CD would never do the job. I have also heard about a CD that produces large amounts of offensive noise on purpose, supposedly to "exercise" (or should we say exorcise!) a sound system and make it sound better. The noise is so unpleasant and is supposed to be played so loudly that users are advised to leave the room during the conditioning process. However, solid-state electronics do not need to be preconditioned to sound better. And playing a CD at super-high volumes through your system could damage it, particularly the speakers.

2. Digital clocks. Perhaps the most outrageous product to hit the audio scene in years is the "Tice Clock," which cost $350 the last time I checked. This is a somewhat cosmetically modified version of the kind of common cheap digital clock (with an alarm, yet) that is widely available in drug, discount, and Radio Shack stores for under $30. However, in its "audio" version, its supplier claims it improves the sound of digital components when it is plugged into the same power line as the sound system.

3. Powerline conditioners. Mr. Tice has also produced a powerline "conditioner" that costs a small fortune and, from what I gather, is no better at protecting components from power surges than run-of-the-mill surge pro-

tectors and no better at providing "conditioned" electricity than what a good power company already does as part of its regular service. There are a number of other powerline filters now available, and, while line filtering is not necessary a bad thing, most power companies do a good job of filtering their power already. You can usually get adequate protection against line surges by merely obtaining a decent surge protector for a few bucks at a hardware store.

4. Harmonic tuning dots. These are small, roughly one-inch-diameter discs that you attach to your components (speakers, electronics, or even the room walls, I suppose) that some audiophiles feel produce a remarkable improvement in sound. However, the only way these dots will influence the sound is if you apply a substantial quantity of them to the moving surfaces of your speaker drivers (especially tweeters), laminate the room walls with thousands of them, or stuff them into your ears. In any case, the change will not be for the better.

BAD THINGS

While the silly things that we discussed would have little effect on anything other than the condition of your wallet, the following items can be counted on to do some serious damage on occasion. All of these apply to CD and digital video players, the latest electronic components to attract the attention of after-market flim-flam artists.

1. Magic-marker treatment of an *entire* disc surface. A number of enthusiasts, pushed further toward insanity by a few dealers and lunatic-fringe magazine writers, have turned their backs on the simple disc-edge treatments discussed previously and have begun coating the entire label surface of their discs with magic marker-type dye to further "clean up" the digital signals. While edge treatment is probably benign (and, of course, ineffectual), this kind of laminating job could damage the lacquer finish and eventually ruin the aluminum layer.

2. Spray-on coatings. While a number of companies produce spray-on or wipe-on coatings that have no effect on CD performance (manufacturer claims notwithstanding), as far as I know these products have no detrimental effects and only bilk you for the cost of the chemicals. However, a number of other substances have become popular as sound improvers, and a few of these will almost certainly cause problems. For instance, coating a disc surface with Armor All (a useful rubber and plastic "protectant" mainly designed for automotive use) was once heavily promoted by an alternative-press writer who was obviously ignorant of chemistry—and no doubt many other subjects. Because of its chemical makeup, Armor All may gradually ruin your discs, and could gradually fog up the laser in the player as well, because of chemical evaporation. In a subsequent issue, the magazine printed instructions for removing the coating (without referring to their previous instructions to put it on!). To its credit, the Armor All company does *not* recommend its product for CD use. Some other folks have recommended using Liquid Wrench to protect your CDs. However, anybody who coats discs with this stuff is going to end up with a mess, because the substance eventually reacts with lacquer. What's more, even if you only apply it to the plastic playing surface, it may be nearly impossible to keep it from spreading to the edges of the label side because of the effects of rotation. Other fad treatments include Rain-X automotive windshield coating, guitar polish, suntan lotion, and even toothpaste—rubbed on and then at least wiped off, I hope. I would not be surprised if some fanatics are using beer, milk, deodorant spray, Sterno—or even RAID. Coating a CD with any chemical is asking for trouble.

3. Freezing. This involves cryogenic temperature reduction of a disc to *very* low levels, combined with gradual thawing. It is said to properly align the CD's molecular structure, thereby improving the sound. Proponents claim to not know *how* freezing helps things but only state that they can hear the difference. While gradual freezing-thawing may not lead to serious degradation problems, I would not be surprised if the molecular condition of the lacquer coating of the nonplaying side of a disc could be harmed by such temperature extremes.

4. Edge stabilizers. These are rubber or plastic rings that fit over the outer circumference of a disc and gyroscopically "stabilize" it as it plays. Often, they are green in color and, consequently, are said to kill two birds with one stone by having an effect on the sound similar to green-ink edge marking! However, tests have shown that they have no audible effect and if anything actually increase the error-correcting load of the player. While the biggest impact of this scam is that it helps to further flatten your wallet, a possible catastrophic effect would be the monumental jam that would occur if one slipped off during play.

5. Weighted stabilizers. CD weights are supposed to be installed on top of a disc just prior to playing it. They are said to reduce digital "jitter." However, a subjective nonproblem like digital jitter can in no way be affected one way or the other by any kind of CD platter weight. What such weights will do for certain is put an extra strain on the mechanical parts of a player and shorten its operating life.

6. Circuit modifications (including nonstandard DACs). These involve both design adjustments by a handful of mainstream manufacturers as well as assorted modifications offered by a number of aftermarket "tweaking"

companies. In some cases nonstandard circuit configurations will be found on very expensive dual-chassis designs from audio cult-oriented high-end manufacturers. In spite of the hype, these mods often consist of nothing more than changes to the analog-output circuitry, involving "recontouring" the frequency response to make the player actually less accurate than it was before. In many cases, the high frequencies are attenuated slightly—making the player a bit more mellow or "tube-like" sounding. In a similar vein, a number of high-end players may feature circuits that supposedly reduce digital "jitter" at the expense of high-frequency flatness.[7] The net result is the same kind of mellow sound that the analog frequency-response modifications produce. While this kind of effect may be pleasant with improperly engineered recordings that display a hot or distorted high end (usually the result of using less-than-ideal microphones), the players themselves are not enhancing playback accuracy. These days we even see expensive CD players that employ tubes instead of transistors in some circuits to supposedly overcome "digital harshness." However, the only things tubes will do for CD player performance is increase distortion, make the player run hotter, and make it more prone to malfunctioning. Some tube-equipped players also roll off the high end a bit, further enhancing their reputations as mellow sounding. For a while, one otherwise straightforward Japanese company installed a non-defeatable circuit in its high-end CD players (and top-line combi players) that produced additional harmonic artifacts above the usual 20-kHz cutoff point. These harmonics, unlike the ultrasonic frequencies that are produced when real music is played, are actually a form of distortion and in no way resemble real-world ultrasonic frequencies. Nevertheless, the company asserted that the circuit allowed its players to better simulate live sound, making them more pleasant to listen to. However, it is unlikely that anybody with normal hearing will notice any difference at all. The only possible artifact that resulted from the circuit was a slight attenuation of the high frequencies in the octave between 8 and 16 kHz. Some listeners might have considered the very slight mellowness to be the result of the additional ultrasonic artifacts, but that was not the case. One American company installed a rather elaborate circuit in some of its CD players that, by means of frequency contouring and phase interactions, supposedly made them sound similar to what is produced by good LP recordings. (Stay calm, it does not include a scratch, rumble, or surface-noise generating circuit!) An LP-simulation circuit may be fine for improving the sound of CD reissues that were mastered from older cutting tapes originally equalized for LP production. However, using it to improve the sound of well-engineered contemporary material would be a mistake and

[7]Jitter is a measure of the instability of the high-frequency clock signals used to drive and synchronize digital circuits. It has no serious audible effects in any decent player, including cheap ones.

contrary to the principles of high fidelity. At least this particular company's players included an on/off switch to bypass the simulation feature.

7. Disc-cleaning machines. As mentioned in Chapter 4, CDs should not require cleaning with normal use; further, to avoid damaging a disc, it is better to spot clean an offending smudged area rather than risking damage by cleaning the entire surface. A compact disc or digital-videodisc cleaning machine cleans the entire disc indiscriminately and therefore may do more harm than good. Keep your discs fingerprint-free, and you should never need to clean them at all.

8. Lens cleaners. Some enthusiasts and reviewers have promoted "lens cleaning" discs to remove surface grime from the plastic lens that directs the laser beam in a CD or LV player. These cleaning discs look like CDs but have a multitude of small brushes covering what would ordinarily be the playing surface. To clean a lens, simply insert the disc and press "play," and built-up grime is quickly scrubbed off. I have two problems with this. First, unless you smoke like the proverbial chimney, indulge in the production of assorted noxious and room-fogging chemicals in your house, or cook with a lot of grease (or gild your discs with Armor All), it is unlikely that a laser lens would ever be layered with enough glop to not function properly. Second, if the little brushes on the disc surface snag something or come loose, they could wreak havoc with the player mechanism. It should be pointed out that virtually every CD-player manufacturer makes it a point to warn people to *never, ever touch the lens with anything.* If you damage your player with a lens cleaner, you will probably void the warranty.

APPENDIX A

Some Dynamite Software

VIDEO AND AUDIO REVIEWS are usually based on a subjective reaction to the content of a recording: Is the film a landmark work or a Z-grade loser? Are the Beatles really the greatest of all rock groups—and is a particular recording their best work? Working from a different assumption—that you already know what you like—I felt that a list dealing only with the *technical quality* of a number of recordings would be useful. There's nothing so disappointing as bringing home a newly remastered CD of a classic album only to discover that it was made from substandard third-generation master tapes, or to sit back to watch a classic film and discover that the transfer was made from a dirty print.

The following lists of both audio and video items should give you a good cross-section of some of the better releases available for playing on your home-theater system. I then list some audio and video test discs that can be helpful in evaluating the pluses and minuses of any home system.

SOME EXCELLENT AUDIO RECORDINGS

In compiling a list of high-quality audio recordings, the following aspects of recorded sound in general were given the greatest weight: articulation and clarity, dynamics, and bandwidth. On a more esoteric level, most of the classical and some of the jazz releases also exhibit what is for some listeners *the* most critical attribute: realistic sound-stage depth and pinpoint imaging.

These two latter characteristics were not major players in determining the value of most of the rock and other pop music on this list—and even some of the jazz. Pop music lovers tend to favor (or are at least used to) heavily manipulated recordings. What's more, when rock fans attend "live" performances, they mostly hear music that is electrified, amorphous, and devoid of stage imaging. Good pop and rock recordings depend more on impact, articulation, and bandwidth than on sound-stage realism, mainly because they are seldom made on a sound stage to begin with.

Not every disc listed will display good articulation and clarity, wide dynamic contrasts, and wide bandwidth all the time, of course. However, every one of them will show off the kind of music recorded as it should sound and will also showcase a fine stereo system for what it is. Not every "great" recording available is itemized here, of course. I could not audition every transcription available, and new material, particularly 5.1-channel releases, will obviously appear after this book is published. However, the list is large enough, up-to-date enough, and eclectic enough to satisfy most tastes.[1]

Alan Morrison, Organist (playing music by Maurice Duruflé, Marcel Dupré, and others). Gothic Records 49083. [Bass]

Abercrombie, John. *Current Events.* ECM 1311. [Bass]

Alwyn, William. *Symphony Number 3*; *Violin Concerto.* Chandos 9187.

An American Panorama (music by Leonard Bernstein, Roy Harris, and Aaron Copland). Dorian 90170.

American Tribute (works by Dan Welcher, John Cheetham, David Sampson, John Stevens, Joseph Schwantner, Donald Erb, and Gunther Schuller). Summit DCD 127. [Bass]

Anderson, Jay. *Next Exit.* Digital Music Products CD-490.

Anonymous 4. *An English Ladymass.* Harmonia Mundi 907080.

Arie Antiche (vocal and harpsichord). Claves 50-9023.

Assad, Badi. *Solo.* Chesky JD-909.

Azahara: Flamenco Guitar Recital. Nimbus NI-5116.

Bach, Johann Christian. *Three Quartets*; *Sextet.* Archiv 423 385.

Bach, Johann Sebastian. *Brandenburgische Konzerte.* Archiv 423 116.

———. *Cantata BWV 63*; *Cantata BWV 65.* Dorian DOR 90113.

———. *The Four Orchestral Suites.* Hyperion 66701/2.

[1]Those wanting more in-depth reviews can look over my book *High-Definition Compact Disc Recordings* (McFarland, 1994), or read my regular record-review column in *The Sensible Sound*, a magazine devoted to quality audio.

Finally, for those of you who are continually looking for items to show off your subwoofer or who want to see if your full-range speaker systems can really reach into the depths, those recordings with notable deep-bass response (sometimes really deep into the below-25- to below-30-Hz range) will have the word "bass" in brackets after the manufacturer's stock number.

For ease of use, classical and popular music are intermixed and alphabetized either by composer, or if no dominant composer is listed, by the most important performer or title.

———. *Organ Works, Volume 4.* Dorian 90151. [Bass]
———. *Organ Works.* Telarc 80049. [Bass]
———. *Organ Works.* Telarc 80088. [Bass]
———. *Suites for Cello Solo.* Channel Classics 1090.
Barber, Billy. *Lighthouse.* Digital Music Products CD-455. [Bass]
———. *Shades of Gray.* Digital Music Products CD-445.
Barock Trompetenmusik. Deutsche Harmonia Mundi 77027.
Bart, Lionel. *Oliver.* Angel 55456. [Bass]
Bartók, Béla. *Concerto for Orchestra.* Delos 3095.
———. *Miraculous Mandarin*; Zoltán Kodály: *Háry János*; *Dances of Galanta.* Delos 3083.
———. *Music for Strings, Percussion and Celesta*; *Divertimento*; *The Miraculous Mandarin.* London 430 352.
Basie, Count (with Joe Turner and Edie Vinson). *Kansas City Shout.* Pablo 2310-920.
Basie, Count (his orchestra, led by Frank Foster). *Live at El Morocco.* Telarc 83312.
Beck, Joe. *The Journey.* Digital Music Products CD-481. [Bass]
Beachcomber: Encores for Band. Reference Recordings RR-62CD. [Bass]
Beethoven, Ludwig van. *Complete Sonatas for Pianoforte and Cello.* Channel Classics 35921.
———. *Missa Solemnis*; Wolfgang A. Mozart: *Mass in C Minor.* Telarc 80150.
———. *Symphony Number 9.* Masters MCD-40.
Berlioz, Hector. *Symphonie Fantastique.* Telarc 80271.
Bernhardt, Warren. *Heat of the Moment.* Digital Music Products CD-468.
Bernstein, Leonard. *Arias and Barcarolles*; George Gershwin: *An American in Paris*; Samuel Barber: *Overture to "The School for Scandal."* Delos 3078.
———. *Chichester Psalms*; Samuel Barber: *Agnus Dei*; Aaron Copland: *In the Beginning*; *Three Motets.* Hyperion 66219. [Bass]
Berry, John. *Standing on the Edge.* Capitol 28495. [Bass]
Billings, William. *Anthems and Fuguing Tunes.* Harmonia Mundi France 907048.
Blazing Redheads. Reference Recordings RR-26CD.
———. *Crazed Women.* Reference Recordings RR-41CD.
Boston Pops. *America, the Dream Goes On.* Philips 412 627.
———. *By Request.* Philips 420 178.
Brahms, Johannes. *Ein Deutsches Requiem.* Telarc 80092. [Bass]
Britten, Benjamin. *Song Cycles.* Chandos 8514.
———. *War Requiem*; *Sinfonia da Requiem*; *Ballad of Heroes.* Chandos 8983/4.
Brock, Jim. *Tropic Affair.* Reference Recordings RR-31CD.
Brown, Greg. *The Poet Game.* Red House CD-68.
Butler, Laverne. *Day Dreamin'. Chesky JD-117.*
———. *No Looking Back.* Chesky JD-91.
Caram, Ana. *The Other Side of Jobim.* Chesky JD-73.
Carmina Burana: The Great Mystery of the Passion. Harmonia Mundi France 901323/24.
Carter, Elliott. *Concerto Pour Hautbois*; *A Mirror on Which to Dwell; Penthode.* Erato 45364.

Catharine Crozier at Grace Cathedral (playing works by Mendelssohn, Schumann, Liszt, and Reubke). Delos 3090. [Bass]

Catlingub, Matt. *Your Friendly, Neighborhood Big Band.* Reference Recordings RR-14CD.

Cephas, Johns, and Phil Wiggins. *Bluesmen.* Chesky JD-89.

Charles, David, and David Friedman. *Junkyard.* Digital Music Products CD-491. [Bass]

Chopin, Frédéric. *The Four Scherzi and Other Works.* Dorian DOR 90140.

Cincinnati Pops. *The Great Fantasy Adventure Album.* Telarc 80342. [Bass]

Classics of the Silver Screen. Telarc 80221. [Bass]

Codex Las Huelgas. Sony 53341.

Cohen, Leonard. *Famous Blue Raincoat, Songs of Leonard Cohen.* Jennifer Warnes, vocals. RCA/Cypress PD90048.

Coil, Pat. *Just Ahead.* Sheffield Lab CD-34.

———. *Steps.* Sheffield Lab CD-31.

Copland, Aaron. *Fanfare for the Common Man*; *Rodeo*; *Appalachian Spring.* Telarc 80078.

———. *Old American Songs*; Charles Ives: *Songs.* Argo 433 027.

———. *Rodeo*; *Billy the Kid*; *El Salón Mexico*; *Danzón Cubano.* Argo 440 639.

———. *Symphony Number 3*; *Music for the Theater.* Telarc 80201. [Bass]

Corelli, Archangelo. *Concerti Grossi Numbers 1–6.* Harmonia Mundi France 907014.

Corigliano, John. *Symphony Number 1.* Erato 45601.

DMP Big Band. *Carved in Stone.* Digital Music Products CD 512.

Däfos (with Mickey Hart, Airto Moreira, and Flora Purim). Rykodisc RCD 10108 (Originally Reference Recordings RR-12CD).

Daniels, Eddie. *Real Time.* Chesky JD-118.

Daniels, Eddie, and Gary Burton. *Benny Rides Again.* GRP 9665.

Davies, Peter Maxwell. *Solstice of Light*; *Five Carols*; *Hymn to the Word of God.* Argo 316 119. [Bass]

———. *Strathclyde Concerto Number 3*; *Strathclyde Concerto Number 4.* Collins 12392.

Dean, Billy. *Fire in the Dark.* Liberty 98947.

Debussy, Claude. *La Mer*; Albert Roussel: *Symphony Number 4*; *Sinfonietta*; Darius Milhaud: *Suite Provençal.* Chandos 9072.

Dello Joio, Norman. *Antiphonal Fantasy on a Theme of Vincenzo Albrici*; Robert Planel: *Concerto for Trumpet and Strings*; Vincent Persichetti: *The Hollow Men*; Charles Ives: *Variations on America.* Summit DCD-145. [Bass]

Dire Straits. *Brothers in Arms.* Warner Bros. 25264.

———. *On Every Street.* Warner Bros. 22680.

D'Rivera, Paquito. *Havana Cafe.* Chesky JD-60.

———. *Tico! Tico!* Chesky JD-34.

Dunlap, Bruce. *The Rhythm of the Wings.* Chesky JD-92.

Dupré, Marcel. *Symphony in G Minor*; Josef Rheinberger: *Organ Concerto Number 1, in E.* Telarc 80136. [Bass]

Duruflé, Maurice. *Organ Music* (complete). Delos 3047. [Bass]

———. *Requiem.* Summit DCD 134. [Bass]
Elgar, Edward. *Symphony Number 1*; *Pomp and Circumstance Marches Numbers One and Two.* Telarc 80310. [Bass]
Ellington, Duke. *Dick Hyman Plays Duke Ellington.* Reference Recordings RR-50CD.
Enya. *Watermark.* Geffen 9 24233. [Bass]
Fagen, Donald. *Kamakiriad.* Reprise 45230.
———. *The Nightfly.* Warner 23696.
Fiesta. Reference Recordings RR-38CD. [Bass]
Flim and the BB's. *Big Notes.* Digital Music Products CD-454. [Bass]
———. *The Further Adventures of Flim and the BB's.* Digital Music Products CD-462.
———. *Neon.* Digital Music Products CD-458.
———. *Tricycle.* Digital Music Products CD-443.
———. *Vintage BB's.* Digital Music Products CD-486.
Frigo, Johnny. *Debut of a Legend.* Chesky JD-119.
———. *Live from Studio A in New York City.* Chesky JD-1.
Gabriel, Peter. *Security.* Geffen 2011.
Garson, Mike. *The Oxnard Sessions, Volume 1.* Reference Recordings RR-37CD.
———. *The Oxnard Sessions, Volume 2.* Reference Recordings RR-53CD.
Gershwin, George. *Concerto in F*; *Variations on an American Theme*; *Doo-Dah Variations.* Chesky CD-98.
———. *Marni Nixon Sings Gershwin.* Reference Recordings RR-19CD.
———. *Piano Improvisations.* Special Music SCD 6039.
The Great Organ of St. Patrick's Cathedral, New York City. Gothic Records 49081. [Bass]
Green, Bunky. *Healing the Pain.* Delos 4020.
Gregorian Chants in a Village Church. Hungaraton 12742.
Grusin, Dave. *The Gershwin Connection.* GRP 2005.
———. *Migration.* GRP 9592. [Bass]
Grusin, Dave, and Don Grusin. *Sticks and Stones.* GRP 9562.
Händel, George Frideric. *Giulio Cesare* (Julius Caesar). Harmonia Mundi France 901385/87.
———. *Messiah.* Telarc 80322.
———. *Water Music.* Harmonia Mundi 907010.
Hanson, Howard. *Symphony Number 5*; *Symphony Number 7*; *Mosaics*; *Piano Concerto in G Major.* Delos 3130.
Harrell, Tom. *Passages.* Chesky JD-64.
Haydn, Joseph. *The Creation.* Telarc 80298.
———. *String Quartets, Op. 54, Numbers 1 and 2.* Deutsche Harmonia Mundi 77028.
———. *Symphonies Numbers 23, 35, and 42.* Dorian 90191.
Hersch, Fred. *Forward Motion.* Chesky JD-55.
———. *The Fred Hersch Trio Plays. . . .* Chesky JD-116.
Higgins, Billy. *The Essence.* Digital Music Products CD-480.
Hindemith, Paul. *Organ Works*; Hugo Distler: *Spielstücke*; Augustinus Kropfreiter: *Toccata Francese.* Argo 417 159. [Bass]
———. *When Lilacs Last in the Dooryard Bloom'd.* Telarc 80132. [Bass]

Hohner, Robert (his Percussion Ensemble). *Different Strokes.* Digital Music Products CD-485.

———. *Lift Off.* Digital Music Products CD-498.

Holst, Gustav. *The Planets.* Telarc 80133. [Bass]

———. *St. Paul's Suite.* Chandos 9270.

———. *Suite Number 1 in E-Flat; Suite Number 2 in F; A Moorside Suite; Hammersmith.* Reference Recordings RR-39CD. [Bass]

Hornsby, Bruce. *Harbor Lights.* RCA 66114.

Hovhaness, Alan. *Magnificat; Psalm 23.* Delos 3176. [Bass]

Howard, James Newton. *James Newton Howard and Friends.* Sheffield Lab CD-23.

Howells, Herbert. *Hymnus Paradisi; An English Mass.* Hyperion 66488. [Bass]

Hyman, Dick. *From the Age of Swing.* Reference Recordings RR-59CD.

Ian, Janis. *Breaking Silence.* Morgan Creek 2959-20023.

Ireland, John. *A London Overture; These Things Shall Be.* Chandos 8879.

Jamal, Ahmad. *Chicago Revisited.* Telarc 83327.

Jarvis, John. *Something Constructive.* MCAD-5963.

Jefferis, Mark. *Talk to Me.* B&W 026. [Bass]

Joan Lippincott and the Philadelphia Brass (playing works by Dupré, Karg-Elert, Widor, and others). Gothic Records 49072. [Bass]

John Weaver Performs (playing organ works by J. S. Bach, Kenneth Leighton, Marcel Dupré, Franz Schubert, and others). Gothic Records 49060. [Bass]

Jones, Ricky Lee. *Traffic from Paradise.* Geffen 24602.

Jones, Vince. *One Day Spent.* Intuition 3087.

Jongen, Joseph. *Symphonie Concertante for Organ and Orchestra*; César Franck: *Fantasie in A*; *Pastorale.* Telarc 80096. [Bass]

Jordan, Clifford, and Ran Blake. *Masters from Different Worlds.* Mapleshade 01732.

K., Sara. *Closer Than They Appear.* Chesky JD-67.

———. *Play on Words.* Chesky JD-105.

Kodo: Heartbeat Drummers of Japan. Sheffield Lab CD-KODO (12222).

Korngold, Erich. *String Sextet in D Major;* Arnold Schoenberg: *Verklärte Night.* Hyperion 66425.

Lauer, Christof. *Bluebells.* CMP Records CD-56. [Bass]

Lennox, Annie. *Diva.* Arista 18704.

Liszt, Franz. *Christus.* Hungaroton 12831-33.

———. *Piano Concertos Numbers 1 and 2*; *Hungarian Fantasy*; *Totentanz.* London 433 075.

———. *Sonata in B Minor* (and other, shorter pieces). Etcetera KTC 2010.

Lloyd, Charles. *The Call.* ECM 21522.

Loeb, Chuck. *Balance.* Digital Music Products CD-484.

———. *Life Colors.* Digital Music Products CD-475.

———. *Mediterranean.* Digital Music Products CD-494.

Loeb, Chuck, and Andy Laverne. *Magic Fingers.* Digital Music Products CD-472. [Bass]

MacLeod, Doug. *Come to Find.* AudioQuest 1027.

Mahler, Gustav. *Symphony Number 3*; *Kindertotenlieder.* Chandos 9117/8.

———. *Symphony Number 5.* Teldec 46152.
———. *Symphony Number 6*; Alexander Zemlinsky: *Six Maeterlinck Songs.* London 430 165.
Manakas, Van, and Jim Brock. *Letters from the Equator.* Reference Recordings RR-56CD. [Bass]
Martino, Donald. *Solo Piano Music.* Centaur 2173.
Mary Preston Plays the Organ Works of Duruflé and Widor. Gothic Records 49079. [Bass]
Mays, Bill, and Ray Drummond. *One to One.* Digital Music Products CD-473.
———. *One to One 2.* Digital Music Products CD-482.
McGriff, Jimmy, and Hank Crawford. *Right Turn on Blue.* Telarc 83366.
Mendelssohn, Felix. *Organ Works.* Argo 414 420. [Bass]
———. *Symphony Number 2.* Delos 3112.
Michael Murray: The Ruffatti Organ in Davies Symphony Hall. Telarc 80097. [Bass]
Mintzer, Bob. *Art of the Big Band.* Digital Music Products CD-479.
———. *One Music.* Digital Music Products CD-488.
———. *Urban Contours.* Digital Music Products CD-467. [Bass]
Monteverdi, Claudio. *Ghirlande Sacre; Ghirlande Profane.* Europa 350-237.
Moreira, Airto. *Fourth World.* B&W 030.
———. *The Other Side of This.* Rykodisc 10207.
Mozart, Wolfgang Amadeus. *String Quartets Numbers 14 and 17.* Titanic 154.
———. *Die Zauberflöte* (complete). Telarc 80302. (Excerpts are available on Telarc 80345.)
Mussorgsky, Modest. *Pictures at an Exhibition* (orch. by Ravel). Chandos 8849.
———. *Pictures at an Exhibition* (orch. by Ravel); Maurice Ravel: *Night on Bald Mountain.* Telarc 80296. [Bass]
———. *Pictures at an Exhibition* (organ transcription). Dorian 90117. [Bass]
Neto, José. *Neto.* B&W 037.
Nielsen, Carl. *Saul and David.* Chandos 8911/12.
Not Drowning, Waving. *Claim.* Reprise 26181.
O Mistress Mine: A Collection of English Lute Songs. Dorian 90136.
Offenbach, Jacques. *Gaîté Parisienne*; Jacques Ibert: *Divertissement for Small Orchestra.* Telarc 80294. [Bass]
Oldfield, Mike. *The Songs of Distant Earth.* Reprise 45933. [Bass]
Organo Deco: Sophisticated American Organ Music, ca. 1915–1950 (including material by Sowerby, Simonds, Bennett, James, Bingham, and Crandell). Delos 3111. [Bass]
Orquesta Nova. Chesky JD-54.
Palestrina, Giovanni. *Masses.* Gimell 020.
Papa Doo Run Run. *California Project.* Telarc 70501.
Peña, Paco. *Misa Flamenca.* Nimbus 5288.
Piston, Walter. *Symphony Number 2*; *Symphony Number 6*; *Sinfonietta.* Delos 3074.
Pomp & Pipes (music by Sigfried Karg-Elert, Alfred Reed, Charles Widor, Marcel Dupré, and others). Reference Recordings RR-58CD. [Bass]
Prokofiev, Sergei. *Alexander Nevsky*; *Scythian Suite.* Chandos 8584.

———. *Alexander Nevsky*; Dmitri Shostakovich: *Symphony Number 9*. Dorian 90169. [Bass]

———. *Alexander Nevsky*. London 410 164. [Bass]

———. *Symphony Number 5*. Chandos 8576. [Bass]

Pugh, Jim, and Dave Taylor. *The Pugh-Taylor Project*. Digital Music Products CD-448.

Rachmaninov, Sergei. *Isle of the Dead*; *Symphonic Dances*. London 410 124.

———. *Symphony Number 2*; *Vocalise*. Telarc 80312. [Bass]

———. *Symphony Number 3*; *Symphonic Dances*. Telarc 80331.

———. *Vespers*. Telarc 80172.

Raitt, Bonnie. *Luck of the Draw*. Capitol 96211.

Rameau, Jean Philippe. *Pièces de Clavecin*; *Suite in A* (harpsichord). Reference Recordings RR-27CD.

Rankin, Kenny. *Because of You*. Chesky JD-63.

Ravel, Maurice. *Daphnis et Chloé* (complete); David Diamond: *Elegy in Memory of Maurice Ravel*. Delos 3110.

———. *Rapsodie Espagnole*; *Ma Mère l'Oye*; *Valses Nobles et Sentimen-tales*; *La Valse*. London 430 413. [Bass]

———. *Shéhérazade*; *Ma Mère l'Oye*; *La Valse*. EMI 54204.

Regner, Max. *Three Sonatas for Unaccompanied Violin*. Dorian 90175.

Respighi, Ottorino. *Church Windows* (*Vetrate di Chiesa*); *Poema Autunnale*. Reference Recordings RR-15CD. [Bass]

———. *Pines of Rome*; *Roman Festivals*; *Fountains of Rome*. Chandos 8989. [Bass]

Reubke, Julius. *Sonata for Organ*; *Sonata for Piano*. Dorian 90106. [Bass]

Rich, Robert. *Propagation*. Hearts of Space 11040. [Bass]

Rimsky-Korsakov, Nikolai. *Scheherazade*; Alexander Glazunov: *Stenka Razin*. Chandos 8479.

———. *Scheherazade*; *Russian Easter Overture*. RCA 61173.

Robert Noehren Premiers the New D.F. Pilzecker Organ at the Church of St. Jude. Delos 3045. [Bass]

Rorem, Ned. *Organ Music* (including *A Quaker Reader*; *Views from the Oldest House*, and *Rain over the Quaker Graveyard*). Delos 3076. [Bass]

Rossini, Gioacchino. *Arias*. (Vocals.) London 425 430.

Rotella, Thom. *Home Again*. Digital Music Products CD-469.

———. *Without Words*. Digital Music Products CD-476.

Ruff, Michael. *Speaking in Melodies*. Sheffield CD-35.

Rutter, John. *Requiem*. Reference Recordings RR-57CD. [Bass]

Saint Saëns, Camille. *Symphony Number 3*; Charles Marie Widor: *Symphony Number 6* (Allegro movement only). Philips 412 619. [Bass]

Sarum Chant: Missa in Gallicantu and Four Hymns. Gimell 017.

Scarlatti, Domenico. *Eighteen Sonatas* (harpsichord). Koch 3-7014.

Schickele, Peter. *P.D.Q. Bach: 1712 Overture*. Telarc 80210. [Bass]

Schnittke, Alfred. *Symphony Number 3*. BIS CD-477. [Bass]

Schoenberg, Arnold. *Pellas und Melisande*. Chandos 8619.

Schonberg, Claude-Michel. *Les Miserables*. Relativity/First Night 1027. [Bass]

Schubert, Franz. *String Quartet in C Major*. Nimbus 5313.

———. *Symphony Number 5; Symphony Number 8*. Virgin 59273.

Schuman, William. *Symphony Number 7*. New World 348.

Schumann, Robert. *Symphony Number 1; Overture, Scherzo and Finale; Konzertstrück for Four Horns and Orchestra*. Delos 3084.

Severinsen, Doc. *Facets*. Amherst 93319.

———. *The Tonight Show Band, Once More . . . with Feeling*. Amherst 94405.

———. *Unforgettably Doc: Music of Love and Romance*. Telarc 80304.

Sevåg Øystein. *Link*. Windham Hill 11123. [Bass]

Sharp, John. *Better Than Dreams*. Reference Recordings RR-54CD. [Bass]

Short, Bobby. *Swing That Music*. Telarc 83317.

Shostakovich, Dmitri. *Symphony Number 5; Ballet Suite Number 5 from the Bolt*. Chandos 8650.

———. *Symphony Number 8*. Collins 12712. [Bass]

———. *Symphony Number 10; Ballet Suite Number 4*. Chandos 8630.

———. *Symphony Number 10; Chamber Symphony*. London 433 028. [Bass]

Smith, Bob. *Bob's Diner*. Digital Music Products CD-471.

———. *Radio Face*. Digital Music Products CD-483.

Sowerby, Leo. *Organ Music* (including *Fantasy for Flute Stops, Requiescat in Pace*, and *Symphony in G Major*). Delos 3075. [Bass]

Spies. *By Way of the World*. Telarc 83305. [Bass]

———. *Music of Espionage*. Telarc 85503. [Bass]

Star of Wonder (Christmas music). Reference Recordings RR-21CD. [Bass]

Sting. *Soul Cages*. A & M 16405.

Strauss, Richard. *Der Rosenkavalier*. EMI 7 54259.

———. *Thus Spake Zarathustra; Dance of the Seven Veils; Four Symphonic Interludes from Intermezzo*. Delos 3052. [Bass]

Stravinsky, Igor. *Firebird Suite* (1919 version); *Petrouchka*. Telarc 80270. [Bass]

———. *Le Sacre Du Printemps*; Sergei Prokofiev: *Scythian Suite*. Dorian DOR 90156.

Tallis, Thomas. *Complete English Anthems*. Gimell CDGM-007.

Tavener, John. *We Shall See Him As He Is*. Chandos 9128. [Bass]

Taverner, John. *Ave Dei Patris Filia*. Nimbus 5360.

———. *Missa Gloria Tibi Trinitas*. Gimell CDGM-004.

Taylor, Samuel Coleridge. *Ballade in A Minor; Symphonic Variations on an African Air*; George Butterworth: *Two English Idylls*. Argo 436 401. [Bass]

Tchaikovsky, Peter. *1812 Overture; Capriccio Italien*. Telarc 80041. [Bass]

———. *Nutcracker Suite; Swan Lake Suite; Sleeping Beauty Suite*. Deutsche Gramophon 437 806.

———. *The Seasons*. Dorian 90102.

———. *Symphony Number 4; Romeo and Juliet Fantasy Overture*. Telarc 80228.

———. *Trio in A Minor*. Chandos 8975.

Terry, Clark. *Having Fun*. Delos 4021.

Terry, Clark, and Frank Wess, with the DePaul University Jazz Ensemble. *Big Band Basie*. Reference Recordings RR-63CD.

Testament: 20th Century American Music for Male Chorus and Band. Reference Recordings RR-49CD.
Three-Way Mirror. Reference Recordings RR-24CD.
The Timeless All-Stars. *Essence.* Delos 4006.
Tippett, Michael. *A Child of Our Time.* Chandos 9123.
Trittico. Reference Recordings RR-52CD.
Turtle Creek Chorale. *Postcards.* Reference Recordings RR-61CD.
Tyner, McCoy (his Quartet). *New York Reunion.* Chesky JD-51. (There is some background rumble on this disc.)
Vaughan Williams, Ralph. *Symphony Number 7: Sinfonia Antarctica*; *Toward an Unknown Region.* Chandos 8796. [Bass]
The Very Best of Erich Kunzel and the Cincinnati Pops. Telarc 80401. [Bass]
Vierne, Louis. *Symphony Number 6*; Max Reger: *Second Sonata in D Minor* (album title: *Organ Music of Reger and Vierne*). Delos 3096. [Bass]
Virtuoso Trumpet Concertos. Nimbus NI 5121.
Vivaldi, Antonio. *Choral Works.* Hungaroton 11695.
———. *Concerti a Due (Six Concertos).* Novalis 150 074.
Vivino, Jerry. *Something Borrowed, Something Blue.* Digital Music Products CD-513.
Vivino Brothers. *Chitlins Parmigiana.* Digital Music Products CD-492.
Waller, Fats. *Dick Hyman Plays Fats Waller.* Reference Recordings RR-33CD.
Walton, Cedar. *Cedar Walton Plays.* Delos 4008.
Warnes, Jennifer. *The Hunter.* Private Music 82089.
Wells, Junior. *Better Off with the Blues.* Telarc 83354.
Widor, Charles-Marie. *Symphony Number 5*; Francis Poulenc: *Concerto for Organ, Strings and Timpani*; Felix Guilmant: *Symphony Number 1.* Chandos 9271. [Bass]
Williams, Joe. *Here's to Life.* Telarc 83357.
Williams, Joe. *I Just Want to Sing.* Delos 4004.
Woods, Phil, and the Little Big Band. *Here's to My Lady.* Chesky JD-3.
———. *Real Life.* Chesky JD-47.
Yellow Jackets. *Four Corners.* MCA 4994.
ZZ Top. *Antenna.* RCA 66317.

A FEW TOP-GRADE VIDEOS

Here are a few examples of video software that will reward the purchase of a world-class subwoofer, potent main and center speakers, a good surround processor, a big amplifier—and a large, top-quality video monitor, particularly one with a wide screen. While the tape versions of some of these performances may be quite good, particularly if they are THX certified, I believe that only the DVD or letterboxed LV versions with digital sound will fully satisfy those with truly fine A/V systems.

If you intend to experience these movies only on prerecorded videotape, you may not need a subwoofer at all if your speaker systems have decent response down to, say, 40 Hz. However, if you play videotaped material loud, or if your A/V room is large, you may need a subwoofer or two just to protect your main-speaker woofers from action-film pyrotechnical damage.

For in-depth reviews of still more current software, I suggest you subscribe to journals like *The Laser Disc Newsletter*, *Widescreen Review*, and *Video.*

The Abyss: THX version. James Cameron feels that the 4:3-ratio LV version is aesthetically superior to the letterboxed one, but the latter is more workable if you have a wide-screen set. The DVD release should allow both options.

Apocalypse Now: The Paramount LV Wide-Screen edition, which is actually an early, uncredited THX release, is impressive, but the AC-3 encoded presentation is even more gripping; the distant B-52 raid on either sounds splendid, and the Pro Logic decoding on the LV version is better than normal.

Batman, *Batman II*, and *Batman Forever*: All three have dynamic musical scores and enough big-screen cartoon impact to satisfy the most-jaded 15-year-old.

Broken Arrow: Nonstop pyrotechnics and impact enough for any action-film enthusiast. Plenty of subwoofer-grade bass.

Clear and Present Danger: The first AC-3 surround video release.

Cliffhanger: There is a remarkable helicopter "flyover" at the beginning that shows what standard Pro Logic can do.

Crimson Tide: Dolby soundtrack was enhanced by the "Spatializer" surround process.

Eraser: Enjoy it as a comedy. Dynamic impact, with plenty of bass punch for subwoofer nuts.

Forrest Gump: The "chopper" scene will jump right out at you, and the audio in the combat sequences will mow you down, figuratively speaking.

The Fugitive: Plenty of sound in the famous train wreck and the dam sequence.

Gettysburg: The extended-length version is no better in terms of technical quality than the earlier, shorter release. Indeed, in some respects the picture of the reissued program is worse in places, and the sound often seems compressed compared with the shorter version. However, the additional footage will be important to Civil War buffs.

The Hunt for Red October: In either the 5.1-channel or DPL version, the film abounds with plenty of deep-bass submarine noises and enveloping surround sound.

Independence Day: Mastered with the Sony SDDS surround process, it entertains with plenty of impact; outstanding visuals.

Jurassic Park: Another THX release. While the Tyrannosaurus footfalls and roarings are impressive, the Brachiosaurus tree-feed sequence near the beginning is also a sonic delight, as are the raptor wheezes.

The Last Action Hero: Excellent AC-3 surround, with wonderful explosion noises.

The Rock: Excellent pyrotechnics.

Stargate: While aesthetically vacuous, this release has very impressive AC-3 surround sound.

Star Trek IV: Experience the shock-wave sequence at the beginning. A Trekkie's delight.

The Star Wars Trilogy: The reissued THX version; the earlier releases were not so hot. The picture quality is "impressive." Also available in a decent THX tape release.

Terminator II: The THX, director's-cut version was only slightly better technically than the earlier release but considerably more workable in terms of story line.

Top Gun: Over ten years old, the Dolby Digital reissue version still has an impressive and very dynamic opening-title sequence. Indeed, some people feel the opening credits are the most exciting part of the movie. The Dolby Digital version has flybys aplenty.

True Lies: Impressive impact in both Dolby Digital Surround and standard Dolby Surround versions.

Twister: Steamroller-grade sound effects give the remarkable visuals even more impact.

Who Framed Roger Rabbit? Listen to the rumble of the "dip" cannon.

Here's a quick, and yet subtle, video-subwoofer "needs" test that you can perform yourself if you have an LV player. During the sequence displaying the action of the "dip" cannon near the end of *Roger Rabbit*, switch back and forth between "digital" and "CX" with your player's remote control. If you hear no difference in extremely low bass strength between these two audio-playback modes, you probably need a subwoofer, at least if you plan on building a collection of dynamite-quality videodiscs.

This test obviously cannot be performed with videotape, and there is no analog sound on the DVD version to use as a comparison. However, even the tape version of this material should have a solid, very low-frequency rumble during this sequence (meaning that if you play your system at fairly high levels, a subwoofer may still be a good investment); the tonal quality and bandwidth of the DVD pressing matches the PCM-digital soundtrack of the LV version. If you cannot detect extremely deep bass in any version, you're missing a lot of what a subwoofer can deliver.

SOME TEST DISCS AND SAMPLERS

I believe that any serious enthusiast should own at least a couple of test discs to use both as tools and as reference standards for good recorded sound and video. Nearly all the discs contain audio tracks that will fully exercise even the most potent stereophonic and surround-sound systems. A few of these are strictly "test" discs, while the rest are "samplers" designed to attract listeners to the complete works by the companies that produce them; some are combinations of both.

Each album contains written material describing the program, but some also devote space to describing the techniques used to make the recordings. While the quality of these assorted treatises is variable (with a few substituting pseudo-technical hype for real engineering information), some of them are miniature but accurate tutorials on recording, sound-system, and even video theory. However, the reader should remember that the people who produced some of these discs often have different (and sometimes opposing) ideas about what constitutes good sound and what kind of playback system is necessary to achieve it. A few of these presentations, usually those which also spout hyperbole somewhere in their printed inserts, may have material of dubious value, such as absolute-polarity tests, "jitter" tests, and digital-converter comparisons.

There are a lot of samplers and digital test discs available these days, and more are appearing all the time. Those below are just a few of the better ones I've personally used. Future test/sampler discs assembled by the people who made these discs probably will also be worthwhile purchases.

Auditory Demonstrations, prepared at the Institute for Perception Research, Eindoven, The Netherlands, in association with the Acoustical Society of America, Philips 1126-061: This is a very interesting disc, in that it contains little that would ordinarily interest a lover of recorded music. It features 39 very interesting psychoacoustic tests that will tax your hearing abilities and teach you a great deal about what it simply means to hear. While the information packet that comes with the disc contains all that is necessary to understand the tests and put them to good use, you will gain further insight if you read "Auditory Demonstrations on Compact Disc," by William Hartmann, in the January 1993 issue of the *Journal of the Acoustical Society of America* (volume 93). Technonuts will fall in love with this disc, but it is important to remember that the material it contains will not be directly related to standard recorded musical sounds.

Bravura (includes pieces by Respighi, Strauss, and Lutoslawski, played by several different orchestras), Delos 3070: This is a showpiece album, with great impact, impressive dynamics, and a wide-band frequency range. The enclosed flyer has important data on how each of the segments was produced. While not a "test disc" in the strict sense, this production is a fine educational tool for those interested in learning about why certain microphone and hall configurations interact the way they do. It should also entice you into buying the complete versions of the assorted samples.

Chesky Jazz Sampler, Volume 1, Chesky JD-37; *Chesky Jazz Sampler, Volume 2*, Chesky JD68; *Chesky Sampler Vol. 3: Best of Chesky Classics & Jazz and Audiophile Test Disc*, Chesky JD-111: A must for enthusiasts who want to evaluate imaging and want good examples of advanced recording techniques. For the most part, the company's fine sound comes mainly from the proper use of good microphones, an adherence to minimalist microphone techniques, and careful attention to mixing. The musical examples are all excellent. Volume 1's vertical imaging demonstration is entertaining and will reveal weaknesses in speaker sound-staging capabilities and placement asymmetries. I suggest you listen to this test section on several different systems to get an idea of what it should sound like.[2] The functional but also notorious "Bonger" test should only be used to evaluate the low-level artifacts of your CD player, since the initial part of the test signal made on the custom processor was over-driven and distorted during the recording process. The absolute-phase comparison is irrelevant if your speaker systems have low asymmetric (even-order) harmonic distortion and you listen well into the *reverberant* field—although it may be meaningful if you listen with headphones or occupy the *direct* listening field of speakers designed for that kind of use. Chesky's use of 128-times oversampling is also pretty much audibly inconsequential, because any good 16-bit converter can do subjectively as well. The second volume offers revealing imaging and sound-stage tests. The tests for D/A converters and digital cables meant nothing to me, but maybe your system can reveal these nuances—although it pays to keep a cool head as you try to spot these things. The third release is somewhat more helpful than the first two volumes, because it carefully outlines a variety of microphone and recording techniques and even includes diagrams to help in understanding what you are hearing. On the other hand, the random-noise track, designed to "break in" your speakers, is a waste of time and could damage your equipment. Additionally, the tracks comparing 20- and 16-bit technology are misleading, and the track dealing with power isolation is so useless that Chesky actually apologizes in the accompanying booklet for including it!

Demonstration of Stereo Microphone Techniques, Performance Recordings PR-6-CD: If you want to hear what a number of different microphone types and placement techniques sound like, this 19-minute disc is for you. The main goal of the producers was to demonstrate the stereo imaging produced by a variety of recording methods; their secondary goal was to demonstrate the tonal qualities available from several microphone types. The first goal was magnificently achieved; the second less so. A weak point in the production is the lack of any data on how microphones in halls behave in terms of their handling of ambience, especially that reflected from the sides and rear. However, dealing with reflected ambience, given the small scale of the sound stage and the tools used, was obviously not a goal of the producers. Another deficiency is the lack of any example of the "Decca Tree" recording technique, which places the center microphone of a three-capsule array considerably out in front of the flanking mikes. There was an example of a standard, three-omni-in-a-row configuration—similar to what is used by Telarc—although the spacing between the capsules was much smaller. Still, this disc is worth every penny of its nearly dollar-per-minute cost.

[2]While I don't know how the Chesky vertical-imaging test was performed, it is known that adding a sharp notch filter centered at about 8 kHz in the frequency spectrum of a playback system or recording will subjectively move the instrument or instruments being reproduced to a position above where they actually are. Adding notches at other frequencies will move the image down, farther up to the side, or even behind the listener. The notches simulate outer-ear pinna effects.

Denon Digital Audio Check CD, Denon 33C39-7441: This is a fine little combination sampler/test disc that contains 12 excerpts from the Denon catalog, plus a series of test sequences on 14 additional tracks that will be helpful in setting up a basic stereo system (speaker phase, balance, centering, etc.). Perhaps the most helpful item is the low-level test sequence, which will help you evaluate the D-A converter in a CD player.

The Digital Domain: A Demonstration, Elektra 9 60303-2: This is a good selection of electronic music and test sequences by the Stanford University Center for Computer Research in Music and Acoustics. The opening jet-plane sequence should be run at low volume at first, to prevent speaker damage. For sound-effects nuts, this piece alone will be worth the price of the disc. Test tones are also available (sine waves, pink noise, etc.), which can help you calculate volume control and power output levels if you have power meters on your amplifier or receiver. They can also assist you in adjusting an equalizer if you have a real-time analyzer (RTA). An important warning about the square-wave test on track 19: this track should *not* be played back through speakers or headphones at all, because even a small increase of the volume control could fry some speaker drivers. An announcement on track 18 warns about possible speaker damage from this track, but it is still possible to punch a programming-sequence button by mistake and immediately land your speakers (or your ears) in a whole lot of trouble. The test tone should have been on the same band as the announcement—directly following it—making an accidental sonic disaster more difficult.

Engineer's Choice (a sampler featuring 22 excerpts of material recorded by engineer John Eargle), Delos 3506: It's hard to beat this disc, featuring masterful examples of engineer Eargle's skills. The insert gives interesting information on the microphone techniques used, as well as the various recording environments encountered. Eargle presented a technical description of this disc in the October 1991 issue of *Audio*.

Gems of Jazz (a sampler featuring Joe Williams, Cedar Walton, Red Holloway, Bunky Green, and others), Delos 3507: This disc features wonderful sound throughout and is an example of what recording engineering is all about. This is not just a sampler to take with you to evaluate audio systems at your local dealer: it is also one to listen to and enjoy over and over. If this disc does not make you want to collect the complete recordings, you are just not into good jazz and superior sound.

HDCD (High Definition Compatible Digital) Sampler, Reference Recordings RR-S3CD: A compendium of excerpts from assorted HDCD presentations, this disc has all the punch, clarity, imaging, and ambience qualities you would expect from a company that specializes in audiophile recordings. The heart and soul of RR is Keith Johnson; over the years, he has set his products off from the competition by employing unique recording hardware and intelligently using microphones. Johnson has done everything from designing his own analog tape recorder heads to employing a new technique he and Michael Pflaumer have contrived and labeled HDCD (High Definition Compatible Digital). Apparently, the full HDCD system is a 20-bit to 16-bit compression-expansion ("companding") process that requires a low-level-signal downward-expanding device to realize its potential.[3] Without the decoding circuitry, the passages at the lowest levels may have a slight tendency to be pumped up and down in loudness (because of the operation of the companding process). Some critics claim this process sometimes adds very-low-level background-noise fluctuations to the signal—although I never felt their presence to any serious degree either in this sampler or in any other RR recording. The undecoded signal may also exhibit a tendency to amplify the fade-to-silence reverb of the hall, subjectively highlighting background ambience. Ironically, because of this, some listeners may actually prefer the undecoded version to the decoded one. Few listeners will be aware of the compression at low levels in any case.[4] Although the system is impressive on paper, its practical use will be limited to recordings with extreme dynamic range. Even those will probably not be able to fully show their stuff unless the playback environment is substantially quieter than a typical home listening room. In the long run, perceptual-coding techniques (particularly 5.1-channel manifestations) will probably render this system, essentially a solution looking for a problem to solve, obsolete before it gets much of a chance to proliferate. It's a case of too little, too late.

Hearts of Space: *Universe Sampler 92* (samplings from the Hearts of Space catalogue of synthesizer music, compiled by Stephen Hill), Hearts of Space HS 11201-2 (available from Reference Recordings): Even though taken from analog masters, the overall sound displays all the digital format's abilities: very wide dynamic range at times, a great and monumental sense of bigness and depth, and a wide bandwidth—with occasionally awesome bass. Although most of these pieces are impressive sounding, many listeners will consider them to be of limited musical value.

Hi-Fi News and Record Review Test Disc II, HFN 015: This is an excellent series of tests and musical passages engineered by Tony Faulkner. However, two warnings are in order. First, some very clean test tones (running from 2 Hz to 20 kHz) are recorded at maximum level (0 dB below maximum output potential), which requires keeping the volume fairly low during playback unless you want to blow a speaker driver. Levels this high may also overload certain signal processors installed in an external processor or tape loop. Second, the central image in the early spoken "walkaround" test sections (both stereo and Ambisonics) are slightly skewed to the left side, an odd error for a test disc! Further into the test sequence, an excellent central-image test copied from an earlier Denon test disc will accurately evaluate the central image.

In Sync: ProJazz Sampler #2 (selections by the Tokyo Union Big Band, the Michel Camilo Trio, Ronnie Cuber, Ronnie Foster and quiet a few others), ProJazz CDJ

[3]Pacific Microsonics, the outfit that designed the system for Reference Recordings, has described its operating principles only vaguely, but it is possibly a subtractive-dithering method that cancels out the sound of the input dither and simulates 20-bit reproduction from a 16-bit system.

[4]One perceptive critic has noted that the HDCD encoding process creates occasionally pleasant-sounding dynamic-range distortions that the decoding process eliminates. In essence, the end result is no better than if no HDCD system had been used at all.

599: This release has plenty of punch when needed, with fine clarity, good depth, and well-etched detail. However, the quality is variable.

Jazz Showcase (excerpts from assorted Telarc jazz albums, with material by the Count Basie Orchestra, Oscar Peterson, Dave Brubeck, and others), Telarc 83342: These excerpts were all made in different environments, with different recording techniques and microphones, and—as you might expect—all sound different. However, they all sound good, reflecting the talents of the near-legendary recording engineer Jack Renner.

King of Instruments (a listener's guide to the art and science of recording the organ, with short excerpts from Bach, Gherardeschi, Buxtehude, Lefébure-Wély, and others, along with written descriptions by John Eargle of the recording techniques employed), Delos 3503: This is a remarkable series of excerpts taken from the Delos catalog. A variety of microphone techniques, which are clearly described in the accompanying insert, were used to make this disc. Needless to say, this is an outstanding educational guide to the art and science of recording organs and an equally outstanding example of state-of-the-art recorded sound.

Kunzel, Erich. *The Very Best of Erich Kunzel and the Cincinnati Pops*, Telarc 80401. This is a sometimes smashing combination of sound effects and music recorded by Kunzel and engineer Jack Renner over a ten-year period—and apparently remastered (and sometimes enhanced) for this sampler. The sound is variable, ranging from excellent to incredible. Some tracks were enhanced by a technique called "Spatializer," delivering a three-dimensional effect with only two front speakers.[5] The dynamics on this recording are spectacularly impressive, so if you have less-than-monumental speakers and amplifiers, you would do well to watch the volume level on the first playthrough. The orchestral sound on each track is typical of Telarc at its best, with excellent clarity, good imaging, and a wide, deep, and spacious soundstage. Deep-bass nuts will definitely want this disc, because there are some genuinely astounding sections—both musical and wildly synthesized—that will tax even the most robust subwoofer systems and big-time amplifiers. There are warnings for the bass-demo tracks given in fine print, which indicate that the bass sometimes plunges to 5 Hz. (Nobody can hear this low.) Consequently, those with bass-reflex speakers—and even individuals owning thoroughly robust acoustic-suspension models—should take care.

An Organ Blaster Sampler (excerpts from Bach, Jongen, Widor, Soler, Franck, and others; Michael Murray, playing several different organs), Telarc 80277: There is variable quality on this sampler disc, ranging from good to great. Some of the stuff here has a tremendous impact and will highlight the potential of any world-class sound system.

[5]While Spatializer technology is also available in outboard processors which are similar to the Carver Sonic Hologram device and to some Lexicon surround decoders (performing much like Polk's SDA speakers), the internal-cross-talk-cancelling, software-encoding process is similar to that used by Q-Sound. Both systems allow for a broader sweet spot than what Carver, Lexicon, and Polk offer, although at dead center the focus is not quite as precise. The Spatializer process is different form the NuReality SRS system, with which it is sometimes confused. The latter shapes extracted ambience signals so that they mimic the side- and rear-sound response contouring provided by the pinna of the outer ear.

Soul of the Machine (The Windham Hill Sampler of New Electronic Music), Windham Hill WD-1700: This is a good cross section of this company's material. Every one of the selections have fine tonality, a good left-right sound-stage spread, and exemplary detail and transparency. Most are not realistic in the live-music sense, of course, but they are certainly impressive as works in themselves.

Stereophile Test CD: STPH 002-2 (available from the magazine): This item was produced by the staff of *Stereophile* magazine and contains interesting and educational excerpts from recordings made over the years by Gordon Holt, John Atkinson, Robert Harley, Peter Mitchell, and Brad Meyer. Some of the recorded selections are very fine indeed, and the descriptions are worthwhile. The disc also offers some *very* interesting comparisons of 18 commonly used microphones, although it is too bad that only one B&K and no highly regarded Schoeps units or the Calrec Soundfield model were included. In addition, the warble-tone test section is not much help, since the device used to produce it was not functioning properly during its production (there has been a lot written in some of the other hobby journals about this goof-up). Both correlated and "uncorrelated" pink-noise tones are also available; the latter will be helpful in setting up certain kinds of L – R surround processors and in doing stereo sound-field measurements with an RTA. Some of the comments about critical A/B testing in the accompanying booklet are misleading, unfortunately.

Stereophile Test CD2. STPH 004-2 (available from the magazine): This is philosophically similar to the first *Stereophile* test disc, although the content is somewhat different. New music example tracks are included, with one of the best recordings of acoustic guitar that I have ever heard. The overall playback level seems excessively low (probably a wasted attempt on the part of the producers to eliminate any chance of digital overload), but the chance of hearing any noise from your amp or digital artifacts from your CD player (as the gain of the former is turned up to compensate) is eliminated by the mostly hissy-sounding AAD or ADD source material. While these recordings are quite good in terms of perspective, depth, clarity, and ambience, they lose a lot because of that hiss. (Background noise of this kind seems to be a "badge of honor" for a few current tweak-oriented CD producers, who appear to be obsessed with analog recording gear and tube electronics; according to the insert that comes with the disc, the producers are aware of the noise and state that most of it is the result of the tube microphones.) The highly touted "EAR" microphones used in some of the recordings also seem—at least to my ears—a bit harsh sounding at times, although in the first *Stereophile* test CD I thought they sounded a bit muffled. The pink-noise sections are useful, particularly the uncorrelated stereophonic sequence, especially if you have a microphone-fed RTA to help in placing and equalizing speakers. This section can also be useful in double-checking surround-speaker levels, although an RTA or sound-level meter is still required for proper tweaking. The harmonic-distortion test-tone comparisons are interesting, but their impact is limited, because steady-state "pure" tones are poor tools for evaluating the subjective effects of distortion on complex musical signals. The tests dealing with digital "jitter" are nonsense.

Surround Spectacular (a matrix-surround-encoded two-disc set containing excerpts from the Delos catalog and a series of test sequences), Delos 3179: With the matrix-encoded musical sequences put together by John Eargle, Stephen Basili, and Al Swanson and some spectacular sound effects engineered by Brad Miller, it is hard to imagine a better array of music and system exercisers than this. Just as impressive are the series of surround-sound and woofer tests engineered by David Ranada, of *Stereo Review* magazine. These are very basic (although some work better if you have a sound-level meter) and are well-explained in the enclosed booklet. Unlike some of the surround-sound tests put together by others, this release, with probably the best DPL tests available, will show you what your matrix surround and synthesized-ambience systems are doing.

The Symphonic Sound Stage, A Listener's Guide to the Art and Science of Recording the Orchestra (eleven excerpts from symphonic performances, all engineered by either John Eargle or Marc Aubort), Delos 3502; *The Symphonic Sound Stage, Volume 2* (subtitled *Second Stage* and consisting of more excerpts similar to those just above), Delos 3504: These short passages can help a knowledgeable listener quickly evaluate important aspects of system performance, particularly the speakers and their placement. These are commendable examples of the current state of the two-channel stereophonic recording art (in spite of all being produced prior to 1987; 1989 in the case of Volume 2), accompanied by worthwhile descriptions of the techniques used. While both volumes are excellent educational tools, those wanting top-grade examples of matrix-surround techniques should look to *Surround Spectacular*, above, also produced by Delos.

A Video Standard (a twelve-inch LaserVideo disc designed to "Optimize your Audio/Video System"), Reference Recordings LD-101: This disc, produced by Joe Kane of the Imaging Science Foundation, is equal to thousands of dollars worth of test hardware and, combined with its encyclopedic instruction manual, is a veritable tutorial in video and audio. The color and resolution charts will help you to align a TV monitor expertly, and the audio sections are nearly as helpful. The package comes with a blue filter for use with some of the color charts. (Note, however, that because there is no pause or warning prior to their appearance during the moving program, many of the video "stills" will automatically flash past rapidly during normal play. To lock into them, you must quickly hit the still button on your player and then move back and forth through the sequence of tests as required, by using the player's frame-advance feature.) The audio tests are workable, although not as to the point as those in Delos's *Surround Spectacular*, and some sections move along rapidly, requiring you to engage the repeating "loopthrough" feature of your player if you want to effectively use them. Be prepared to spend time reading the manual carefully. Kane has recently produced a more user-friendly video test disc, *Video Essentials*, that has been favorably reviewed by several knowledgeable critics. Although the test sections may not be as useful to those with elaborate measurement gear as is this disc, it may be a more practical buy for ordinary consumers.

APPENDIX B

Publications

AUDIO, VIDEO, AND MUSIC JOURNALS

WHILE SCIENTIFIC PUBLICATIONS like the *Journal of the Audio Engineering Society* and the *Journal of the Acoustical Society of America* are useful mainly to hard-core tecchies, many consumer A/V magazines are embarrassingly inadequate when dealing with audio and video theory or evaluating the hardware needed to build a fine system. This is particularly true when they analyze room acoustics, loudspeakers, and surround-sound processors. Most of the limited-circulation, "fringe" publications (some of which are nothing more than mimeographed and stapled-together tracts) are even worse, being notoriously inept when it comes to evaluating amplifiers, CD players, and other electronic hardware. However, on occasion even the worst of the fringe publications will contain something worthwhile (especially when it concerns trends and industry gossip), and nearly all of them have workable software reviews of audio recordings, prerecorded videotapes, and laser discs.

Consumer-oriented audio journals for the most part can be divided into two categories: equipment magazines that assign record reviews to secondary status at the back of the magazine and music-oriented or record-review magazines that may or may not have a column or two on audio in each issue. My experience has been that the publications that stress record reviews over hardware reports have the most scholarly software reviews and also often take a surprisingly down-to-earth approach to hardware. The equipment-oriented publications are variable, with some being very measurement oriented and hard-headed and others taking a less-grounded approach. Some, like *Audio* and *The Sensible Sound*, split the difference—with variable results.

Consumer-oriented video and home-theater publications span the same philosophical extremes as those devoted to audio. While some are level-headed and useful, a fair number, particularly those promoting high-end equipment, are less reliable, particularly when it comes to audio. This is especially true of some high-end home-theater publications, many of which also feature picture essays on assorted high-end systems installed by various "certified" experts, which stress visual aesthetics over performance.

Nonetheless, staying current with the better publications can help you keep up to date with the latest trends in equipment and software. Here are some of the audio, video, home-theater, and record-review journals that I've seen. Note that editorial policies (and ownership) can change, so do not be surprised if the journal you pick up on the newsstand does not exactly resemble what I have described. Also included are some music-oriented journals that regularly contain at least a few record reviews.

The Absolute Sound. (Irregular.) An alternative-press bible for audio-equipment extremists and diehard subjectivist buffs. Interestingly, while often out in left field in its equipment evaluations, *TAS* usually has pretty good record reviews that stress artistic qualities as well as sound.

American Record Guide. (Bimonthly.) A record-collector's journal, the *ARG* traditionally has reviews that are impeccable for evaluations of aesthetic standards but less revealing when it comes to sound quality. When *High Fidelity* magazine folded a few years back, its companion journal, *Musical America*, was nearly lost. It has now merged with the *ARG*, and the combination has become a must for serious record collectors and serious lovers of classical music. They occasionally run technical articles as well.

Andrew Marshall's Audio Ideas Guide. (Quarterly.) This Canadian-based journal is mainly equipment oriented, but a few record reviews are included in each issue. There is a lot of pseudoscience in the equipment commentaries, particularly those dealing with CD players and amplifiers, and the speaker-system reviews cater to the "direct-sound" crowd rather than to those who listen to more distantly placed speakers in typical, non-acoustically treated living rooms. In spite of this, the record reviews are usually quite to the point in terms of aesthetics, and the reviewers also spend some time discussing sound quality.

Audio. (Monthly.) For years, this magazine was an audio-enthusiast journal that attempted to bridge the gap between consumer-oriented literature and strictly technical material. While this was certainly a worthwhile goal, they did not always achieved the proper balance, sometimes being too technical, other times applauding current trends that failed to pan out. Now under new editorial control, the philosophical position of the magazine has improved somewhat. The magazine has a few record reviews in each issue, and while the aesthetic competence of the critics is variable, the reviews nearly always include comments, which are usually accurate, on sound quality.

The Audio Amateur. (Quarterly). A well-known "must read" for serious do-it-yourselfers who can read schematics and drawings and like to build things, but who also have a true-believer mindset about audio. Because the articles are not edited, the magazine occasionally combines technically accurate material with articles of lesser merit.

The Audio Critic. (Irregular.) Mainly hardware oriented, with a desire, like *Audio,* to span the gap between technical-only and consumer-oriented journals, the magazine is an often continuous polemic against the lunatic fringe in audio—and sometimes video. Most of the technical essays are very accurate and informative, although the stress on "upscale" gear, instead of lower-priced equipment that would often perform as well, is a weakness. The record reviews are sound-quality oriented and are informative and accurate.

Audio-Video Interiors. (Monthly.) Touted on the cover as "the home theater authority," this journal is very much home-decor oriented, with many picture essays dealing with expensive and photogenic systems belonging to the rich and (sometimes) famous and lots of impressive ads and equipment picture spreads. Even so, it still has space for some interesting and informative essays on new and existing technology, interviews with significant personalities, a few software reviews, and even an occasional equipment test. The home-decor spreads are often pretty, but some of the systems, in spite of their often being expensive and installed by individuals with CEDIA (Custom Electronic Design and Installation Association) credentials, are unsatisfactory for good A/V performance.

Car Stereo Review. (Monthly.) Strictly for car audio buffs, the magazine includes equipment reviews by individuals who definitely know their stuff, as well as construction tips and in-depth analyses of assorted elaborate and comprehensive installations. If you like car or truck sound systems, you will want to subscribe to this magazine.

CD Review. (Monthly; once called *Digital Audio.*) Although always a software-oriented journal, at one time the equipment reviews by the likes of Ken Pohlmann (on digital gear) and Dave Moran (on speakers) were very helpful. However, recent issues have little in the way of helpful information on hardware, and most of the so-called technical essays have been nothing more than product description, "catalogue-with-commentary" pieces. CDs are reviewed for both performance and sound quality, each rated on a dual 10-point system. The magazine strives to look hip, but for me the result is a hard-to-read format.

The Cinema Laser. (Irregular.) Small in scale and dedicated to videodisc reviews and industry information (including occasionally analyses of manufacturer philosophies and pricing structures), this little newsletter is similar in layout to the *Laser Disc Newsletter*, but with fewer reviews, particularly of musical material, and no classified ads for disc and A/V equipment and no regularly appearing general index. However, the subscription price is substantially lower. There is little in the way of equipment commentary, but the reviews are to the point and deal with both technology and aesthetics. Some of the information can be found on the journal's Web site.

Classic CD. (Monthly.) This magazine rarely has much material dealing with audio hardware. However, each issue comes with a "sampler" CD that contains a dozen or so excerpts from the most significant recordings reviewed in the issue. In addition, many other discs are reviewed, with the total usually topping 200. As with most such journals, the comments on sound quality are something that must be searched for as the review is read. It also contains many outstanding essays on the musical scene as well as interviews with notables in the industry. The layout is sometimes a bit confusing, with "glitz" often taking precedence over readability.

Consumer Reports Magazine. (Monthly.) The most visible general consumer-products testing organization in the USA, Consumer's Union often tests audio and video gear in the low-to-medium price category and usually does so with all the expertise that their well-equipped and well-staffed laboratories allow. While occasionally superficial, ostensibly to keep from confusing their less technically educated readers, the tests on most electronic hardware usually tell you most of what is required to make an intelligent purchase. However, I have never felt that they did an adequate job of testing surround-sound equipment—although when they do test it, they will usually have an informative essay or two on design and performance principles. Their speaker test ratings, based on probably the best sound-power measurement procedures in the industry, work to best effect as starting points. If they rate a system low, it is certainly something to avoid. However, if they rate a system high, the prospective shopper should make it a point to very carefully compare it with others rated similarly, because two systems with substantially different radiation patterns can still score close to each other in a sound-power test (see Chapter 2). If there is any philosophical weakness in the magazine's editorial policies, it stems from their sometimes apparent pleasure in downgrading products that have received rave reviews from assorted hobby-oriented journals.

Consumer's Research Magazine. (Monthly.) Oriented towards consumer products of all kinds, this outfit often tests medium- and low-priced audio and video gear. While workable as a starting point when shopping, I have found that the tests are often less than rigorous and tend to leave the reader in the dark as to why a specific piece of gear has been given a certain rating. Notably, each issue usually has a short section of record reviews, which are typically incisive and accurate and always rate the performance and sound quality.

Downbeat. (Monthly.) No home-playback hardware is dealt with here at all, but there are a few record reviews in each issue and those, although not heavy on technical analysis, are artistically informative. In addition, there is a wealth of material on the live-music jazz scene, and the magazine contains many excellent interviews and essays.

Fanfare. (Bimonthly.) Although primarily a journal for serious record collectors, for some time this magazine had several equipment-oriented columns that were written by individuals who set new standards for irrational opinions on audio. However, hardware is stressed very little now (the publisher has learned his lesson), and the current equipment critic usually seems levelheaded. The record reviews are similar in format and expertly written style to those in the *American*

Record Guide. The comments about recorded sound vary from none to substantial, depending on which of the many reviewers does the write-up. A normal issue of *Fanfare* has hundreds of reviews.

Gramophone. (Monthly.) This British journal caters to record collectors and is full of reviews that are excellent in terms of performance aesthetics. The sound-quality comments must be hunted for, but they are usually quite cogent and meaningful. There are occasional essays on audio principles, but the sections on specific audio hardware are limited and mainly of use to British buffs, although US- and Japanese-made products are also reviewed.

Hi-Fi News & Record Review. (Monthly.) Another British journal, *HFNRR* is more equipment oriented than *Gramophone* and has record reviews that are along the lines of what is found in the journals *Audio* and *Stereo Review.* The magazine has even produced an interesting test disc (see Appendix A). While the equipment-evaluation reports done are often less rigorous than they might be (with some wandering deep into the realm of speculation), it is a fairly good buy for music lovers wanting information on sound recordings.

High Performance Review. (Quarterly.) This journal contains excellent record reviews by musically knowledgeable critics (usually over 100 per issue), combined with equipment reviews of limited value. Hardware is subjectively evaluated on the basis of supposedly easy-to-identify differences heard when the same recordings are played with different CD players or amplifiers, even though the magazine's own published test results show that these pieces of equipment are functionally close to identical. Their speaker tests cater to the "direct-field" school.

Home Theater. (Monthly.) Touted on the cover as the "ultimate audio/video experience," this magazine caters to those who want to build or improve upon a home-theater setup. While it has the usual product-list articles, it also often has very informative technical pieces and test reports that I recommend highly. Software reviews are usually limited in scope, with the stress on videodiscs, but there will occasionally be compilations and/or reviews of music videos available on both disc and tape.

Home Theater Technology. (Monthly.) Bigger and glitzier than *Home Theater* (but no more expensive), this journal is also a bit more comprehensive, with essays on equipment and new technology (both audio and video), interviews, picture stories on various custom installations, and equipment evaluations—although little in the way of software reviews. While some of the technical material is very informative, particularly when dealing with new technology, the equipment reviews are variable. The video gear is well handled, but there are often notable problems in dealing with the subjective performance of amplifiers, receivers, and CD players and the audio performance of videodisc players.

International Audio Review Hotline. (Irregular.) Basically a vanity-press religious-like tract that is even more extremist than *Absolute Sound.* For diehard subjectivists only.

Jazz Times. (Monthly.) Not hardware oriented at all, this journal surveys the world of current live-music jazz. Excellent essays and personal interviews are common.

The recorded-music review section is not particularly large, but the writings are to the point and expertly handled. Needless to say, little is mentioned about sound quality.

Laser Disc Newsletter. (Monthly.) If you collect videodiscs, or even if you simply rent them, seriously consider subscribing to this straightforward, non-glitzy publication. Nearly every decent videodisc that is produced gets a review, and a cumulative index makes it easy to find reviews of older releases. While picture and sound quality is dealt with expertly (program content usually gets less coverage), there is none of the usual pontificating about the high-end equipment "required" to enjoy the presentation. You will not be confronted with long-winded editorials designed to make you ashamed of your modest system—or shamed into purchasing something more elaborate and expensive.

Musician. (Monthly.) Not hardware oriented, this publication contains information about the current pop and jazz music scene, with fine essays, interviews, and announcements. The record reviews, although musically comprehensive, typically stress the artistic over the technical.

Notes: The Quarterly Journal of the Music Library Association. (Bimonthly.) This academically oriented journal is mainly found in music libraries. It features summaries of the record reviews by many of the other magazines listed here, serving as an excellent reference guide; sound quality is sometimes noted in the short summaries. The magazine also contains many fine reviews of books on music and music-related topics.

Opera. (Monthly.) This fine British journal deals mainly with information about the world of live-performance opera. However, it does contain a few record reviews, which are written by individuals with outstanding credentials for evaluating the artistic merits of a recording. Unfortunately for sound buffs, the stress is rarely (if ever) on the technical aspects. If a recording sounds "OK," that is good enough. No hardware is reviewed.

Opera News. (Monthly or biweekly, depending on the time of year.) This is another opera specialty journal, which usually has only a couple of pages of lengthy recording reviews in each issue. The few that are done are written by experts on opera who ordinarily have only a passing interest in the sound quality of what they review, provided it meets rather minimal standards. Sound, when mentioned, usually gets only a brief "fine sound" comment, and home-audio equipment is never dealt with. While the stress here is mostly on opera, there are also frequent reviews of material by the likes of Gershwin, Porter, etc.

Opera Quarterly. A fine music journal, containing basic and detailed information about the world of live-performance opera. In addition to reviews of performances, interviews with opera personalities, historical discographies, and a few in-depth scholarly reviews of opera recordings are featured. Like the other opera journals, the stress is on performance and not on sound quality; home-audio hardware is not to be found.

Perfect Vision. (Discontinued, but still possibly available in some high-end audio-video stores.) Originally produced by the same people who publish *The Absolute*

Sound, this magazine was aimed at home-theater enthusiasts with a decidedly upscale outlook. Nearly always informative in the video realm, mainly because of guest essays by video guru Joe Kane, the magazine was firmly entrenched in the subjective extremist camp, particularly when analyzing the audio performance of video gear.

Request. (Monthly.) A pop-music magazine (stressing rock and country) dealing mostly with the current live-music scene via interviews and essays. The record reviews are strictly pop oriented, and although they cover the music's artistic merits (or lack thereof) quite well, they deal very little with sound quality. Music videos are also reviewed. The occasional essays on equipment are shopping oriented and not of much use to those wanting hard information.

The Sensible Sound. (Bimonthly.) Like most small audio-enthusiast magazines, the stress here is on equipment. While the magazine presents itself as a balanced alternative for frugal high-end readers, closer reading shows that it is often nearly as subjectively anti-measurement as *Stereophile.* While hard-headed material does get printed, much of the time the stress is on mystic audio—with amplifiers, CD players, and even cables said by some of the magazine's more imaginative writers to have a major influence on sound-staging, imaging (both horizontal *and* vertical), and the reproduction of ambience, among other things. Careful reading of the more rational material can pay off, however. The magazine does have a few record reviews, and those are pretty good from an aesthetic standpoint and also deal quite well with technical qualities.

Sound and Image. (Previously bimonthly; merged with *Video* in 1995.) Although no longer currently available, back issues of this journal can be found in some libraries. It has consistently been one of the best consumer-oriented publications of its type available, with technically excellent, well-written material on both video and audio hardware and home-entertainment technology. The merger with the larger-circulation *Video* magazine was a productive one, because the two publications were somewhat alike in their reader orientation—although *S&I* always seemed superior in terms the technical competence of most of its writers.

The Speaker: The Magazine of the Boston Audio Society. (Irregular.) Mostly concerned with information on the audio scene and the industry itself, this fine, little, alternative-press magazine often has very important data on current events and audio (and occasionally video) equipment, along with interviews and personal observations. A sizable percentage of each issue is set aside to summarize the monthly meetings of the Society. The meetings themselves often have interesting guests, summaries of assorted audio shows by sharp-witted members, and equipment evaluations or comparisons. Many consider the BAS to be a group of hard-headed skeptics with a low tolerance for the mumbo-jumbo propagated by tweak-magazine journalists, and for the most part this is true—although a fair number of true-believer types are also members of the Society. The magazine itself (and the Society) claims to have no particular point of view and was set up to be a forum for all the members, no matter what their prejudices. However, because most members are engineering oriented and often quite well-educated, the rationalist viewpoint tends to prevail. When record reviews appear, they are usually done as one-time write-ups by a guest writer or staff member and stress engineering analyses over aesthetic qualities.

Speaker Builder. (Bimonthly.) Published by the same organization that gives us *Audio Amateur*, this how-to magazine is generally less technical than its parent publication, simply because of the nature of loudspeakers as opposed to electronics. Its content varies from eccentric to remarkably good, so the reader must be able to pick the wheat from the chaff. There are many obtuse "parts-measure-the-same-but-sound-different" commentaries, along with less-than-enlightened original speaker designs, but these are often combined with rewarding and interesting material on renovating older speakers and building workable new ones. There will occasionally be a useful technical piece, and decent articles on woodworking are a regular feature. The magazine is also a good source for parts. Overall, an interesting read for knowledgeable enthusiasts with a penchant for working with their hands.

Stereo Review. (Monthly.) Like *Audio*, this is primarily an upscale enthusiasts' magazine, although nowhere near so much so as most of the fringe and so-called elitist journals. However, where *Audio* tries to be fairly technical at times, *SR* puts the stress on consumer needs and basic principles. Consequently, the technical articles tend to be fairly rudimentary, and even people just getting started in the hobby should have no trouble understanding them. The review and essay sections on audio hardware are usually quite good—although somewhat limited in the depth of analysis at times, no doubt to keep from scaring off neophyte readers. The regular editorial comments by the equipment tester Julian Hirsch, as well as the summaries on surround-sound hardware by David Ranada, should be required reading by all audio enthusiasts, particularly those just getting started. Each issue will have a fairly good crop of record reviews in the back pages, and sound, as well as artistic, quality is always at least mentioned. This is the largest-circulation audio magazine.

Stereophile. (Monthly.) Doubtless, this is *the* major "audio subjectivist" magazine. While its record reviews are first-rate and quite helpful, the technical material usually leaves a bit to be desired. The stress is on subjective evaluations, as opposed to the more instrument-oriented measurement and double-blind listening tests I prefer. Frequently, test reviews of equipment will feature elaborate instrument-oriented testing procedures that look both professional and impressive. However, the subjective interpretations of those measurements, especially when dealing with amplifier and CD-player performance, often appear to have no relationship to the measured data at all. Nevertheless, the industry information features and newsletter material are a must for enthusiasts wanting to keep up with the latest gossip, and the "letters" section, which has material from both the fawning and technically ignorant at one extreme and nonplussed engineers and skeptical college professors at the other, is usually more entertaining than most TV sitcoms.

Stereophile Guide to Home Theater. (Quarterly.) An outgrowth of the parent magazine, but mainly addressing video and home theater. Lots of good software reviews are a highlight, as are interesting and often very worthwhile essays on video and surround-sound hardware and satellite and videodisc systems. The magazine also has some very good introductory and advanced pieces on surround-sound technology and room-system interactions, as well as summaries of seminars on video technology. The area of weakness, as you might expect, involves audio-hardware performance.

Stevenson Classical Compact Disc Guide. (Quarterly.) Although lacking any printed material on audio or video equipment, this large compendium of excerpted reviews is jam-packed with the latest word on audio recordings, with each issue listing both summarized performance and sound-quality information. In addition, short original reviews by the magazine's staff are included, as is a simplified list of discs that received outstanding ratings. This journal is a must for individuals who are looking for an index to longer reviews of classical audio recordings (and some videos) as well as a quick-reference guide for shopping.

The Strad. (Monthly.) Another fine British magazine that stresses the world of classical-music stringed instruments. It contains scholarly essays on the subject, plus discographies, interviews, and a smattering of lengthy and well-written record reviews—but no information on hardware. Like most music-oriented journals, the emphasis is on performance and not on engineering. The reviews deal mainly with recordings that feature the sound of stringed instruments.

Tempo. (Quarterly.) Still another fine British music magazine that contains serious essays and interviews dealing with contemporary music theory, but no write-ups on A/V hardware. The record reviews are limited, but they are artistically revealing. Unfortunately, little is said about sound quality in those reviews.

Video. (Monthly.) This is probably the major video magazine now in print and the video "counterpart" to *Stereo Review.* For some time it was conspicuous for the decent quality of its video-hardware reviews and for the subjective sloppiness of its audio-hardware reviews—including the audio performance of the video hardware. With the incorporation of *Sound and Image,* the magazine appears to have adopted some of the latter magazine's balanced outlook, and current issues have been very satisfactory. While the stress is on video, audio is treated in depth, with a strong emphasis on surround sound. In addition to often excellent equipment tests, there are regular pieces on video and audio theory, along with the usual material on new and current products available. Each issue also has a few disc and tape reviews, with a good analysis of their technical merits.

Widescreen Review: The Journal of the Widescreen Laserdisc Home Theater Experience. (Bimonthly.) Its title notwithstanding, this a basically a high-end audio journal (containing the usual eye-opening theories about hardware and psychoacoustics) that includes material about video hardware and video performance, especially wide-screen monitors and wide-screen videodisc formats. Each issue has plenty on the picture and audio quality possible with current and upcoming technology, industry gossip and news, and detailed software reviews. The letters section is often huge (sometimes funny, sometimes caustic), another tweak-journal characteristic. The sections on high-end video technology and performance requirements, written on occasion by video guru Joe Kane, are frequently revealing. However, sporadic pieces deal with such hair-splitting topics as viewing-room color schemes or the best color temperature for behind-TV-set lights. Hardware eccentricities notwithstanding, this colorful journal is a worthwhile item if you want to keep up with trends and industry gossip—and the software reviews are quite informative.

Glossary

PERHAPS ONE OF THE MOST INFURIATING THINGS about conversing with audiophiles is their use of seemingly obscure terminology. In order to better understand the material in this book and in any articles that you might read in both popular and technical journals (and to better understand the often-arcane conversations of dedicated enthusiasts and not a few salesfolk), a glossary has been provided.

Most of these terms and abbreviations have been analyzed in the book fairly well. This additional information offers brief summaries for purposes of clarification. In addition, there are also some general-purpose audio and video terms listed that may be of interest to individuals who want to know more about the basics. Common acronyms are given their full-form designations and sometimes further defined on the spot. Less common acronyms are referenced to their more common full-form entries.

AAD Analog/Analog/Digital. A designation found on some compact discs, indicating that the program was recorded with an analog recorder, edited with an analog mixer, and transferred to the digital medium for playback.

ABX Comparator A proprietary switching device, invented by Arny Krueger, designed to make rigorous double-blind tests easier to do.

AC Alternating Current. The standard electrical power available from typical U.S. power outlets. The current flow alternates direction, usually 60 times per second. AC can be affected by capacitance and inductance, depending on frequency, and is also affected by resistance. *See also* DC.

AC-3 *See* Dolby Digital.

A/D Analog to Digital. Refers to the conversion of analog sound or video to digital during storage, manipulation, or recording.

ADD Analog/Digital/Digital. A designation found on some compact discs, indicating that the program was recorded with an analog recorder, edited in the digital domain, and transferred to the digital medium for playback.

AES The Audio Engineering Society. A professional audio society with members throughout the professional, manufacturing, and educational community. They publish the *JAES* (*Journal of the Audio Engineering Society*).

AFM Audio Frequency Modulated.

AM Amplitude Modulation. A radio transmission technique that conveys data by varying the strength of a high-frequency carrier signal. AM radio, while capable of being high-fidelity, is rarely configured that way.

ATRAC Adaptive Transform Acoustic Coding. The low-bit-rate data-reduction encoding process used in the Sony-developed MiniDisc system. *See also* **Data reduction.**

ATV Advanced Television. The new digital-video spectrum assigned to handle HDTV and standard-resolution formats. The space allocated for one HDTV signal can also be used to handle several standard-resolution programs.

A/V receiver *See* Receiver.

Acoustic-suspension speaker A sealed-box system that makes use of the air behind a woofer to control cone movement. Originally conceived by Harry Olson many decades ago, this woofer system design was refined and put into use by Edgar Villchur, making it the foundation of his company, Acoustic Research, in the 1950s and '60s. *See also* **Infinite baffle.**

Active crossover A powered electronic network that divides up the frequency constituents of an audio signal (bass, midrange, and treble) before it is amplified and sent to the various drivers in a speaker system. While active crossovers are often contained within subwoofer enclosures along with the bass driver(s), those that work with multiway systems may also be outboard mounted.

Ambience As a general audio term, ambience is the background-sound quality of a listening room, surround processor, and/or recording. The ambience of a recording is what gives it space and a sense of realism. It is the sound of the "hall" or recording studio itself. Ambience is often synthetically added by the recording engineer if the recording environment was not reverberant enough to do the job naturally. *See also* **Sound field.**

Ambience extraction The use of left-minus-right dematrixing and signal rerouting and (sometimes) delaying techniques to send hall ambience or reverberation already present on a stereophonic audio or video recording to surround speakers to simulate a concert-hall effect. *See also* **Matrixing; Dolby Surround.**

Ambience synthesis The routing of delayed, processor-modified or processor-created ambience signals, in addition to those already on a recording (even one that is monophonic), to surround speakers to simulate a concert-hall effect. *See also* **DSP.**

Ambisonic recording A surround-sound recording technique practiced by Nimbus, Hyperion, and a few other companies. *See also* **Calrec Soundfield microphone.**

Amp Ampere. A measurement of electrical current. This term is also sometimes used as an abbreviation for amplifier.

Amplifier A device (sometimes called an "amp") for boosting the amplitude of a given electrical signal; ideally, without affecting its quality.

Analog signal The exact electrical or mechanical replica of any particular audio or video input to a system. Any signal originally produced by nondigital recording equipment, even though the finished item may be a digital audio disc or a digitally compressed video signal. Note that no matter what the recording medium, the sound or picture we ultimately experience is analog. We live in a subjectively analog world.

Analog videodisc *See* LV.

Anamorphic In video and film, a wide-screen process of recording images so that each frame is horizontally compressed ("squeezed") on a videodisc or strip of film. During playback via a theater projector (by means of a special lens), from a disc player (done electronically), or within a TV set (also done electronically), the image is reciprocally expanded to restore its shape to normal. Anamorphic expansion can best be accomplished in the video realm if the playback monitor is a wide-screen model. The best-known anamorphic film process is CinemaScope, which applies an approximate 2:1 compression–2:1 expansion.

Anechoic Without echo. An anechoic situation exists when acoustic signals produced by a source are not reflected back to it or anywhere else. Room reverberation does not exist under anechoic conditions. Most recordings are not recorded anechoically, because the sonic signature of the environment is a part of what the engineer will want to record. Note that because the ground is reflective, a true anechoic condition would only exist fairly high up off the ground and outdoors. A skydiver experiences anechoic conditions. *See also* **Ambience**.

Antenna A device for receiving radio-frequency (RF) signals from a source and making them strong enough to be handled by a tuner, television set, satellite receiver, etc.

Articulation As commonly used to describe recordings, articulation refers to the clarity and inner detail of the assorted instruments of a recorded ensemble. Regarding hardware, it refers to the ability to delineate the material on recordings.

Aspect ratio The width-to-height ratio of a television screen, letterboxed image on that screen, or motion-picture theater screen. Typical TV sets have a 1.33:1 (4:3) ratio, while wide-screen versions have a 1.77:1 (16:9) ratio. Modern motion-picture ratios run the gamut from 1.66:1 to 2.76:1, and these will often be the ratios used when images are letterboxed to a conventional-ratio TV screen.

Attenuate To reduce in amplitude.

Audiophile A person who has an enduring interest in audio.

Azimuth In audio, the angle between the magnetic gap of a tape head and the direction of travel of the tape, ideally 90 degrees. In video (VCR) use, it involves the angle at which a tape-head gap intersects the scan movement. This angle between the direction of the rapidly moving head and the slowly moving tape will vary, depending on the nature of the signals and the positioning of the various video and hi-fi audio heads.

Bandwidth A range of frequencies. With audio recordings, bandwidth refers to a sound system's or recording's ability to capture the frequency-response range of the ensemble and soloists. With regard to a home playback system, it refers to the "audible" bandwidth the system should be able to reproduce, usually from 20 or 30 Hz up to 15 or 20 kHz. *See also* **Frequency**.

Bass The low-frequency range of the audible spectrum, running from 20 Hz (or a bit lower) up to anywhere from 200 to 500 Hz, a total of four octaves or more.

Bass reflex A speaker-box design that makes use of a port or drone cone that, according to parameters outlined by Thiele and Small more than twenty years ago, allows the rear radiation of a woofer cone to reinforce the output of the front, extending and smoothing low-range response. At frequencies below the reinforcement range, there will be a sharp cutoff as the port signal goes back out of phase with the front.

Betamax Also called **Beta**, this is the original home video recorder pioneered by Sony in 1975. More advanced versions are SuperBeta and ED Beta. (The latter is not record/playback-compatible with SuperBeta or any of the earlier versions.) While some Beta recorders are still available, the JVC-promoted VHS system has become the dominant format for everyday home video recording. *See also* **ED Beta; VHS**.

Biamping Using separate amplifiers to power the crossover-separated drivers in a speaker system. When a powered subwoofer is added to a system, the latter automatically becomes biamped, with the satellites separately amplified from the subwoofer. With the right speakers, biamping can boost the output capabilities of a sound system considerably.

Bias An inaudible, high-frequency signal combined with an audio signal recorded on analog tape to magnetize it properly and reduce distortion. The factors that determine a particular bias level and frequency are the tape-head gap, the tape formulation, and the recording speed. Ordinarily, increasing the bias level will lower distortion at the expense of a bit more noise and reduced high-frequency response. Reducing the bias level will lower the noise floor and flatten out the high end a bit, but at the expense of higher distortion. This tradeoff does not exist with digital tape recorders.

Bipole loudspeaker A speaker system with drivers facing front and rear that are wired in phase. Because of this, their signals do not generate out-of-phase cancellation effects, and side radiation is not radically attenuated. Bipoles should be placed away from the front wall so that their rear-facing signals can be properly reflected. *See also* **Dipole**.

Bit An abbreviation of "binary digit." A bit is a single digit in a binary number. *See also* **Byte**.

Bitstream processing This form of digital processing is used in most of the new compact-disc and laser-video players and involves sampling at extremely high rates (also called single bit,

MASH, pulse-width modulation, pulse-density modulation, etc.). While bitstream processing reduces low-level distortion, its main advantage is cost savings for the manufacturer—and hopefully you.

Blend When used in reference to audio recordings or playback systems, blend relates to the smooth interaction of assorted instruments or singers within a recorded ensemble.

Boston Audio Society The oldest national audio hobby society in the United States, with membership from around the country and world; they publish an influential journal.

Boundary effects The wave cancellation and reinforcement effects that exist when audio signals interact with a room, its furnishings, and even the speaker cabinet itself. In a recording studio, boundary effects will color the sound that is received by the microphones. Sometimes this enhances the sound; sometimes it does not. *See also* **Comb filtering**.

Bright A subjective term to describe a recording that has a lot of audible high-frequency energy.

Byte A byte is the number of bits necessary to encode one character of information in any given computer system, including digital video and audio systems.

CAV Constant Angular Velocity. CAV discs rotate at the same speed throughout their playing time. A feature of "standard-play" laser videodiscs, CAV allows sharp and steady freeze-frame and slow- and fast-motion video (but not audio) playback with standard laser-video players. A major disadvantage of CAV videodiscs is their short playback time. *Compare* **CLV**.

CD Compact Disc.

CD+G Compact Disc, plus Graphics. This format stores still images, graphics, and textual material in addition to audio. A special player, decoder, and TV monitor are required to enjoy this format.

CD-I Compact Disc, Interactive. This format stores video, graphics, text, and audio, with the user in control of the way this material is displayed. A special player/decoder and TV monitor are required to enjoy this format.

CEDIA Custom Electronic Design and Installation Association. A national dealer organization that requires its members to have at least two years experience and be licensed and insured. While not exactly a degree in home audio-video, CEDIA certification at least means that a dealer has some basic knowledge about audio and video. However, I have seen installations by CEDIA members that were much less effective than they could have been, probably because the customer was more interested in visual aesthetics than performance.

CLV Constant Linear Velocity. CLV discs rotate at different speeds during their playback time, running fast at the beginning and slowing down as they play along. "Extended-play" videodiscs, CDs, and DVDs are all made in this format. The primary advantage of CLV discs is their extended playback time, because full use is made of the available linear space on the disc. The main disadvantage of CLV analog laser videodiscs is their inability to display sharp and steady freeze-frame and slow- and fast-motion video with conventional laser-video players. *Compare* **CAV**.

CRT Cathode Ray Tube. The picture-producing part of a television set. *See also* **Direct-view television set; Projection television set; LCD.**

CX An audio noise reduction process developed by Columbia for use in LP records and FM radio. Not particularly successful at first, it was later successfully used with LV discs.

Cables The shielded copper or fiber-optic interconnecting wires used to connect audio or video components, although unshielded speaker wire is sometimes included in this category.

Calrec Soundfield microphone A specialized, four-capsule, four-channel, coincident-pickup microphone that was specifically designed for Ambisonic recording. The Calrec unit is also a superb stereo microphone and has the additional advantage of being remotely adjustable for pickup pattern, making it easier for a recording technician to adjust for best frequency response and sound-stage imaging.

Capacitance In active or passive AC circuits, a form of frequency-dependent resistance produced by a capacitor. A capacitor will block DC and will, depending on its design (its capacitance), let higher frequencies pass through at differing levels of attenuation, with very high frequencies often not affected at all.

Capstan A rotating, usually metal, shaft in a tape recorder which, in conjunction with the rubberized pinch roller, pulls the tape across the heads. A dual-capstan recorder has capstans at each end of the head block for more uniform tape movement.

Cardioid microphone A microphone designed for picking up sounds mainly from the front and sides, with little sensitivity to sounds toward the rear. The pickup pattern is heart-shaped—thus the name. Cardioid pickup patterns tend to be frequency dependent, making it necessary to carefully place and aim them for good balance. Design variants include the hyper-cardioid and super-cardioid, which have less sensitivity to the side and somewhat more sensitivity to the rear.

Cartridge In a phonograph, the device that converts the mechanical output of the stylus to an electrical signal for the preamplifier.

Cassette A self-contained tape storage and playback device, designed to be used with an audio- or videocassette tape recorder.

Center channel In A/V systems, this is the so-called "dialogue channel" that is located between the left and right main speakers. However, in most video applications, it does much more than reproduce dialogue. In audio-only recordings—which are given Dolby encoding—this channel can add central focus, particularly when you are sitting away from the central axis. While in the Pro Logic version it is "derived" from the identical left and right signals, with Dolby Digital and DTS Digital Surround, the center channel is a discrete source. *See also* **Dolby Digital; Dolby Pro Logic; DTS; Sweet spot.**

Channel In audio, a distinct path for a signal that is being recorded or played back. Standard stereo has two channels. Pro Logic–decoded audio still has two, but they carry two additional "matrixed" channels. Dolby Digital and DTS audio have five full-range channels and a subwoofer channel. In video, a signal transmitted at a particular frequency.

Channel block A feature on some television sets that allows parents to make it impossible for children to watch undesirable programs.

Channel separation In audio, a measurement of the amount of leakage between the various channels in a multichannel installation, specified in dB. While a higher number is better, anything greater than 20 dB (a ratio of 100:1) will be adequate for full stereo separation. *See also* Cross-talk.

Chroma-differential gain In video, a measure of how color saturation varies with scene brightness.

Chroma-differential phase In video, a measure of how color hue varies with scene brightness.

Chroma level In video, a measure of color saturation.

Chroma phase In video, a measure of color hue, usually adjustable with the tint control on a TV set.

Chrominance The color component of a modern television signal.

Class-A amplifier A design in which the output devices of the amplifier conduct current all of the time. These amps have very low distortion but also tend to run hot and normally have fairly low maximum power outputs.

Class-AB amplifier Much more common—and cheaper—than the Class-A type, the output devices of this amp design are set to conduct current only part of the time. While exhibiting more measurable distortion than the Class-A design, the Class-AB amplifier's distortion will still be inaudible, and the amp will run cooler, produce more power, and cost far less.

Clipping In audio, the result of an analog signal's being overdriven to the extent that its peak levels cannot be accommodated, and therefore are "clipped" off from the audible signal. Typical in smaller amplifiers, it is the most audible of common electronic distortions.

Closed-loop drive A tape-recorder drive system in which the tape is pulled by dual capstans on either side of the heads. The result is a very uniform tension and less wow, flutter, and scrape-induced distortion.

Coincident-microphone recording A technique whereby two directional microphones (one for each channel) are located very close to each other and aimed at specific sections of the ensemble to be recorded. This is said to keep timing differences as well as phase and comb-filtering effects to a minimum. Also called Intensity stereo.

Coloration In audio, a subjective term to describe levels of audible distortion.

Color noise The irregular, grainy characteristic that appears in large color areas on all video pictures. The level of noise will vary, depending on the quality of the TV set, the quality of the playback device, and the quality of the source material.

Color temperature A measure of the relative warmth or coolness of a television picture; most often stated in degrees Kelvin. Warm pictures display a reddish cast; cool pictures, bluish. While NTSC specifications call for a certain standard, individual viewers (and manufacturers) often have ideas of their own regarding what looks right.

Comb filter: video A circuit that separates chrominance and luminance signals in a television set or laser-video player to control interference. In many sets, it is digitally implemented. It is superior to the simple "notch" filters found in older and cheaper sets.

Comb filtering: audio The result of two audio signals interacting in such a way that their combined outputs cause the frequency response to become more irregular and choppy appearing—like the teeth of a comb. This can happen when the outputs of two speaker systems (or even speaker drivers with overlapping outputs within the same system) reach the listener's ears at slightly different times. The effect is rarely detrimental unless the alternating peaks and dips are widely spaced. Wall reflections combining with the main signals also cause comb-filtering effects, although the result here is usually an enhanced sense of spaciousness. Indeed, at higher frequencies, comb-filtering effects are usually not unpleasant if the speaker systems are wide-dispersion models and listening is done in the reverberant field. During recording, the comb-filtering effects of widely spaced microphones can be measurably similar to what is reproduced by speakers, but the result may be subjectively more disturbing. Microphone comb filtering is similar to what is sometimes intentionally applied electrically to a monophonic signal to create a pseudo-stereo effect. *See also* **Diffraction**.

Combi player An LV or DVD player that can play a variety of audio and video recordings.

Component input/output High-grade video connection found on some data-grade and high-end monitors and line doublers. These allow suitable input sources to deliver even better video performance than an S-Video hookup. *See also* **RGB input**.

Compression In radio transmissions, the process of making the louder passages a bit quieter (and sometimes, making the quiet ones a bit louder) in order to reduce background noise and increase the effective range of the station. In tape recording, compression is used to mask background noise during the recording process. During playback, a mirror-image expansion of the signal will result in the original dynamics being reproduced—minus the background noise. *See also* **cx; Dolby; dbx**.

Crossover network The circuit that routes the proper electrical signals (highs, midrange, bass) to the various drivers in a loudspeaker system (if it is a passive design) and to the various amplifiers in a biamped system (if it is an active design). *See also* **Active crossover; Passive crossover**.

Cross-talk In audio, the leakage of a signal from one channel of a system to another. A system with low cross-talk will have good separation between channels. In a stereo audio program, a separation of 20 dB (100:1) should be adequate, although in some professional applications a level of up to 60 dB may be required. In video recorders and disc players, the leakage of a signal from one track to an adjacent track.

Current The flow of electricity through a conductor. *See also* AC; DC.

Curve In audio, the representation of frequency over a given range, in relation to a fixed standard of amplitude.

D/A Digital to Analog. Refers to conversion of digital material back to analog during the playback process.

DAC Digital-to-Analog Converter. The circuit that changes binary digital data back to an equivalent analog form so that it can be handled by conventional amplifiers, speakers, or TV monitors.

DAT Digital Audio Tape. DAT recorders, which use a magnetically coded PCM system rather than an optically read one like the compact disc, are divided into two types: RDAT, which has its tape heads attached to a rotating drum to keep linear tape speeds low (the heads are similar to but smaller than those used on video recorders), and SDAT, which uses stationary heads and requires great quantities of tape running at high speed.

dB One-tenth of a Bel. Named in part after Alexander Graham Bell (hence the capital B) and used in both audio and video applications, the number of Bels is the common logarithm of the ratio of two powers. If two powers differ by 1 Bel, the greater one will be 10 times the other. A 100-watt amplifier is 1 Bel, or 10 dB, higher in output than a 10-watt unit. Decibels are ratios, not fixed quantities. While used to describe both video and audio phenomenon, the more common popular use involves the latter. For example, it is said that an individual can usually hear volume changes in the neighborhood of 1 dB, depending on the bandwidth of the manipulated signal. When measuring audio signal-to-noise ratios, the difference between the quietest and loudest sounds is stated in dB. With some kinds of equipment, such as microphones, analog tape recorders, or LP playback systems, the measurement is "weighted" as to audibility, because the ear is more sensitive to some frequencies than to others. Two common corrections for hearing characteristics are the A-weighted and the somewhat more rigorous C-weighted scales, indicated as dBA or dBC, respectively. *See also* **Signal-to-noise ratio**.

DBS Direct-Broadcast Satellite.

dbx noise reduction A system making use of complementary compression and expansion techniques to reduce background noise in analog tape and MTS video systems. It was also used for a limited time in some LP recordings and FM radio transmissions.

DC Direct Current. Electrical energy that flows in one direction only. DC is blocked by capacitance, restrained by resistance, and unaffected by inductance.

DCC *See* **Digital Compact Cassette**.

DDD Digital/Digital/Digital. A designation found on some CDs, indicating that the program was recorded and edited digitally, before being transferred to the final digital format.

DPL *See* **Dolby Pro Logic.**

DSP Digital Signal Processing. Used in both audio and video. In audio playback systems, it is most often used with surround-sound synthesizers to simulate hall, club, or studio ambience. However, it is also used in equalizers and filters, and versions of it are also employed to enhance material produced by Dolby Surround decoders. In video, DSP is used in everything from comb filters to MPEG data compression to line-doubling circuits—with the goal of enhancing picture quality.

DTS Digital Theater Systems. A discrete, 5.1-channel format designed originally for motion-picture use. It is the main competitor of Dolby Digital. *See also* **Dolby Digital.**

DVD Digital Video (or Versatile) Disc. The CD-sized, digital laser-video format that is replacing the old analog laser-video system and may replace the CD as an advanced surround-sound audio-only format.

DVT Digital Video Tape.

Damping: electrical Also called "damping factor," a measurement of a power amplifier's ability to control the motion of a speaker diaphragm after the signal drops to zero. Directly related to the amplifier's output impedance.

Damping: mechanical The mechanical resistance that is applied to a speaker diaphragm to keep it from resonating after the input signal drops to zero. Also applicable to a phonograph stylus.

D'Appolito speaker configuration In this arrangement, three speaker drivers are stacked vertically, with the tweeter sandwiched between two woofer and/or midrange units. This controls vertical dispersion and crossover lobbing for less ceiling and floor bounce, and it often improves focus and clarity. Most THX speakers follow a variant of this design.

Data reduction In digital video and audio transmission or storage systems, a process that eliminates nonvisible or nonaudible aspects of pictures or sound that are not ordinarily perceived because of "masking," allowing a much higher storage density. Data reduction—sometimes called lossy compression—is not the same as data compression. The latter allows the compressed information to be restored to its original status; the former permanently eliminates material that cannot be detected by eye or ear. *See also* **Compression; Masking; PCM.**

De-emphasis A form of equalization used in both analog FM tuners and CD players to reduce noise and distortion in program material that has received pre-emphasis.

Delay Line An electrical circuit designed to delay the output of a given input signal a fixed amount, usually for the purpose of adding a synthesized ambience to the program. This is done in the studio to add artificial reverb to a program and can also be done at home with a surround-sound processor. Dolby Surround also adds delay to help separate the surround-channel sound from that of the main channels.

Dematrixing *See* Matrixing.

Depth In the context of sound reproduction, depth refers mostly to the ability of a recording or sound system to project a sense of front-to-back distance within an ensemble or the sound stage. It may also refer to a sense of depth within the recording environment itself, especially with Dolby-encoded material. *See also* **Envelopment**.

Derived center channel *See* **Dolby Surround; Matrixing**.

Diffraction The deflection of a sound wave by an obstacle in its path. Its wavelength must be short in relation to the size of the obstacle if the effect is to be significant. With loudspeaker playback situations, diffraction effects often manifest themselves as comb-filtering or phase anomalies, most of which are inaudible at normal listening distances.

Diffuse sounding An undesirable quality in a recording or improperly positioned speakers that results in an unrealistically spread-out sound, particularly with centered, solo instruments. *See also* **Phasiness**.

Digital Compact Cassette Philips's not particularly successful, data-reduced, digital-tape format. Audibly equal in quality to the CD, but less convenient to work with.

Digital output On all DAT decks, as well as some DCC decks and CD, LV, and DVD players, this is the coaxial or fiber-optic output that can pass digital signals to outboard D/A converters or surround processors or other digital recorders. While it may be useful as a way to transfer digital data to another recording device for dubbing purposes or to an AC-3 decoder, connecting a digital output to an outboard converter to "improve" ordinary playback sound quality beyond what a typical (even cheap) unit's built-in D/A converter can deliver is pointless and may actually reduce sound quality.

Dipole With regard to loudspeakers, the sound-radiating pattern produced by all flat-panel designs and some surround speakers, including all THX-certified models. The sound is radiated equally from the front and rear, with the two wave fronts out of phase with each other and with the energy radiated to the sides attenuated because of cancellation effects. *See also* **Bipole loudspeaker**. With regard to microphones, another name for the figure-eight design that picks up sound front and rear, with the two signals recorded out of phase with each other and with little energy picked up from the sides. With regard to antennas, a type that receives signals mainly from two opposite directions, with little sensitivity to the sides. Most wire-lead antennas sold with receivers and tuners are dipole types.

Direct field The listening position in a room where the direct sound from a speaker, set of speakers, or live performer(s) is louder than the sound reflected from nearby boundaries. Normally, you would have to be very close to the sound source for this to occur at all audible frequencies. *See also* **Reverberant field**.

Direct-view television set A TV that employs a single picture tube that projects the image upon the inner surface of its flared end. The end of the tube is specially treated, faces the viewer, is rectangular in shape, and ranges in diagonal size from a few inches on up to 40 inches.

Dispersion The ability of a loudspeaker to radiate sound over a given angle. In a microphone, it is the ability of that device to receive sound over a given angle. *See also* **Radiation pattern; Polar response.**

Distortion Any changes made to an original, "clean" audio or video signal, either at the recording end or at the playback end.

Distribution amplifier A powered video splitter that divides an incoming video RF signal for several pieces of equipment (TVs, VCRs) while at the same time amplifying it enough to compensate for losses incurred during the process.

Dither A very low-level amount of random noise that, when added during the digital recording process, decorrelates quantization error by spreading the quantization noise across the audio spectrum, reducing distortion and the sometimes abrupt and unrealistic silence that occurs when PCM digital-audio signals drop to very low levels. Dither allows engineers to record at levels below the least significant bit and the apparent noise floor of the recording system, allowing for better very low-level ambience pickup and a higher subjective dynamic range. Dither can be audible, but it is possible to shape its spectrum so that it is less intrusive. This is a feature of the "Sony Super Bit Mapping" recording process, for example, and a number of other recording companies have similar "20-bit" designs. Done well, these really can give us true 19- or 20-bit performance from the 16-bit PCM system employed with the CD, although with nearly all music the subjective improvement is marginal.

Dolby Digital The discrete 6-channel (OK, 5.1: five, plus subwoofer) digital surround-sound system designed by Dolby and employing its AC-3 digital coding. While primarily a theater and video-sound format, the process is also workable for audio-only programs. *See also* **DTS Digital Sound; Dolby Surround Sound.**

Dolby HX Pro A special circuit in analog tape recorders that uses the recorded signal's high frequencies to simulate high-frequency bias. This feature automatically lowers the recorder-generated bias to reduce distortion and improve headroom at high frequencies. Unlike Dolby B, C, or S, this system is not complementary and does not require special decoding during playback.

Dolby noise reduction A noise-attenuating system that makes use of complementary compression and expansion techniques over specific frequency bands to reduce background noise in analog tape systems. Dolby A and SR are wide-band systems for professional use. Dolby B offers about 10 dB of noise attenuation above 4 kHz. Dolby C works above 1 kHz and increases the attenuation to about 20 dB. Dolby S gives about 24 dB of noise reduction.

Dolby Pro Logic Sometimes abbreviated DPL, an enhanced version of Dolby Surround Sound that employs analog or digital "steering" circuitry to enhance surround effects and also provide a signal for a center-channel speaker. *See also* **Steering.**

Dolby SR-D Identifies 35-mm film releases that incorporate both a standard 4:2:4 Dolby matrix soundtrack (in analog form, as compared with the PCM digital version used with some LV discs) and the AC-3, Dolby Digital soundtrack.

Dolby Surround Sound Four-channel ambience-extraction, derived-center-channel system used in theaters and home audio-video systems to provide three-dimensional effects. *See also* **Hafler circuit; Matrixing; Dolby Digital.**

Dome driver A common design for tweeters and occasionally midranges that uses a hemispherical radiating surface instead of a conventional cone. Its advantages are low mass, rigid structure, high power handling, and wide dispersion, given the voice-coil size.

Doppler distortion The frequency shift caused when a high-frequency signal is being reproduced by the same speaker driver that is also reproducing a signal at a lower frequency. Doppler (sometimes called FM) distortion may be audible with certain test tones but is rarely heard with musical material.

Driver An individual speaker element in a loudspeaker system.

Drone cone *See* **Passive radiator**.

Dropout In audio or video tape recording, the result of a coating defect or a dirt deposit on the tape. This creates a momentary discontinuity in the played-back signal. These effects are more audible or visible at lower tape speeds.

Dry sounding In a recording, this refers to a lack of hall reverberation and ambience. Under some conditions, and with some kinds of music, this may not be bad. Under most conditions, especially when large-scale ensembles are performing, it is not a desirable quality. *See also* **Ambience; Reverberation.**

Dubbing Copying a recording from one audio or video recorder to another.

Dynamic range The relationship between the loudest and quietest parts of a live- or recorded-music program. The technical definition is the total harmonic distortion, plus 60 dB, when a device reproduces a 1-kHz signal recorded at –60 dB below maximum. (Example: THD + N of –25 dB plus 60 dB = a dynamic range of 85 dB.)

ED Beta The professional-grade Beta format produced by Sony that is similar in concept to S-VHS but somewhat higher in quality. Unlike the latter, which is partially compatible with standard VHS, ED Beta is not adaptable to the older Beta or SuperBeta systems. *See also* **Betamax; VHS.**

EIA Electronic Industries Association.

EP In VCR parlance, Extended Play. Sometimes called **SLP** (Super Long Play).

EPL External Processor Loop. Essentially a relabeled tape loop within a preamplifier, integrated amplifier, or audio receiver. Its function is to allow the easy installation of outboard-mounted signal processors.

Early reflections With regard to room acoustics, the reflections that arrive within a few milliseconds of the original sound. Depending on the direction from which they are coming, they can either add spaciousness to the sound or muddy the detail. With regard to DSP, they are the electrically delayed signals that a processor creates to simulate smaller concert halls (or the reflections close to an ensemble in a larger hall).

Echo These are reverberation artifacts so spread out in time (especially the initial reflection) that the reflected signal is perceived as a distinct sound. A distinct echo is usually not desirable, unless a recording was made in a reverberant space, such as a very large church.

Efficiency The ability of an audio device to turn mechanical energy to electrical (microphones, phonograph cartridges) or vice versa (loudspeakers, amplifiers). For example, the more efficient a loudspeaker is, the louder it will play with a given input. A typical acoustic-suspension speaker may be anywhere from 0.5% to 2% efficient; some horn speaker systems surpass 20%. The leftover energy is dissipated as heat. Under most conditions, efficiency has little to do with sound quality, but with speakers, high efficiency allows one to use a lower-powered amplifier.

Electrostatic speaker A design that uses the attractive and repulsive forces of electrostatic charges between fixed surfaces and a lightweight, typically large, movable diaphragm. The prime advantage of this design is the uniform distribution of force on the moving mass. Its main drawbacks are poor dispersion at high frequencies, limited movement (output), and the lack of an enclosure. The latter two restrict deep-bass output. *See also* **Planar-magnetic loudspeakers.**

Envelopment In the context of sound reproduction, envelopment mainly refers to the ability of a recording or audio-video system to impart a sense of space, depth, and ambience to the sound. With regard to playback system hardware, the term deals with the ability to *re*create, or possibly synthesize, that same sense of space. In most cases, a system will do this better if a surround-sound feature is employed.

Equalizer There are many types of equalizers, but most use discrete controls to vary rather narrow sections of the response range of a sound system to reduce speaker, room, or recording anomalies. Tone controls are wide-band equalizers, as are low-bass "subwoofers." Equalizers are also used in recording studios to deal with the same problems as home units.

Expander In home audio, a device that increases the dynamic range of an incoming signal by making the loud passages louder and the quiet ones quieter. Rarely required with modern digital program material, expanders can make older recordings and video soundtracks that were compressed to accommodate analog-playback-medium limitations more realistic sounding. Some expansion circuits, like Dolby B, C, and S, as well as the dbx system still used in MTS video sound systems, are designed to work with signals that were previously compressed in a specific manner.

Extraction processors These are surround-sound devices for home use that "extract" a left-minus-right component from the sound of a recording and send it to specially placed effects speakers for additional ambience and reverberation. The technique works best with material that has been encoded with the necessary matrixed signals (such as Dolby), but it also works well with standard recordings that have a substantial amount of noncoherent reverberation on them. The extraction processor routes a lot of that reverb to the ambience speakers. *See also* **Synthesizing processors; Surround sound; Dolby Surround Sound; Hafler circuit; Matrixing.**

FCC Federal Communications Commission. The regulating body for radio and television transmission in the USA.

FET Field-Effect Transistor.

FFT (Fast Fourier Transform) analyzer A mathematical operation used to test a variety of audio (and other) components. One notable use is in speaker testing, where it is used to simulate anechoic testing conditions without employing a special chamber. It is best at plotting frequency and phase characteristics, but must be used with care if real-world listening conditions in more reverberant environments are to be considered.

FM Frequency Modulation. A radio-transmission technique that conveys data by encoding audio signals as variations in the frequency of the carrier signal.

Far field A listener is in the far field when each doubling of the distance from the source results in a reduction of 6 dB in sound level, due to the inverse-square law. The far field exists between the near field and the reverberant field and is typically from 2 to 3 times the distance between the most separated points of a speaker system that are radiating at the same frequency. For practical purposes, consider it to be the point where the reflected energy begins to dominate over the direct sound energy. Therefore, its location may vary with frequency, becoming closer to the speaker system at low frequencies or at higher frequencies if the system has excellent high-frequency dispersion, and farther away at crossover points or at frequencies where the system has limited dispersion.

Ferrofluid™ Originally developed as a rotary seal, this substance is a magnetically attracted liquid that works well at cooling tweeter and midrange driver voice coils. This is because it stays suspended in the magnetic gaps between the coil and the heavy magnet assembly, conducting heat to the latter. It also can affect speaker damping. In some tweeter designs, silicone grease will work better than Ferrofluid.

Fiber optics A form of signal transmission that allows digital data to be transmitted as pulses of light, normally through special cable. The main advantage is a reduction of noise and distortion.

Figure-eight microphone A microphone with a dumbbell-shaped pickup pattern that is sensitive to sound from the front and rear but not from the sides. Variants include super- and hypercardioid models, which have less sensitivity to the rear and somewhat more to the sides. *See also* **Dipole.**

Flat response In audio, a condition whereby a signal is not boosted or attenuated at specific frequencies over its operating range.

Flat-screen picture tube Direct-view television picture tube with a flatter front surface (and usually shallower depth) than older (or cheaper) designs, allowing for a more undistorted picture, particularly when viewed from off to the side somewhat.

Flutter A variation from exact speed, normally found in analog recording and playback devices. Called "wow" when the variations are slow, flutter is not a problem with digital record-playback systems, because their outputs are controlled by internal clock mechanisms. *See also* **Wow.**

Flying erase head An erase head built into a rotating VCR head drum. Most decks have the erase head mounted in a fixed position, limiting editing flexibility. The flying head makes it possible to do clean edits.

Focus In two-channel audio, focus relates to the ability of a recording or pair of speakers to keep sound-stage images—especially those in the central area—properly sized and positioned. *See also* **Center channel; Imaging.**

Frame In video, one complete image on a TV screen that has been formed by progressive or interlaced scan lines. *See also* **Progressive scan; Interlaced scan.**

Franssen effect *See* **Precedence effect.**

Free field The condition whereby a sound reaches the listener without having been reflected from any surface. Often misnamed the near field, although the latter exists at distances much closer to the sound source.

Frequency A rate of vibration or signal oscillation. In audio, it normally involves the audible bandwidth. In video, it most typically involves the bandwidth of the sharpness component of the video signal, although the bandwidth of the color component of the signal is often of greater importance.

Frequency response Sometimes called magnitude response, it is the measurement of the amplitude linearity of a component over a given frequency range. Frequency response is probably the most important aspect of audio system performance but there are different opinions about what is to be measured. A big problem with many recordings, even some contemporary ones, is that the microphones chosen to make them often have poor frequency response. The biggest problem with some of the studio-monitor speakers that are still being used to edit those recordings is their limited or ragged frequency response, particularly in the reverberant field. Using studio-located equalizers to compensate for those deficiencies results in recordings that are improperly equalized for playback on high-quality home speakers. *See also* **Direct field; Reverberant field.**

Front-to-back (F/B) ratio The F/B ratio relates to an antenna's sensitivity to signals from the front compared with its sensitivity to signals from the rear. In TV and FM radio reception, assuming the antenna is aimed properly, a higher ratio will help prevent ghosts and multipath distortion.

Gain The amount of amplification developed by an amplifier, preamplifier, etc.

Ghosts *See* **Multipath distortion.**

Golden ear A term describing audio buffs who have the (real or imagined) ability to hear subtle differences in recorded sound.

Gray-scale linearity In video, an indication of how accurately a VCR, disc player, or TV set handles subtle variations of gray—from lighter shades down to near black.

Ground The zero-voltage reference used to signify a negative connection.

Group delay The frequency-dependent variation in signal delivery time from an audio component. In loudspeaker systems, this can result from crossover anomalies or differences in listening distances. Many speaker manufacturers "time align" the drivers to compensate for group-delay problems, but these solutions only work if the listener is locked into a specific location and the system is auditioned from very close up. At normal listening distances, the group delay from any decent loudspeaker system is inaudible.

Haas effect *See* Precedence effect.

HDTV High-Definition Television.

Hafler circuit An ambience-recovery circuit designed by David Hafler in the 1960s. The L-minus-R matrix principle that is its basis is similar to what is employed in the Dolby Surround version, but without the need for extra amplifiers.

Hard matte *See* Matting.

Harmonic distortion The most common form of audio distortion, it shows up as additional unwanted signals at multiples of the original frequency. Thus, a 1-kHz tone may have second-order harmonic distortion at 2 kHz, third-order at 3 kHz, etc. These can continue upward to beyond the seventh or eighth order. The percentage total of *all* these measurements is called total harmonic distortion (THD) and is commonly used in audio test reports. However, different components generate different ratios of odd and even orders, making some sound better than others—even though their THD measurements may be the same.

Heads The parts of an audio or video tape recorder that lay down or pick up the magnetic signal on the tape.

Hi-fi video The videotape medium that makes use of specially encoded signals to carry the audio part of the video program. The Beta version uses the video heads on the rotating tape drum. The VHS system has separate audio heads—also on the tape drum—in addition to the video heads. Newer formats employ digital audio.

High-pass filter Within an audio crossover network, the electronic or passive circuitry that allows the high frequencies to go to a speaker system or amplifier. *See also* **Low-pass filter**.

Home-theater system An audio-video system that is high enough in quality to simulate a theatrical experience in the home. While most video components, especially television sets, may be inadequate to achieve near perfection, good results in the audio realm can be had for a reasonable amount of money.

Horizontal resolution In video, one of the more common specifications listed by manufacturers. It is the ability of a component (television, VCR, disc player) to resolve detail sideways across a television screen. Many NTSC television monitors have horizontal-resolution capabilities well beyond any source material they might have to reproduce, making the one-upmanship battle of resolution specifications more of an advertiser's tool than something significant. This measurement can be calculated either from a component's video frequency response or by means of a resolution chart on a monitor.

Hz Hertz, or cycles per second (cps), or pitch. The name comes from Heinrich Hertz, a German physicist.

IC Integrated Circuit. A miniature electrical circuit.

IDTV Improved-Definition Television. IDTV sets employ digital line-doubling circuitry, which allows NTSC-spec interlaced scan lines to simulate a more artifact-free progressive scan. *See also* **Interlaced scan; Progressive scan**.

IEC International Electrotechnical Commission.

IEEE Institute of Electrical and Electronic Engineers.

IHF Institute of High Fidelity.

IPS Inches per second.

ISO International Standards Organization.

Imaging The ability of a component (usually a loudspeaker pair) or recording to form a realistic sound stage with precise instrumental and/or vocal localization. In fact, imaging is often more dependent upon recording techniques than speaker-system design. A few speaker systems, when reproducing certain recordings, perform imaging feats that even live music cannot duplicate.

Imaging Science Foundation (ISF) An organization founded by Joe Kane to promote the correct alignment of existing TV picture systems and improve the quality of future systems.

Impedance In a DC (direct current) circuit, the same thing as resistance. In an AC (alternating current) circuit, impedance is the complex interaction of inductive and capacitive forces—in addition to resistance. In such a circuit, impedance is dependent upon frequency. *See also* **Resistance; Input impedance; Output impedance**.

Indexing With audio (DCC and MiniDisc) and video recorders, this is the ability to electrically mark a point on a tape or disc for later access. Some CD, LV, and DVD players also have an indexing playback function, but it will only work with discs that are specially encoded with indexing points.

Infinite baffle If the front of a loudspeaker driver is acoustically isolated from its back, it is said to be operating in an infinite baffle. Practical limitations result in enclosures behind drivers that still isolate the rear from the front. *See also* **Dipole; Acoustic-suspension speakers; Electrostatic speakers; Planar-magnetic speakers**.

Infrared A part of the electromagnetic spectrum that is just below the frequency range of visible light. Most remote controls work with infrared light.

Infrasonic filter A type of high-pass filter that attenuates frequencies below the audible range—reducing the work that woofers and amplifiers must do when reproducing signals that contain very low-frequency, but audible, sounds or even subsonic energy. Useful with LP-record playback to limit annoying record-player and cutting-lathe rumble and the studio or hall noise present on some CD recordings, particularly those made in churches and older halls.

Input impedance The "load" actually seen by a source connected to an input. In audio, the input impedance should be considerably larger than the connected component's output impedance to avoid signal losses and frequency-response irregularities. In video, the input and output impedances should nearly match.

Integrated amplifier Sometimes called a **control amplifier**, this is a receiver minus a tuner or, if you like, a power amplifier plus a preamplifier. Some integrated amplifiers are very elaborate and contain A/V switching and even surround-sound processing.

Intensity stereo *See* **Coincident-microphone recording**.

Inter-aural cross-talk An effect created when the signals from a pair of stereo speakers are heard as individual events, rather than a coherent, single one. This effect can muddy stereo imaging and sound-stage realism. *See also* **Cross-talk.**

Inter-aural cross-talk cancellation By emitting out-of-phase cancellation signals to null inter-aural cross-talk, this process can improve focus and sound-stage imaging. To work properly, it requires the listener to sit exactly in the sweet spot, out in front of and exactly between the speakers. The effect can be influenced by early room reflections, and some listeners think the process adds substantial sound-stage phasiness. It is available on some recordings in the form of the Q-Sound, Spatializer, or Roland RSS systems; variant designs that work with conventional recordings are also available in home processors made by Spatializer, Carver, and Lexicon, as well as in Polk Audio's SDA speaker systems. *See also* **Early reflections; Phasiness; Sweet spot.**

Interlaced scan The process of imaging a television picture by having the numerous scan lines that form the picture laid down at two intervals, with each positioned adjacent to the other. Done correctly, it allows for a sharper picture at any given transmission bandwidth. *See also* **Progressive scan.**

Intermodulation distortion (IMD) Electronically similar to mechanical Doppler distortion in that it results from a higher-frequency signal distorting as it rides on one of lower frequency. Unlike Doppler distortion, the one-dimensional nature of IMD within an electronic component can make it quite audible. *See also* **Doppler distortion**.

Jog/shuttle dial A control dial found on some VCRs and laser-video players that allows the user to more easily execute forward and reverse picture searches.

Kbps (sometimes kb/s) Kilobits per second.

kHz Kilohertz, or thousands of cycles per second. *See also* **Hz**.

Kell factor A psychovisual phenomenon that determines how much the eye can resolve on a TV screen.

LCD Liquid Crystal Display. Instead of cathode ray tube (CRT) displays or CRT projection tubes, some television sets employ the LCD, which is cooler running, lighter in weight, and smaller in size. Some very small sets have direct-view displays. A few others are front-projection models that focus light through several LCD panels, with the resultant image projected to an external screen. *See also* **CRT; Direct-view television set; Projection television set.**

LED Light Emitting Diode. Often used as an indicator on A/V components.

LEDE Live-End–Dead-End room. A room designed to attenuate speaker reflections from adjacent walls, while highlighting the more delayed, scattered reflections from the far end, behind the listeners. This keeps the recorded signals from being strongly modified by the front of the listening room and allows the longer delays from the rear to place room-generated ambience where it belongs. An LEDE room will be heavily padded at the speaker end and lined with diffusing panels at the other end. The absorptive characteristics of this room may result in recordings that are overly bright when played back on wide-dispersion speaker systems located in typical, somewhat more reflective, home-listening rooms.

LP Long Play. In VCR parlance, the middle recording and playback speed available on some units. In audio, the short term for the analog, long-play, 33⅓-rpm, vinyl disc.

LSB Least Significant Bit.

LV LaserVideo. This is the earliest laser-read videodisc system; sometimes called the analog videodisc.

Late reflections In room acoustics, the sounds that arrive at the listening position after being reflected from multiple room surfaces. They are the aural clues to the size of the listening space. In DSP, they are the electrically delayed signals that a home or studio processor creates to simulate larger room spaces.

Learning remote A remote control that is designed to learn commands from a variety of other (dedicated) controls, simplifying user control of multiple components.

Letterboxing Video reproduction of a film that places the entire, uncropped picture on the TV screen—eliminating the pan-and-scan problems that result when a wide format is cropped to fit a 4:3-ratio (or even, in the case of extremely wide originals, a 16:9-ratio) screen. While often helpful in capturing all the action on the screen, the smaller size of the individuals within the picture (even when the TV monitor is a fairly big one) and the loss of detail involved in not using all the vertical scan lines may be counterproductive. *See also* **Matting; Pan and scan.**

Line doubling *See* **IDTV.**

Line level Low-voltage output signals available at the shielded (RCA, XLR) connections of preamplifiers, CD players, tape recorders, etc., designed to interface with the line-level inputs of amplifiers, subwoofers, tape inputs, etc. *See also* **Speaker level.**

Line-source loudspeaker A line source is a tall, vertically oriented, narrow driver or line of drivers. Because of this design, the "driver" will behave like a very large source over the vertical dimension and like a smaller one over the horizontal dimension. When very tall, the resultant erratic vertical dispersion and phase cancellations will affect performance in both the direct and reverberant fields.

Linear audio track The monophonic analog sound track that runs down one side of a videotape. Far inferior in sound quality to what can be obtained with hi-fi videotape, this is what you will hear if you plug your VCR (even a hi-fi model) directly into the RF input of a TV set.

Linear stereo tracks These are the non-hi-fi stereo tracks that are available on some prerecorded videotapes. They usually employ Dolby Noise Reduction to improve the S/N ratio that is sacrificed when going from a mono linear audio design to stereo, but they are still far inferior to the stereo tracks that are standard on any hi-fi-audio-equipped video recorder. This feature has fallen into disuse but may be found on some used models.

Liquid-cooled speaker *See* Ferrofluid.

Liquid-cooled tube These are found on CRT-type projection television sets and involve a liquid solution hermetically sealed between the projection tubes and the lens assembly. The coolant prolongs the life of the tubes and keeps heat expansion from distorting the picture.

Listening distance In home audio, the subjective distance of the listener from the performers on a recording. The distance can be somewhat altered by careful use of the volume control, but the recorded sense of space around the instruments and the depth of the sound stage that result from good minimalist microphone techniques will also play a large part in determining it.

Lossy compression *See* Data reduction.

Loudness compensation A circuit available on many preamplifiers, integrated amplifiers, and receivers that attempts to compensate for the loss in low-frequency hearing sensitivity at lower volume control settings. While simplified switched versions usually are crudely effective at best and certainly less workable than simple bass tone controls, some of the more sophisticated, continuous-control versions may work well—although still no better than the bass controls.

Low-pass filter Within an audio crossover network, the electronic or passive circuitry that allows the low frequencies to go to a speaker system or amplifier. *See also* High-pass filter.

Luminance The brightness component of a television signal. *See also* Chrominance.

Mbps (sometimes mp/s) Megabits per second.

MD *See* MiniDisc.

MOL Maximum Output Level.

MOSFET Metal-Oxide-Semiconductor-Field-Effect Transistor. A special, high-peak-current output transistor used in some power amplifiers.

MPEG Motion Pictures Experts Group. A group that meets under the auspices of the International Standards Organization in order to generate standards for digital-video and video-audio data compression/reduction.

MSB Most Significant Bit. The first bit in a binary number. In 16-bit digital-audio playback systems, it contributes 32,000 times more to the output signal than the 16th (least significant) bit. Thus, errors in MSB circuitry occurring at very low levels can cause audible distortion and nonlinearities.

MTS Multichannel TV Sound. The standard stereophonic audio reception and noise-reduction process used in all true stereo television receivers not using satellite or digital decoders.

Macrovision A jamming signal encoded into most prerecorded videotapes that makes it difficult to do tape-to-tape copies.

Masking Under ordinary conditions, the process by which the threshold of hearing of one sound is raised by the presence of another. In both digital video and digital audio, a technique that allows a system to delete superfluous (inaudible or invisible) artifacts from a data stream by means of data reduction or data compression, enabling the system to transmit or store wide-bandwidth information within a much smaller bandwidth. Four notable uses of masking involve Dolby AC-3 Digital Surround Sound, MPEG video, DCC cassettes, and the MiniDisc. *See also* **Data reduction.**

Matrixing In audio, the electrical mixing of two or more channels of sound down to one or more new ones. The latter can later be "dematrixed" back to the original number. With two-channel stereo, this will involve both left-plus-right (derived center) and left-minus-right (extracted ambience) processing. Dematrixing can also be applied to two-channel stereophonic signals that were not consciously matrixed from multiple originals, with variable results. While used in FM-signal transmissions and processes to receive stereo audio signals, its most notable use is in surround-sound processors. *See also* **Dolby Surround**; **Hafler circuit**; **Extraction processors**.

Matting The application of a mask to a film or video program to remove information from the top and/or bottom of a picture. Used extensively in both theater presentations and video letterboxing. A *hard matte* is applied to the camera during the filming or videotaping process and, like anamorphic manipulation, delivers a true wide-screen image. A *soft matte* is a postproduction process that is done digitally when a film is transferred to videodisc or by means of projection gates in a theater.

Microphone An electroacoustic device that turns the acoustic signals that come in contact with it into electrical signals for recording. Its behavior is just the opposite of that of a loudspeaker.

Midbass The part of the bass frequency range between roughly 100 and 300 Hz.

Midrange The middle range of the audible spectrum, running anywhere from 300 to 500 Hz on up to 3 or 4 kHz, a total of four octaves or more. The speaker component that handles this area is called the *midrange driver.*

MiniDisc (MD) Sony's new data-reduced, small-disc, digital format.

Minimalist technique The use of very few microphones, combined with very little editing, to achieve a natural sound on an audio recording, particularly with classical or jazz music. Also called *purist technique.*

Mixing console The piece of equipment that recording engineers use to edit the material they recorded or are in the process of recording.

Monaural *See* Monophonic.

Monitor: audio With regard to recording, this refers to the listening the recording engineer does while "recording" and editing a program (usually music). With regard to audio playback, it refers to the speaker systems used in the monitoring and mixing room, which may be commercial models but can also be models designed for consumer use that are often better than the commercial models.

Monitor: video Refers to TV sets without a tuner, which thereby require connection to a video source of some kind to produce a picture. However, many monitors designed for home use have both monitoring connections and a tuner.

Monophonic A recording or sound system that has only one channel, usually with all the sound (in most cases, music) coming from just one speaker system.

Moving-coil cartridge A phonograph cartridge that makes use of a moving coil attached to the internal end of the stylus assembly to excite a magnetic field in a fixed-magnet structure, producing an electrical output for amplification. Rarely available these days, except as high-end audio items, and electroacoustically no better than the moving-magnet design.

Moving-magnet cartridge A phonograph cartridge that makes use of a moving magnet attached to its stylus assembly to excite a magnetic field in a fixed-coil structure, producing an electrical output for amplification.

Multipath distortion In FM radio transmissions, this effect occurs when a signal, because of being reflected from some surface (building, hill, etc.), arrives slightly later than the signals arriving directly from the transmitter. Because of the nature of FM-stereo matrixing, it can cause audible problems if the antenna and/or tuner is not well designed. In video, the effect causes ghost images.

NAB The National Association of Broadcasters.

NR Noise Reduction.

NTSC The National Television System Committee. The body responsible for the color television broadcast standards in the USA. The term NTSC is often applied to the performance parameters of pre-HDTV video hardware and software in this country.

Near field Technically, the region where the particle velocity is mostly out of phase with the sound pressure—meaning that it can be very close to the listener at higher frequencies. Popularly—and incorrectly—it is often considered to be any point where the direct sound is significantly louder than the reflected sound.

Negative feedback In all amplifiers, a part of the output signal that is fed back and added to the input signal out of phase, somewhat reducing the gain, limiting distortion, and imparting stability. Negative feedback, when used properly, can also improve frequency response. At higher frequencies, the feedback may not be fast enough, and the result will be increased transient intermodulation distortion. Under most conditions, this will not be audible. Feedback may be used "locally," in sections of an amplifier, or "generally," to control the response of the whole unit. *See also* **TIM.**

Noise floor The noise generated by an audio device in the absence of any input signal.

Noise reduction: audio A blanket term to describe a variety of background-noise-suppressing systems (Dolby, dbx, CX, etc.), which are employed in audio and video sound systems. Even hi-fi video recorders have proprietary audio noise-reduction circuitry. Most digital-audio systems do not require it.

Noise reduction: video On some VCRs and laser-video players, digital noise reduction is used to improve picture quality, especially as it relates to video grain and snow in dark areas.

Noise shaping Digital recording techniques that take advantage of the ear's reduced sensitivity at high frequencies.

Notch filter In video systems, this removes a small part of the TV signal where color information is most concentrated, reducing unwanted artifacts from less-than-perfect signals. *See also* **Comb filter.**

Objective testing The proper use of instrumentation or rigorously managed listening comparisons, rather than casual or uncontrolled techniques, to evaluate audio or video equipment. *See also* **Subjective testing.**

Octave A pitch interval or frequency ratio of two to one. Thus, a jump from 50 Hz to 100 Hz is one octave, as is a jump from 5,000 Hz to 10,000 Hz (5 kHz to 10 kHz). In listening to musical programs, the interaction of ear and brain makes it difficult to resolve minor frequency-response anomalies narrower than about a third of an octave.

Off-axis Any listening, viewing, measuring, or recording position that is not directly in front of the forward axis of a TV set, loudspeaker, or microphone.

Ohm A basic unit of electrical resistance. *See also* **Resistance; Impedance; Reactance.**

Omni-directional microphone A microphone that picks up wide-bandwidth sound equally well from all directions. A variant is the subcardioid, which has somewhat less sensitivity in one direction.

On-axis Any listening, viewing, measuring, or recording position that is directly in front of a TV set, loudspeaker, or microphone.

Open-reel recorder A tape recorder that holds its tape in individual reels rather than cassettes. Reels vary in diameter from 5 to 10 inches.

Output impedance The impedance seen by an electrical load attached to the output terminals of an audio or video device. For practical purposes, the output impedance of any audio amplifying equipment should be low in comparison to what it is connected to. It should not only be low at low frequencies, where it will affect bass damping, but should also be low at higher frequencies to insure a flat frequency response. In video systems, output and input impedances should closely match.

Oversampling In most digital playback equipment, the sampling frequency is increased two, four, or, in the case of bitstream devices, even hundreds of times. However, the new samples are artificially included between the originals and will not actually affect the 16-bit information. What this digital filtering technique does is reduce the need for steep analog filters to remove ultrasonic hash, saving the manufacturer and hopefully the purchaser money. Although nearly all modern CD players use this technique in one form or another, there is no evidence that oversampling markedly improves playback sound.

PAC Perceptual Audio Encoder. A 5.1-channel surround-sound system developed by Bell Laboratories, and designed originally to compete in the broadcast realm with Europe's Musicam and Dolby's AC-3 systems.

PASC Precision Adaptive Sub-Band Coding. The low-bit-rate, digital data-reduction coding process used in the Philips-developed DCC tape-recording system. *See also* **Data reduction**.

PCM Pulse Code Modulation. The standard playback or recording system employed by the CD and most professional-grade digital recorders, including DAT. In contrast to digital data-reduction systems, PCM recording systems allow 100 percent of the material recorded to be played back.

PIP Picture in Picture. A TV set that can place a smaller picture derived from a different signal within the larger main picture. In most sets, this requires the addition of another tuner, usually from a VCR.

P-mount cartridge A plug-in phono cartridge originally designed by Technics but now used by a number of other companies. Its main advantage is ease of alignment.

PWM Pulse Width Modulation. *See* **Bitstream processing**.

Pan and Scan A method of transferring wide-screen films to smaller-ratio TV screens, whereby the full image is not shown at all times. For example, an original wide-screen shot might show two people talking to each other; in a pan-and-scan version, each person might be shown individually, with the camera moving (panning) between them as they speak. *See also* **Letterboxing**.

Pan potting The individual level controls for each channel in a multitrack recording mixer are called pan pots. Pan potting is used to adjust each of those tracks for acceptable balance.

Passive crossover A nonpowered electrical network that divides the frequency constituents of an audio signal (bass, midrange, and treble) after it has been amplified and then routes them to the various drivers in a speaker system. In most situations, it is enclosed within the same box as the speaker drivers.

Passive radiator Also called **drone cone**. A nonpowered bass driver. Passive-radiator drivers are often employed and behave as independent bass speakers below the resonance of the active drivers. *See also* **Bass reflex**.

Perceptual coding *See* **Masking**.

Phantom-center channel The image that is formed between two front-center speakers when they combine their identical outputs. Such an image cannot usually be properly formed unless the listener is sitting in the "sweet spot." *See also* **Dolby Surround; Matrixing**.

Phase distortion Also called **phase shift** and sometimes **group delay**, it results when one part of the frequency spectrum is delayed more than another. Phase shifts can cause test waveforms viewed on an oscilloscope to distort but must be fairly extreme if they are to be audible when listening to music under normal home-playback conditions, at least with loudspeakers.

Phasiness An overly spacious characteristic that may be imparted to solo instruments or singers if they are improperly recorded. To get an idea of extreme phasiness, temporarily reverse the leads of one of your loudspeakers and notice how any centralized images become quite diffuse (there may also be a loss in bass power, but that is not what we are dealing with here). On some recordings, the left and right channels will appear to be in phase and solidly imaged, while the center will appear to be slightly out of phase and ill-defined. *See also* **Spaced-array microphones**; **Focus**; **Imaging**.

Pinch roller *See* **Capstan**.

Pink noise Random noise (hiss) that has equal energy in each octave.

Pinna The projected, curved parts of the outer ear that contour the frequency response and phase characteristics of the sounds going to the inner ear, allowing the brain to determine from which direction they emanate.

Planar-magnetic loudspeaker A flat, panel-type speaker that radiates sound from both front and back. This design looks similar to some electrostatic designs but uses a widely dispersed variant of the magnet-and-coil system found in typical dynamic models. Because of this, there is less electrical load on the amplifier, and thus these speakers are less likely to cause erratic amplifier behavior.

Polar response A plot of output amplitude of a single frequency vs. the angle off-axis. In other words, the variation in radiated or received energy with the angle relative to the axis of the radiator or receiver. The measurement can be used with either speakers or microphones. *See also* **Radiation pattern**.

Power response In loudspeakers, the integrated output in all directions. In most rooms, the overall level of the power response swamps the tonal effects of the direct signal. *See also* **Room response**.

Preamplifier Strictly speaking, the stage of an audio circuit that amplifies the very small output of a phonograph cartridge, allowing it to be successfully further amplified by a power amplifier. The term is often applied to the entire control section of a receiver, integrated amplifier, or stand-alone "preamplifier." Some stand-alone preamps also contain surround-sound processing circuitry and A/V switching.

Precedence effect When identical sounds come from two different speaker systems, if the distance is great enough, the ear tends to attribute all the sound to the near one. This phenomenon is one reason that the surround-channel sound in a DPL system is delayed relative to the main channels. Similar to the Franssen effect, where percussive bass signals have their localization determined by the position of higher-frequency drivers in a speaker system. Also known as the **Haas effect**. *See also* **Direct field; Power response.**

Pre-emphasis A deliberate change in the frequency response of a recording system for the purpose of reducing distortion or improving the signal-to-noise ratio.

Pro Logic The proprietary system of center-channel steering licensed by Dolby Corporation. Its function is to "steer" center-channel information to a center speaker in Dolby-encoded audio programs. On nonencoded material, the steering may still offer an improvement over standard two-front-channel playback. *See also* **Phantom-center channel.**

Progressive scan The process of imaging a picture by having the numerous scan lines that form it laid down continuously, eliminating artifacts that result from interlacing. Commonly used in computer monitors and high-definition television sets. *See also* **Interlaced scan.**

Projection television set A TV that employs either three CRT tubes or an LCD arrangement to project an image on a special screen. The most common are *rear*-projection models, which use lenses and mirrors within a large box to project the image to the inside of a translucent screen, the outside of which faces the viewer. Less common are *front*-projection models, which mount the projector across the room from a conventional screen.

Psychoacoustics The study of the relationship between human hearing perception and stimulus; in other words, the study of how we hear.

Punch A strictly subjective term that refers to the ability of a recording to deliver dynamic snap and impact.

Push-pull woofer system A bass loudspeaker that makes use of two woofer drivers mounted in the same cabinet but facing in opposite directions. Wired out of phase from each other, this mounting technique allows the two to move in and out together, reducing even-order distortion products. The system is used in both full-range systems and subwoofers.

Q In loudspeakers, a measure of directionality. At low frequencies, the Q will always be low. At higher frequencies, it gets larger, depending on the size of the drivers involved. Thus, Q is a measurement of frequency-dependent radiation pattern and polar characteristics. Q is also a measurement of the slope of any peaks in loudspeaker, equalizer, or microphone frequency-response curves.

Quadraphonic sound The term used to describe any of several surround-sound systems developed in the 1970s. These days, the term surround sound is more popular.

Quantization In a digital-audio signal, the number of possible values available to represent various levels of amplitude.

RCA plug The standard audio line level and video connecting plug found on amateur-grade equipment in the USA.

RF Radio Frequency. A signal used to transmit audio and video information through the air or through cable. While virtually all receiver-equipped TV sets and VCRs can receive RF signals, all VCRs and some laser-video players can also transmit them through a cable to a TV set. The latter function results in picture and sound that is inferior to what is possible with direct video and audio hookups.

RFI Radio Frequency Interference.

RGB input Red/Green/Blue input. The separate-color professional-grade interface that some TV monitors employ to receive data-grade video. The result is a picture much improved over that delivered by regular direct-video or even S-Video inputs.

RIAA Recording Industry Association of America. This group develops standards for recordings in this country. The RIAA "curve" is a record/playback compensation curve applied to LP records that allows them to have flat response with minimum distortion.

RMS Root Mean Square. A common measurement of average power output in audio amplifiers.

RPM Revolutions per minute.

RTA *See* **Real-time analyzer.**

Radiation pattern (R-P) The polar response characteristics of a loudspeaker system at all frequencies. Along with power response, the R-P is what mainly determines the subjective impression of a loudspeaker. In a microphone, this might be called its radiation-pickup pattern. *See also* **Polar response; Dispersion.**

Random noise Any kind of hiss-like noise produced by special noise generators. Similar noise can also be heard when a TV or radio tuner is tuned to a channel that has no station transmitting. *See also* **Pink noise; White noise.**

Reactance In passive or active AC circuits, a form of frequency-dependent resistance produced by an inductor. An inductor will let DC current pass through unaltered and will attenuate higher frequencies, depending on its reactance.

Real-time analyzer (RTA) A device for measuring the amplitude of specific signals in the audio bandwidth. An RTA presents a continual readout of the signal amplitude in evenly divided spectral bands, with either music or test signals as a source. *See also* **Frequency response.**

Real-time counter On VCRs, DCC decks, and MiniDisc recorders, a device that measures play and record time in actual seconds, minutes, and hours instead of arbitrary numbers.

Receiver In audio, a component combining a tuner, preamplifier, and amplifier into one chassis.

Most modern audio receivers also contain A/V switching abilities and surround-sound circuitry. In video, any component that can receive antenna or cable video signals.

Resistance Commonly, the non-frequency-dependent resistance of current flow within an electrical circuit. *See also* **Impedance**.

Resonance The tendency for a mechanical or electrical system to vibrate at specific frequencies. The most common problems with resonances in modern audio hardware involve loudspeaker systems and microphones.

Reverberant field A technical term that defines the sound field that exists when the reflected sound in a listening or monitoring room predominates over the direct sound from the source (be it a loudspeaker or performers). Obviously, it is strongly effected by room layout, reflectivity, and size. *See also* **Direct field**.

Reverberant sound (reverb) The amount of ambience and hall reflections captured during the recording process. Reverb can be recorded naturally, but many engineers add it synthetically to compensate for deficiencies in the recording environment.

Reverberation The multiple sound reflections that result when sound is produced in an enclosed space. *See also* **Ambience; Early reflections; Late reflections**.

Ribbon speaker A design that uses a long, very thin narrow metal conductor suspended in a magnetic field. Ribbons are usually employed as tweeters or tweeter-midranges, because their design does not allow for good performance in the bass range. Ribbons usually have good horizontal and limited vertical dispersion. *See also* **Line-source loudspeaker**.

Rolloff Commonly, a gradual reduction of audio output above and below specific frequencies. Usually applied to loudspeaker or microphone performance, it can also be used to describe the sound of recordings at their frequency extremes.

Room response The power response of a loudspeaker as measured in a given room. The measurement includes both the direct signal and the reflections from the room boundaries, minus the sound absorbed by the furnishings. *See also* **Power response**.

Rumble The low-frequency mechanical noise that appears on some recordings, which can be caused by any number of things, including mechanical or stage noise at the recording source. In the old days, rumble was also caused by LP turntables feeding through to the speakers or from the sound made by the cutting lathe that made the record master.

SAP Second Audio Program. In video systems, the SAP channel can be used to provide an alternate soundtrack—especially helpful when there is a need to broadcast dialogue in a language different from what is being delivered by the main channels.

S-VHS *See* Super-VHS.

S-connector The video hookup employed by S-VHS, ED Beta, and some laser-video players to keep the Y (luminance) and C (chrominance) signals separate. This hookup is sometimes called a **Y/C connection**. *See also* **Super-VHS**.

SLP In VCR parlance, Super Long Play, the slowest play and record speed. Sometimes called **EP**, or Extended Play.

SMPTE Society of Motion Picture and Television Engineers.

S/N ratio *See* **Signal-to-noise ratio.**

SP In VCR parlance, Standard Play, the fastest play and record speed.

SPL Sound-Pressure Level. *See also* **dB.**

Sampling rate In digital systems, the rate in Hz at which the circuitry determines the signal amplitude. For CDs, this is 44.1 kHz; for RDAT recorders, it can be either 48, 44.1, or 32 kHz.

Saturation A magnetic-recording term used to describe a condition whereby recording tape or tape heads are carrying all the signals that they can handle. Any additional input results in no additional storage or recording output levels.

Scan-velocity modulation A feature on some TV sets that adjusts the rate of horizontal movement of the electron beam as it scans the picture. This results in a sharper picture.

Sensitivity A standardized speaker measurement that determines how loud a system will sound under controlled conditions. The standard procedure agreed upon by the industry is output, in dB at 1 meter with 2.83 volts applied, which will amount to 1 watt at 8 ohms (2 watts at 4 ohms). While sensitivity has little bearing on overall sound quality, it will be a factor in determining the required amplifier power.

Shadow mask On a direct-view television picture tube, this is the perforated screen that is bonded behind the front glass surface, which limits color distortion (or blooming) and also improves contrast. *See also* **CRT.**

Signal-to-noise ratio (S/N ratio) Often arbitrarily assigned, the S/N ratio should be the difference, in dB, between the noise floor of a playback component or sound recording and the loudest level it can achieve with inaudible distortion. The measurement is sometimes weighted as to audibility, because the ear is more sensitive to some frequencies than others. The most generous scale is dBA (A-weighted). In any case, the larger the S/N number, the better. *See also* **dB; Noise floor.**

Slew rate *See* **TIM.**

Slope In audio, the rate of change that a frequency-response curve displays, normally stated in dB per octave. Among other things, slope can relate to crossover-point attenuation rates, woofer low-end rolloff rates, or equalizer control functions.

Software Another term for audio or video recordings.

Solid state Electronic circuits whose active elements are transistors and integrated circuits, rather than tubes.

Sound field In audio-video circles, this term relates to the "totality" of the sound presented by the sound-system–recording combination. In audio-only recordings it will involve the direct sound of the players, the sense of envelopment, the reverberation and ambience of the studio or hall—and even the interaction of the recording with the playback system and its environment. In A/V performance, it will involve how well the sound of the system interacts with the material on the TV screen. While the quality of the source material is critical, the sound field will be greatly influenced by the quality of the playback system, its arrangement within the listening-viewing room, and whether it incorporates surround-sound hardware.

Sound power The amount of energy radiated by an audio source, measured in joules per second, or watts. Its most common use is with loudspeakers, where power response is measured by how sound power varies with frequency.

Sound stage In audio or video sound, this often vaguely defined term refers mostly to the left-right spread of the sound between the speakers in a playback system. It can also be used to define a sense of front-to-back depth. While the sound system layout can be critical, recording quality is also of great importance in influencing the sound stage.

Source The signal that is played through an audio or video system. It may be something received over an antenna or cable system or be from an installed component like a VCR, videodisc player, CD player, or audiocassette deck.

Spaced-array microphones A technique whereby the microphones recording a stereo program are placed some distance apart and in front of the ensemble or individual. This allows timing, as well as intensity, cues to be reproduced. Some critics think the technique results in a sense of direct-sound phasiness or diffuseness in the center image. Adding an additional center microphone may alleviate some of these negative characteristics. *See also* **Coincident-microphone recording; Comb filtering; Phasiness; Monophonic**.

Spatial averaging An energy average over a given *space* around a loudspeaker system. Spatial averaging measures the effects of the speaker's radiation pattern but may include only small segments of the front hemisphere or sphere around a speaker. Its advantage as a measurement technique is that it limits the effects of acoustical interference while taking into account the amplitude irregularities caused by resonances within the speaker itself. *See also* **Spectral averaging; Power response**.

Speaker level The moderate-voltage outputs of an amplifier or amplifier section of a receiver or integrated amplifier. While these are mainly designed to power loudspeaker systems, some subwoofers have speaker-level inputs to their built-in active or passive crossover networks. *See also* **Line level**.

Spectral averaging An energy average over a given *band of frequencies* produced by a loudspeaker system. Thus, it will measure frequency response at specific, fixed angles around a system, as well as the power response. *See also* **Spatial averaging**.

Spectral balance Relates to the ability of a speaker system to integrate its direct and reflected sound so that it sounds balanced, smooth, and transparent in a typical listening room.

Square wave A waveform consisting of a fundamental and all the odd-numbered harmonics it produces. Because it consists of energy to at least the 20th harmonic, it can be used for frequency-response evaluation with electronic components. Any amplifier that can reproduce an exact 1-kHz fundamental square wave cleanly will be clean to 20 kHz.

Standing waves These are irregularities (quite audible and unwanted in the bass range) that result when sounds reflected back and forth between the walls of a room interact with each other and with the direct sounds from the speaker systems that produced them to form alternate reinforcements (peaks) and nulls. The effect is dependent upon the size and shape of the room and the listening position and, to a smaller degree, on the positioning of the speakers. Standing waves can be detrimental to sound reproduction at lower frequencies in small and/or badly proportioned rooms, where their effects are often extreme.

StarSight A proprietary subscription-activated menu system built into some TV sets and VCRs to aid in program selection and recording.

Steering Most notably used in Dolby Pro Logic systems, the electronic manipulation of recorded audio signals from two-channel sources allowing encoded center-channel material that would ordinarily only be vaguely imaged to be positively routed to a center speaker and surround material to be similarly routed to the surround speakers. The goal of steering up front is to simulate three discrete-channel sources, with surround steering normally simulating a broad sense of space around the viewer.

Stereo From the Greek for "solid." In audio, it ordinarily refers to a recorded program that uses two speakers in front to recreate the left-right sound-stage image of a live performance. If done right, stereophonic reproduction can also lend a certain degree of depth to the sound. Surround- and ambient-effect sound systems, making use of more than the standard two "front" speakers, are also an advanced form of stereo. The latest incarnation for home audio-video is Dolby AC-3.

Stridency A nontechnical term that usually refers to violin sound that is too close up, edgy, bright-sounding, or metallic. Many recording engineers get in close to those instruments with accent microphones, in order to capture their detail and allow them to compete with the much-louder brass section of the orchestra. However, even minimalist recording techniques may result in stridency, because they are often produced when the hall is nearly empty and there is no audience to absorb some of the excess high-frequency energy. *See also* **Minimalist technique**.

Stylus The external moving part of a phonograph cartridge. Usually, it will included a jeweled tip (nearly always a diamond) and a cantilever or shank connecting the tip to the magnets or coils within the cartridge body.

Subjective testing Judging audio or video gear by listening or viewing without using any measurement instruments. While some people can be quite sensitive to differences in audio or video quality, many are misled either by environmental factors or personal predispositions. *See also* **Objective testing**.

Subsonic filter *See* **Infrasonic filter**.

Subwoofer An electronic or mechanical device that extends the deep-bass response of an audio system. The most common are add-on, large, conventional woofers, which must be carefully aligned to work properly. Electronic-type "subwoofers" are actually equalizers that are dedicated to standard woofer systems and electrically boost the low-bass range to achieve smooth, flat low-bass response. Many add-on subwoofers incorporate electronic equalizers to flatten out the bottom of their ranges. *See also* **Equalizer; Woofer.**

Suckout Bass-range reflections from nearby floor or wall boundaries that partially null the primary signal coming from the speaker itself. The suckout phenomenon differs from standing waves or higher-frequency reflections in that relocating the listening position or padding the walls is not a cure. Suckout involves only the bass, particularly the midbass (although it can also cause interactions in the low bass between two widely spaced woofer systems), and requires very careful speaker placement to correct. *See also* **Standing waves.**

Super VHS Also called S-VHS; the high-band, sharper-picture upgrade to standard VHS.

Supertweeter A tweeter designed to reproduce the very highest frequencies above the 2–15 kHz range normally handled well by a good standard tweeter. Supertweeters are usually found in four- or five-way systems and are sometimes placed on the back of a cabinet, facing the wall behind the system. Note that a decent conventional tweeter should be capable of doing everything important that a supertweeter should do, because the highest frequency most people can hear distinctly (particularly if they are past the teen years) is about 15 kHz, and most music and film sound does not have significant energy past 12 to 13 kHz. The only way a supertweeter would offer an advantage would be if its radiating-surface diameter was very small—say one-half inch or less. This would result in improved dispersion above 10 kHz, compared with that of a typical 1-inch dome tweeter. Some supertweeters are said to have strong response to well above 20 kHz, but CDs, videodiscs, and videotapes do not reproduce that range, and nobody can hear up that high, anyway.

Surround sound The matrixed, synthesized or discrete rear-, side-, or center-channel outputs that are integrated with the main channels of a stereophonic audio or audio-video system to enhance realism and ambience. Most modern versions have separate amplification for those channels. *See also* **Dolby Surround; Ambisonic; Hafler circuit; Dolby Digital; DTS; Center Channel; DSP; Extraction processors; Synthesizing processors.**

Surround speakers The usually small speakers that are placed toward the sides or toward the rear in a surround-sound playback system and handle the decoded, extracted, or synthesized ambience signals. Some manufacturers refer to them as "rear-channel" speakers, a misnomer.

Sweet spot The so-called "best" listening (or viewing) position for enjoying an audio (or audio-video) system. Usually, it is centered between the main speakers and about as far from their connecting axis as they are from each other. Sweet-spot listening is mandatory for good imaging with systems that employ only two speakers up front. *See also* **Center channel.**

Synthesizing processors These are surround-sound devices for home use, such as those produced by Yamaha, Onkyo, and Lexicon, that add their own preprogrammed hall ambience and reverberation to the sound of a recording. This "overlay" of ambience can greatly benefit some recordings, particularly those that are fairly dry sounding. However, the effect can muddy the sound of recordings that have a fairly large amount of reverberation to begin with. Recording engineers often employ synthesizing devices to add ambience to the recordings themselves. *See also* **Ambience; Surround sound; Extraction processors.**

THD *See* **Harmonic distortion.**

THX A LucasFilm Corporation performance certification program for A/V software and hardware, particularly dealing with Dolby Pro Logic and Dolby Digital behavior but also involving TV monitor and laser-video picture and sound quality. *See also* **Dolby Pro Logic; Dolby Digital.**

TIM Transient Intermodulation Distortion. The intermodulation distortion caused by time lags in amplifiers operating at very high frequencies that have very high levels of negative feedback. Also called *slew-induced distortion;* an amplifier that is relatively immune to it is said to have a *high slew rate.* TIM can be controlled by an input that is rolled off above the audible frequencies so that signals too "fast" for feedback to handle will be nonexistent. *See also* **Intermodulation distortion; Negative feedback.**

Tape loop On most preamplifiers, integrated amplifiers, and receivers, the switch-operated hookup that allows a tape deck to be properly integrated into the system. A tape loop will have an input for tape playback and an output for tape recording. *See also* **EPL; Tape monitor.**

Tape monitor The switch that inserts a tape loop into a circuit. With some recorders, this allows you to do an A/B comparison between the source material and the recording as it is being made.

Timbre The quality given to a sound, particularly a musical sound, by its overtones. In audio, a popular term describing the basic tonal quality of a sound system, particularly the speakers.

They-are-here sound A recording technique that tries to simulate the effect of a performer or performers on a recording actually being in the listening room, rather than having the listener subjectively transported to the hall itself. This is only viable when small-scale sound is being recorded, particularly that of solo instruments with limited volume capabilities, such as guitar, harp, and violin—or maybe a string quartet. In many cases, the effect is probably the accidental result of the engineer and/or performer simply trying to reproduce a small-hall effect. *See also* **You-are-there sound.**

Three-way speaker A loudspeaker system that uses separate drivers for the high frequencies, midrange, and bass. Certain designs may have more than three speaker drivers, but because some are paired together to handle the same frequencies, they will still be three-way designs. *See also* **Two-way speaker.**

Time-base corrector A circuit found in all analog LV players (advanced versions employ digital circuitry) and some VCRs that electrically corrects for small mechanical speed errors.

Time shifting Setting a VCR to record a program for later viewing.

Tonality A subjective term that refers to the clarity and accuracy of the sound of an instrument or group of instruments. It can also refer to those qualities in vocal reproduction.

Tone arm The mechanism on an LP record player that holds the cartridge in proper position over the record.

Tone burst A momentary sine-wave signal that is used to test the transient response of an audio component. While a tone burst can theoretically measure the tendency of a component, particularly a loudspeaker, to continue to oscillate after the input signal is cut off, a proper frequency-response sweep will do the same thing, because these resonances will show up as peaks or dips in the sweep curve. *See also* **Frequency response; Transient response.**

Tone control The control on a preamplifier, integrated amplifier, or receiver that boosts or cuts certain segments of the audible bandwidth. Bass and treble controls are the most common versions, but some units have midrange controls also. *See also* **Equalizer.**

Tracking The ability of a CD player, LV player, VCR, or LP phonograph stylus to follow the mechanical or electrical pattern on a tape or disc.

Tracking control On a videotape recorder, the control that adjusts the "head-switching" network contained within the deck's electronic circuitry. Many modern decks have automatic tracking controls, an excellent idea.

Transient response The ability of an audio component to quickly respond to the signal being fed to it. Transient response is more critical in mechanical components like speakers, phono cartridges, and microphones. *See also* **Tone burst.**

Transparency In audiophile circles, the ability of a sound system or recording to achieve a realistic sense of imaging, space, and clarity. Most commonly used to describe the capabilities of speaker systems.

Transponder Used in video satellites to receive program material from ground-station uplinks and then retransmit it to properly aimed dish receivers back on the ground.

Transport The mechanical part of an LP turntable, audio tape deck, CD player, LV player, or VCR that moves the disc or tape so that the signal can be reproduced.

Treble The high-frequency range of the audible spectrum, running from 3 or 4 kHz on up to 15 or 20 kHz (less than three octaves).

Tuner The component that receives the RF signals (radio, video, satellite) from an antenna or cable system. Audio tuners are sometimes stand-alone units but are usually configured as part of an audio or audio-video receiver. Video tuners are usually included as part of a TV set or VCR. A satellite receiver is a tuner designed to receive either analog or digital satellite-transmitted signals that are received by a dish antenna.

Tweak To adjust an audio or video component so that it works at its very best. Also, a slang term for an audio extremist who dwells on the more mythical aspects of audio in preference to more rational beliefs that are substantiated by objective testing procedures.

Tweeter The individual speaker unit (driver) designed to handle the treble range. *See also* **Treble**.

Two-way speaker A speaker system that uses separate drivers for the high and low frequencies; the midrange frequencies are split between them. Two-way systems usually suffer from midrange dispersion problems, because the woofer, which must be robust enough to do decent work down low, is usually not small enough in diameter. Some two-way systems employ a nonpowered passive radiator to augment the deep bass. *See also* **Three-way speaker; Passive radiator**.

Universal remote control A remote control that has been preprogrammed by its maker to operate a variety of components.

VBI *See* **Vertical blanking interval.**

VCR Videocassette recorder. *See also* **Video recorder**.

VCR Plus A time-shift control system installed in some VCRs to simplify recording. Also available in handset-type controllers for older video recorders.

VHS Video Home System. The now-dominant home-video tape-recording system developed by JVC as competition to the Sony-designed Beta format.

Vertical blanking interval The horizontal black bar visible on a TV picture when the vertical-hold control is adjusted so that the picture "rolls" off center. The VBI consists of 21 lines and, because each line arrives at two 1⁄60-second intervals, totals 42 of the 525 lines available with the NTSC system. The VBI allows a video-picture scan line to return to the starting point at the next picture frame (the first nine lines contain the signal pulses that synchronize picture transmission) and also is useful for carrying specially encoded data to multiple recipients in the video distribution chain.

Vertical resolution The ability of a television component (VCR, laser-video player, or TV set) to resolve detail vertically on a television screen. Usually stated in lines of detail from screen top to screen bottom, vertical-resolution limits of consumer-grade television sets are set by the FCC.

Video enhancer A circuit designed to boost picture detail. Sometimes useful in dubbing material from one VCR to another.

Video recorder A tape recorder designed to record and play back video and audio signals received via cable or antenna. VCRs can also copy material from another recorder or laser-video player. Hi-fi versions produce higher-quality audio performance than standard models.

VideoGuide A proprietary subscription-activated menu system designed to aid in television program selection and recording. *See also* **StarSight**.

Videophile A person who has an enduring interest in video, particularly video hardware and home theater.

Voice coil The wire coil surrounded by the magnet assembly in a moving-coil, dynamic loudspeaker. The coil is attached to a diaphragm (which may be a cone, dome, or some kind of hybrid air mover) of the driver and causes it to move when excited by a signal from an amplifier. Most voice coils are made from copper wire, although a few are made of aluminum wire.

W/ch Watts per channel.

Watt A unit of power. Amplifiers do not deliver watts (they deliver voltage), nor do speakers create them. When presented to a specific load (speaker impedance), current flows and the power dissipated is rated in watts. Wattage produced may be calculated by multiplying voltage times current or by squaring the voltage and dividing it by the impedance.

White noise Similar to pink noise, except that white noise contains equal energy at each frequency point. *See also* **Pink noise**.

Woofer The individual speaker unit (driver) designed to handle the bass range. Some speaker enclosures contain multiple woofer drivers to increase bass power.

Wow The speed variation of a mechanical playback device such as an LP record player or analog tape recorder. Short-term speed variations are sometimes called **flutter**. Digital recording and playback devices do not produce wow, because their outputs are controlled (and slightly delayed) by an internal clock mechanism. Wow created by warped LP records is called *warp wow.*

XS Stereo Sound A proprietary stereophonic video system designed by Thomson electronics for budget-grade TV sets. This is not a true stereo system like the dbx version used in MTS; instead it uses small amounts of negative cross-feed to simulate stereo. *See also* **dbx; MTS**.

Y/C connector *See* **S-connector**.

You-are-there sound A recording technique that attempts to make it sound as though you are in the hall with the performers. Most recordings of any kind of music strive for this effect. *See also* **They-are-here sound**.

Bibliography

THERE ARE SCADS OF BOOKS and a multitude of articles about audio, video, and surround sound—with more appearing all the time. Some manufacturers, such as Dolby, Lexicon, Carver, and Bose, also supply often very informative "white papers" on their hardware and design philosophies. In this list, I've tried to select some books and articles that deal in greater depth with some of the topics I've just touched on in this book. Some are quite basic and aimed at the novice or hobby-oriented reader, and others are technically quite advanced. A few are repair oriented, but even if you don't intend to "do it yourself," they will be helpful in understanding how your equipment works and in coming to grips with the jargon used by some repairpeople. All are worth a serious look-see. Audio and video texts are intermixed.

Books

Ando, Yoichi. *Concert-Hall Acoustics.* N.Y.: Springer-Verlag, 1985.

Baert, Luc (ed.). *Digital Audio and Compact Disc Technology* (3rd ed.). Newton, Mass.: Butterworth-Heinemann, 1995.

Ballou, Glen (ed.). *Handbook for Sound Engineers.* Indianapolis, Ind.: Sams, 1987.

Bartlett, Bruce. *Introduction to Professional Recording Techniques.* Indianapolis, Ind. Sams, 1987.

Beer, Nick. *Servicing Audio and Hi-Fi Equipment* (2nd ed.). Newton, Mass.: Butterworth-Heinemann, 1995.

Békésy, Georg von. *Experiments in Hearing* (trans. by E. G. Wever). N.Y.: McGraw-Hill, 1960.

Belton, John. *Widescreen Cinema.* Cambridge, Mass.: Harvard University Press, 1992.

Benson, J. Ernest *Theory and Design of Loudspeaker Enclosures.* Synergetic Audio Concepts, 1993.

Benson, K. Blair (ed.). *Audio Engineering Handbook.* N.Y.: McGraw-Hill, 1988.

Benson, K. Blair, and Donald Fink. *Advanced Television for the 1990's.* N.Y.: McGraw-Hill, 1991.

Beranek, Leo. *Acoustics.* N.Y.: American Institute of Physics, 1986.

———. *Music, Acoustics and Architecture.* N.Y.: Wiley, 1962.

Berg, Richard E. *The Physics of Sound.* Englewood Cliffs, N.J.: Prentice-Hall, 1982.

Berger, Ivan, and Hans Fantel. *The New Sound of Stereo.* N.Y.: New American Library, 1986.

Blauert, Jens. *Spatial Hearing: the Psychophysics of Human Sound Localization.* Cambridge, Mass.: MIT Press, 1996.

Borwick, John (ed.). *Loudspeaker and Headphone Handbook.* Newton, Mass.: Butterworth-Heinemann, 1988.

———. *Microphones: Technology and Techniques.* Stoneham, Mass.: Focal Press, 1990.

———. *Sound Recording Practice.* N.Y.: Oxford University Press, 1994.

Bullock, Robert. *Bullock on Boxes* (speaker enclosures). Peterborough, N.H.: Audio Amateur Press, 1991.

Camras, Marvin. *Magnetic Recording Handbook.* N.Y.: Van Nostrand Reinhold, 1988.

Capel, Vivian. *Newnes Audio and Hi-Fi Engineers Pocket Book.* Newton, Mass.: Butterworth-Heinemann, 1992.

Capelo, Gregory. *VCR Troubleshooting and Repair.* Indianapolis, Ind.: Sams, 1991.

CasaBianca, Lou (ed.). *The New TV: A Comprehensive Survey of High Definition Television.* Westport, Conn.: Meckler, 1992.

Clifford, Martin. *Microphones* (3rd ed.). Summit, Pa.: Tab Books, 1986.

———. *Modern Audio Technology.* Englewood Cliffs, N.J.: Prentice Hall, 1992.

Colloms, Martin. *High Performance Loudspeakers.* N.Y.: Halsted Press, 1985.

Davidson, Homer. *Troubleshooting and Repairing Audio and Video Cassette Players and Recorders.* N.Y.: McGraw-Hill, 1992.

———. *Troubleshooting and Repairing Audio Equipment* (2nd ed.). Summit, Pa.: TAB Books, 1993.

———. *Troubleshooting and Repairing Compact Disc Players* (2nd ed.). Summit, Pa.: TAB Books, 1994.

Davis, Don. *Sound System Engineering* (2nd ed.). Indianapolis, Ind.: Sams, 1987.

Davis, Gary, and Ralph Jones. *The Sound Reinforcement Handbook.* Milwaukee, Wis.: Yamaha/Hal Leonard, 1987.

Dickason, Vance. *The Loudspeaker Design Cookbook* (4th ed.). Peterborough, N.H.: Audio Amateur Press, 1991.

———. *Loudspeaker Recipes.* Peterborough, N.H.: Audio Amateur Press, 1994.

Dickreiter, Michael. *Tonmeister Technology.* N.Y.: Temmer Enterprises, 1989.

Drucker, David. *Billboard's Complete Book of Audio.* N.Y.: Billboard Books, 1991.

Eargle, John. *Electroacoustical Reference Data.* N.Y.: Van Nostrand Reinhold, 1994.

———. *Handbook of Recording Engineering.* N.Y.: Van Nostrand Reinhold, 1992.

———. *Handbook of Sound System Design.* Comack, N.Y.: ELAR, 1989.

———. *The Microphone Handbook.* Comack, N.Y.: ELAR, 1981.

———. *Music, Sound and Technology*, 2nd ed. N.Y.: Van Nostrand Reinhold, 1995.

———. *Stereophonic Techniques.* N.Y.: Audio Engineering Society, 1986.

Erickson, Robert. *Sound Structure in Music.* Berkeley: Univ. California, 1975.

Everest, F. Alton. *Acoustic Techniques for Home and Studio* (2nd ed.). Summit, Pa.: TAB Books, 1984.

———. *The New Stereo Soundbook.* Summit, Pa.: TAB Books, 1992.

Ferstler, Howard W. *High Fidelity Audio-Video Systems: A Critical Guide for Owners.* Jefferson, N.C.: McFarland, 1991.

———. *High Definition Compact Disc Recordings: Sound Quality Evaluations of over 1,400 of the Most Technically Excellent Digital Recordings.* Jefferson, N.C.: McFarland, 1994.

Forlenza, Jeff, and Terri Stone (eds.). *Sound for Picture: an Inside Look at Audio Production for Film and Video.* Milwaukee, Wis.: Hal Leonard, 1993.

Gelfand, Stanley. *Hearing: An Introduction to Psychological and Physiological Acoustics.* N.Y.: Marcel Dekker, 1990.

Goodman, Robert. *Maintaining and Repairing VCR's.* N.Y.: McGraw-Hill, 1993.

Hack, David. *Digital Audio Tape Recording.* Washington, D.C.: Congressional Research Service, 1990

Hall, Donald. *Musical Acoustics* (2nd ed.). Belmont, Calif.: Wadsworth, 1991.

Hayes, Robert. *Wide Screen Movies.* Jefferson, N.C.: McFarland, 1988.

Hedgecoe, John. *John Hedgecoe's Complete Guide to Video.* N.Y.: Sterling, 1992.

Heller, Neil. *Compact Disc Troubleshooting and Repair.* Indianapolis, Ind.: Sams, 1987.

Holt, Jim. *The Complete Idiot's Guide to VCR's.* Indianapolis, Ind.: Alpha Books, 1993.

Hood, John. *Audio Electronics.* Newton, Mass.: Butterworth-Heinemann, 1995.

Horn, Delton. *The Complete Guide to Digital Audio Tape.* Summit, Pa.: TAB Books, 1991.

Huber, David. *Microphone Manual: Design and Application.* Indianapolis, Ind.: Sams, 1988.

———. *Modern Recording Techniques* (4th ed.). N.Y.: Macmillan, 1995.

Inglis, Andrew. *Behind the Tube: A History of Broadcasting Technology and Business.* Stoneham, Mass.: Focal Press, 1990.

Jones, Steve. *Rock Formation: Music, Technology and Mass Communication.* Newbury Park, Calif.: Sage Publications, 1992.

Jorgensen, Finn. *The Complete Handbook of Magnetic Recorders* (3rd ed.). Summit, Pa.: TAB Books, 1988.

Josephs, Jess. *The Physics of Musical Sound.* N.Y.: Van Nostrand Reinhold, 1967.

Kallenberger, Richard. *Film into Video: A Guide to Merging the Technologies.* Newton, Mass.: Butterworth-Heinemann, 1994.

Kaufman, Richard. *Enhanced Sound: 22 Electronics Projects for the Audiophile.* Summit, Pa.: TAB Books, 1988.

Keith, M. C., and J. M. Krause. *The Radio Station.* Stoneham, Mass.: Focal Press, 1993.

Langman, Larry, and Joseph Molinari. *The New Video Encyclopedia.* Westport, Conn.: Garland, 1990.

Leduc, Jean-Pierre. *Digital Motion Pictures.* N.Y.: Elsevier, 1994.

Lowery, H. *A Guide to Musical Acoustics.* London: Dobson, 1969.

Maddox, Richard. *DAT Technical Servicing Handbook*, N.Y.: Van Nostrand Reinhold, 1994.

Marco, Guy A., ed. *The Encyclopedia of Recorded Sound in the United States.* Westport, Conn.: Garland, 1993.

McComb, Gordon. *The Compact Disc Player Maintenance and Repair Manual.* Summit, Pa.: TAB Books, 1987.

———. *Troubleshooting and Repairing VCR's.* N.Y.: McGraw-Hill, 1991.

McWilliams, Jerry. *The Preservation and Restoration of Sound Recordings.* Nashville, Tenn.: American Association of State and Local History, 1979.

Milazzo, Dino. *Getting the Most Out of Your VCR and Planning Your Home Entertainment System.* Inamar, 1992.

Millard, A. J. *America on Record: A History of Recorded Sound.* N.Y.: Cambridge University Press, 1995.

Miller, Allen Wayne. Choral Recordings as History (a study of the recording techniques of five choral ensembles), Ph.D Thesis, Florida State University, 1992.

Moravcsik, Michael. *Musical Sound: An Introduction to the Physics of Music.* N.Y.: Paragon House, 1987.

Moylan, William. *The Art of Recording.* N.Y.: Van Nostrand Reinhold, 1992.

Nakajima, Heitaro. *Compact Disc Technology.* Tokyo: Ohmsha, 1991.

———. *Digital Audio Technology.* Summit, Pa.: TAB Books, 1983.

Nardantonio, Dennis. *Sound Studio Production Techniques.* N.Y.: McGraw-Hill, 1990.

Nisbett, Alec. *The Use of Microphones* (4th ed.). Stoneham, Mass.: Focal Press, 1993.

Olson, Harry F. *Music, Physics and Engineering.* N.Y.: Dover, 1967.

Pickett, A. G, and M. M. Lemcoe. *The Preservation and Storage of Sound Recordings.* Washington, D.C.: Library of Congress, 1959.

Pohlmann, Kenneth. *The Compact Disc Handbook.* N.Y.: Oxford University Press, 1992.

———. *Principles of Digital Audio.* Indianapolis, Ind.: Sams, 1989.

Pratt, Douglas. *The Laser Video Disc Companion.* N.Y.: New York Zoetrope, 1992.

———. *The Laser Video Disc Companion.* N.Y.: Baseline Books, 1995.

Rand, Herbert C. The Effect of the Absorptivity of Adjacent Walls on Loudspeaker Directivity, Ph.D Thesis, Florida State University, 1987.

Reed, Robert. *Dictionary of Television Cable and Video.* N.Y.: Facts on File, 1994.

Riggs, Michael. *Understanding Audio and Video.* Tustin, Calif.: Pioneer, 1989.

Ritchie, Michael. *Please Stand By: The Prehistory of Television.* Woodstock, N.Y.: Overlook Press, 1994.

Roederer, John. *The Physics and Psychophysics of Music: An Introduction.* N.Y.: Springer-Verlag, 1995.

Rossing, Thomas. *The Science of Sound.* Reading, Mass.: Addison-Wesley, 1982.

Rovin, Jeff. *The Laserdisc Film Guide.* N.Y.: St. Martin's Press, 1993.

Rumsey, Francis, and Tim McCormick. *Sound and Recording: An Introduction.* Stoneham, Mass.: Focal Press, 1994.

Runstein, Robert. *Modern Recording Techniques.* Indianapolis, Ind.: Sams, 1986.

Schetina, Erik. *The Compact Disc.* Englewood Cliffs, N.J.: Prentice-Hall, 1989.

———. *The Complete Guide to Digital Audio Tape Recorders.* Englewood Cliffs, N.J.: Prentice-Hall, 1993.

Sinclair, Ian. *Introducing Digital Audio.* London: PC Publishing, 1992.

———. *Newnes Audio and Hi-Fi Handbook.* Newton, Mass.: Butterworth-Heinemann, 1993.

Solari, Stephen. *Digital Video and Audio Compression.* N.Y.: McGraw-Hill, 1995.

Steiner, John. *Technicians Handbook of VCR Repair.* Englewood Cliffs, N.J.: Prentice-Hall, 1990.

Thomas, Steve. *How to Keep Your VCR Alive: VCR Repairs Anyone Can Do.* Tampa, Fla.: Worthington, 1990.

Villchur, Edgar. *The Reproduction of Sound in High Fidelity and Stereophonic Phonographs.* N.Y.: Dover, 1965.

Watkinson, John. *The Art of Digital Audio* (2nd ed.). Stoneham, Mass.: Focal Press, 1994.

———. *The Art of Digital Video* (2nd ed.). Stoneham, Mass.: Focal Press, 1994.

———. *Compression in Video and Audio.* Stoneham, Mass.: Focal Press, 1995.

———. *An Introduction to Digital Audio.* Stoneham, Mass.: Focal Press, 1994.

———. *Introduction to Digital Video.* Stoneham, Mass.: Focal Press, 1994.

———. *R-DAT.* Stoneham, Mass.: Focal Press, 1991.

Wayne, Victor. *Operating Your VCR.* Chicago: Consumer's Press, 1992.

Weems, David. *Great Sound Stereo Speaker Manual.* Summit, Pa.: TAB Books, 1990.

White, Glenn. *The Audio Dictionary* (2nd ed.). Seattle: University of Washington Press, 1991.

Wilkins, Richard. *Home VCR Repair Illustrated.* N.Y.: McGraw-Hill, 1991.

Wolenik, Robert. *Build Your Own Home Theater.* Indianapolis, Ind.: Sams, 1993.

Woram, John M. *Sound Recording Handbook.* Indianapolis, Ind.: Sams, 1989.

Wysotsky, Michael. *Wide-Screen Cinema and Stereophonic Sound.* N.Y.: Hastings, 1974.

Zaza, Antony *Mechanics of Sound Recording.* Englewood Cliffs, N.J.: Prentice-Hall, 1991.

Articles

Excellent essays on A/V appear regularly in both technical and popular journals. These were selected because of their relevance to the themes covered in this book. Most of the articles deal with generalized subject matter, setup procedures, shopping tips, interviews with noted experts, and/or A/V theory. These are sometimes fairly basic, sometimes quite technical, and possibly beyond the abilities of some lay readers, as with, for example, quite a few *JASA*, (*Journal of the Acoustical Society of America*), *JAES* (*Journal of the Audio Engineering Society*), *IEEE Proceedings* (*Proceedings of the Institute of Electrical and Electronics Engineers*), *SMPTE Journal* (*Society of Motion Picture and Television Engineers Journal*) documents. A few involve equipment test reports and were included because they also included relevant information on operational principles. A very few also deal with the social impact of audio, video, and home theater.

I should point out that a few mainstream A/V journalists, most notably Julian Hirsch, Ian Masters, and David Ranada in *Stereo Review* (Ranada also writes for a number of other journals, including some that deal primarily with video), Michael Riggs and Joseph Giovanelli in *Audio* (his "Audio Clinic" column), Larry Klein in *Electronics Now*, and the often acerbic Peter Aczel in *The Audio Critic*, will regularly highlight the more worthwhile aspects of audio and video system performance in their writings. This is also true for the freelance equipment-testers and/or essayists Brad Meyer, Tom Nousaine, Ed Foster, Joe Kane, Marc Wielage, Ron Goldberg, Gordon McComb, and David Moran. The regular material written for *Audio* by John Eargle is also uniformly informative, and the opinions on video sound regularly published by Tom Holman in a variety of journals are mandatory reading for an understanding of the subject. Finally, any article, anywhere, by Roy Allison, Fred Davis, Stanley Lipshitz, Floyd Toole, or R. A. Greiner will be on the mark. Because of the quantity of what these people have written, only a few of their more relevant essays are included—even though I urge everyone who cares about the subject to make a point of searching out and reading their latest, as well as previously published, material.

Note that when essay titles are cryptic, I have included short explanations of the main themes. You may have to visit a formidably equipped library or contact a publisher to obtain some of these articles. A few appeared as AES preprints—usually available from the Society—and at least one is a private study sent to your author, posted here to substantiate findings within this book.

Aarts, R. M. "Enlarging the Sweet-Spot for Stereophony by Time/Intensity Trading" (paper presented at the 94th Audio Engineering Society Convention, March 1993; available as preprint 3473 [C-1]).

Aczel, Peter. "Seminar 1989, Pt. 1," *The Audio Critic* 13, Winter-Spring 1989; "Seminar 1989, Pt. 2," *The Audio Critic* 14, Summer-Winter 1989–990; "Seminar 1989, Pt. 3," *The Audio Critic* 15, Winter 1990–1991. A three-part series of interviews with the audio experts Bob Carver, Dave Clark, John Eargle, Stanley Lipshitz, and Peter McGrath.

———. "The Wire and Cable Scene: Facts, Fictions, and Frauds, Part 1." *The Audio Critic* 15, Winter 1990–1991.

———. "The Wire and Cable Scene: Facts, Fictions, and Frauds, Part 2." *The Audio Critic* 16, Spring-Fall 1991.

Adams, Robert. "Clock Jitter, D/A Converters, and Sample-Rate Conversion, *The Audio Critic* 21, Spring 1994. An informative but complex analysis of digital-sound artifacts.

Allen, John. "Digital Sound for Motion Picture Theatres??? A Reality Check," *Boxoffice*, July 1993; "Digital Stereo Demonstration: An Invitation to the Industry," *Boxoffice*, November 1993. Two-part analysis of motion-picture sound, including material on Dolby SR-D and DTS.

———. "Video Projection," *The Boston Audio Society Speaker* 20 (6) March 1997.

Allison, Roy. "The Best Place for Your Speakers? Your Computer Knows." *Audio*, August 1994.

———. "The Delicate Question of Speaker Placement," *Stereo Review*, August 1975.

———. "Imaging and Loudspeaker Directivity: To Beam or Not To Beam" (Paper delivered at the 99th convention of the Audio Engineering Society, October 1995; available as preprint 4095 [K-4]).

———. "The Influence of Room Boundaries on Loudspeaker Power Output," *JAES* 22, June 1974. The original, definitive, discussion of the "Allison Effect."

———. "Loudspeakers and Real Rooms," *Hi-Fi News and Record Review* (British), December 1989; "Room for Improvement," *Hi-Fi News and Record Review* (British), February 1990; "Marking the Boundaries," *Hi-Fi News and Record Review* (British), April 1990. Three-part analysis of room-loudspeaker interactions.

———. "The Sound Field in Home Listening Rooms," *JAES* 20, July-August 1972. Must reading for fans of the classic AR-3a loudspeaker system.

———. "The Sound Field in Home Listening Rooms, II," *JAES* 24, January-February 1976.

Allison, Roy, and Edgar Villchur. "On the Magnitude and Audibility of FM Distortion in Loudspeakers," *JAES* 30, October 1982.

Anastassiou, Dimitris. "Digital Television," *Proceedings of the IEEE* 82, April 1994.

Angus, Robert. "Celestial Broadcasting," *Audio Video Interiors*, July 1995. Satellite TV systems.

———. "Shocking News about Surge Protection," *High Performance Review*, Fall 1991.

Ardito, Maurizio. "Studies of the Influence of Display Size and Picture Brightness on the Preferred Viewing Distance for HDTV Programs," *SMPTE Journal* 103, August 1994.

Attewell, Trevor. "Ambisonics—The Future of Surround Sound?" *Hi-Fi News and Record Review*, September 1982.

Ballagh, K. O. "Optimum Loudspeaker Placement Near Reflecting Planes," *JAES* 31, December 1983. See also Roy Allison, "Comments on Optimum Loudspeaker Placement Near Reflecting Planes," *JAES* 32, September 1984.

Bamford, J. S. "Ambisonic Sound for the Masses," *Canadian Acoustics* 22, September 1994.

Barron, Michael. "Bass Sound in Concert Auditoria," *JASA* 97, February 1995.

Barron, Michael, and L. Lee. "Energy Relations in Concert Auditoriums," *JASA* 84, August 1988.

Barry, Jim. "Small Dish, Big Picture," *Video*, August 1993. A good review of small- and large-dish satellite technology.

Barry, Peter. "Making the Scene," *Video*, November 1995. Several experts discuss their reference-standard video recordings.

Bartlett, Bruce. "Choosing the Right Microphone by Understanding Design Trade-offs," *JAES* 35, November 1987.

———. "Good Sound: What Is It?" *High Performance Review*, Spring 1987.

———. "Stereo Imaging," *High Performance Review*, Winter 1992–1993.

Bartlett, Bruce, and Michael Billingsley. "An Improved Microphone Array Using Boundary Technology: Theoretical Aspects," *JAES* 38, July-August 1990. Audio recording techniques.

Baxandall, Peter. "A Technique of Displaying the Current and Voltage Output Capability of Amplifiers and Relating This to the Demands of Loudspeakers," *JAES* 36, January-February 1988.

Beacham, Frank. "An Unexpected Challenge for HDTV," *Video* 1994.

Bech, Søren. "Perception of Timbre of Reproduced Sound in Small Rooms: Influence of Room and Loudspeaker Position," *JAES* 42, December 1994.

Begault, Durand. "Perceptual Effects of Synthetic Reverberation on Three-Dimensional Audio Systems," *JAES* 40, November 1992.

Benjamin, Eric. "Audio Power Amplifiers for Loudspeaker Loads," *JAES* 42, September 1994.

Bently, Helen (U.S. Representative, Maryland). "The Picture Is Coming Clear," *Congressional Record*, 25 May 1989. The decline of the U.S. consumer-electronics industry.

Berger, Ivan. "The Evolution of Stereo," *Video*, September 1988.

Berkovitz, Robert, and Bjórn Edvardsen. "Phase Sensitivity in Music Reproduction" (Paper delivered at the 58th AES convention, November 1977; available as preprint 1294 [J-4]).

Berlant, Bert. "Loudspeaker Directionality and the Perception of Reality," *JAES* 33, May 1985.

Bernard, Josef. "All about Surround Sound," *Radio-Electronics*, June 1990.

Berriman, David. "The Bass Race," *Electronics World and Wireless World*, February 1994. Bass system performance.

Betley, Marge. "Sounding Out," *Opera News* 57, August 1992. Opera-house modification for sound reinforcement.

Blauert, Jens, and P. Laws. "Group Delay Distortion in Electroacoustical Systems," *JASA* 68, May 1978.

Bolt, Richard, and Philip Morse. "Sound Waves in Rooms," *Reviews in Modern Physics*, April 1994.

Booth, Stephen. "Stereo or Doubletalk," *Video Review*, February 1992. A discussion of MTS stereo and Thomson XS Stereo for video.

Borish, Jeffrey. "The Basics of Concert-Hall Acoustics," *High Fidelity*, July 1989.

Bowles, G. H. "The Future of Music Databases," *Fontes Artis Musicae*, January 1987.

Box, Alan (President, EZ Communications). "Digital Audio Broadcasting," Washington, DC: Government Printing Office, 1992. Hearing of the House Committee on Energy and Commerce dealing with the impact of the compact disc and other recorded mediums.

Bradley, J. S. "Architectural Acoustics Research Today," *Sound and Vibration* 25, January 1991. An analysis of new developments in architectural research, measurements, and noise control in concert halls and auditoriums.

———. "Comparison of Concert Hall Measurements of Spatial Impression," *JASA* 96, December 1994.

———. "A Comparison of Three Classical Concert Halls," *JASA* 99, March 1991.

Brockhouse, Gordon. "The CD Bit Wars," *High Fidelity*, April 1989.

———. "Head Monster: An Interview with Noel Lee," *Andrew Marshall's Audio Ideas Guide*, Winter 1990.

Bugliera, Vito. "History of Compatibility between Cable Systems and Receivers," *IEEE International Conference on Consumer Electronics, 13th*, IEEE Service Center 1994.

Burr, Ty. "The Letterbox Dilemma," *Video Review*, February 1990. Wide-screen laserdiscs: pro and con.

Burwen, Richard. "20,000-Watt Hi-Fi Gets Digital EQ," *Audio*, April 1995. An interesting analysis of the use of top-calibre equalization to improve the sound of nearly any recording.

Butterworth, Brent. "Why Widescreen?" *Video*, May 1993.

Cabot, Richard. "Audio Measurements," *JAES* 35, June 1987.

Canby, Edward Tatnall. "Looking for Mr. Good Mike," *Audio*, December 1991. Microphones and their use during live performances.

Carver, Robert. "Results of an Informal Test Project on the Audibility of Amplifier Distortion," *Stereo Review*, May 1973.

Charles, Jeff. "Acoustics," *The Architectural Review* 194, June 1994. A sound analysis of the Glyndebourne Opera House, Sussex, England.

Chiarella, Chris. "Copy Right," *Video*, April 1996. How film is transferred to videotape.

Chisholm, Malcom. "The Acoustical Treatment of Recording Studios," *Mix*, August 1990.

Ciapura, Dennis. "FM Fidelity: Is the Promise Lost?" *Audio*, March 1985.

Clark, David. "ABXing DCC," *Audio*, April 1992. A/B tests of DCC recorders.

———. "High-Resolution Subjective Testing Using a Double-Blind Comparator," *JAES* 30, May 1982. Scientific A/B testing.

———. "To the Max: Better Sub-System Design Parameters," *Car Stereo Review*, November-December 1995.

Cordesman, Anthony. "RCA's Digital Satellite System: A Gourmet Dish or a Pot of Trouble?" *Audio*, August 1995.

Cremer, Lothar. "Early Lateral Reflections in Some Modern Concert Halls," *JASA* 85, March 1989.

D'Appolito, Joseph. "A Geometric Approach to Eliminating Lobing Error in Multiway Loudspeakers (paper presented at the 74th convention of the Audio Engineering Society, October 1982; available as preprint 2000).

Davis, Don. "The LEDE Concept," *Audio*, August 1987. An analysis of live-end–dead-end control and listening rooms by their chief advocate.

Davis, Fred. "Effects of Cable, Loudspeaker, and Amplifier Interactions" *JAES* 39, June 1991. A seminal analysis of speaker-wire performance.

———. "Hi-Fi Audio Pseudoscience," *The Skeptical Inquirer* 15, Spring 1991.

———. "More on Hi-Fi Audio Claims," *The Skeptical Inquirer* 16, Fall 1991.

———. "Speaker Cables: Testing for Audibility," *Audio*, July 1993.

Davis, Mark. "Audio Specifications and Human Hearing," *Stereo Review*, May 1982. A benchmark essay on psychoacoustics for the layman.

———. "Loudspeaker Systems with Optimized Wide-Listening-Area Imaging," *JAES* 35, November 1987. A discussion of the principles behind the original dbx loudspeaker.

———. "What's Really Important in Loudspeaker Performance?" *High Fidelity*, June 1978. Perhaps the best short lay-oriented essay on loudspeaker room behavior ever written.

Deiss, Michael. "A DBS Digital Television System Using MPEG Compression and High-Power Transponders," *IEEE International Conference on Consumer Electronics, 13th*, IEEE Service Center, 1994.

DiAntonio, Peter. "Acoustic Room Treatments," *Recording Engineer/Producer*, July 1989.

Dolman, Peter. "Surrounded by Sound," *Electronics World and Wireless World*, January 1990.

Drury, G. M. "Broadcasting by Satellite," *International Journal of Satellite Communications* 12, July-August 1994.

Eargle, John. "Do CD's Sound Different?" *Audio*, November 1987.

———. "The First Step to Great Sound," *Stereo Review*, December 1993. A top recording engineer discusses his techniques.

———. "On the Processing of Two- and Three-Channel Program Material for Four-Channel Playback," *JAES*, April 1971. Surround-sound matrixing techniques.

———. "Stereo Microphone Techniques," *db* magazine, June 1981.

———. "Testing, 1, 2, Testing," *Audio*, February 1991. A discussion of recording techniques, plus a review of two test discs.

———. "Video Data Reduction," *Audio*, November 1994.

Edeko, F. O. "Image Movement in Stereophonic Sound Systems," *Electronics & Wireless World* 94, May 1988.

———. "Improving Stereophonic Image Sharpness," *Electronics & Wireless World* 94, January 1988.

Eder, Bruce. "Widescreen Fever," *Video*, September 1991. A good analysis of letterboxing.

Eickmeier, Gary. "An Image Model Theory for Stereophonic Sound." (Paper presented at the 87th Audio Engineering Society convention, October 1989. Available from the society as preprint 2869, A-3).

Elen, Richard. "Ambisonic Mixing: an Introduction," *db* magazine, July 1984.

Ellis, Leslie. "The Buzz about Cable," *Video Review*, December 1991.

Elrich, David. "The Big Picture" *Video Review*, February-March 1993. 16:9-aspect-ratio TV sets.

Elson, Mark. "Ghosts in the Machine: How to Recognize and Eliminate Video Distortion," *Video*, October 1995.

———. "It's Unpacked, Now What? Setting Up Your System for Great Sound," *Stereo Review*, May 1994.

Erlichman, James. "Will CD's Live Long Enough to Become Golden Oldies?" *The Guardian*, July 2 1988.

Everest, F. Alton. "Coloration of Room Sound by Reflections," *Audio*, March 1993.

———. "The Uneasy Truce between Music and the Room," *Audio*, February 1993.

Everitt, David. "You Say You Want a Resolution to the Unresolved Question," *Video Review* 15, Winter 1994. A discussion of horizontal-luminance resolution.

Fantel, Hans. "Sight and Sound: Dirty Tricks," *Opera News*, August 1991. Sales behavior and ethics.

Farrar, K. "The Soundfield Microphone," *Wireless World*, October 1979.

———. "The Soundfield Microphone, Part 2," *Wireless World*, November 1979.

Feldman, Leonard. "Dynaco QD-1 Series II Surround-Sound Processor," *Audio*, September 1992. Includes data on the Hafler circuit.

Feldman, Leonard, and John Sunier: "Surround Sound without the Pictures," *Audio*, December 1991.

Fellgett, Peter. "Ambisonics," *New Electronics*, May 1981.

———. "Directional Information in Reproduced Sound," *Wireless World*, September 1972.

Ferstler, Howard W. "The Compact Disc," *American Record Guide*, May-June 1988.

———. "Critical Listening: A Comprehensive Guide," *Stereo Review*, December 1995.

———. "The Great Amp Debate," *Digital Audio*, January 1988. Audible distortion in amplifiers.

———. "Loudspeaker Design: What Matters and What Doesn't," *Stereo Review*, April 1995.

———. "Loudspeaker Testing: CU vs. Who? and Why?" *BAS Speaker*, June-July 1987.

———. "Sound Dispersion," *Stereo Review*, October 1990. A discussion of speaker performance in real-world listening rooms.

———. "Subwoofers," *American Record Guide*, Fall 1987.

———. "Useful Noise," *Stereo Review*, December 1988. Using pink noise to test system hardware.

Fielder, Louis, and Eric Benjamin: "Subwoofer Performance for Accurate Reproduction of Music," *JAES* 36, June 1988.

Fincham, L. R. "A Bandpass Loudspeaker Enclosure" (this is a reprint of a paper delivered at the 63rd AES convention, May 1979, and relates to certain KEF designs; available as preprint 1512 [D-4]).

Finger, Robert. "Video CD: A Coding Challenge," *Audio*, December 1994. A performance analysis of the new CD-sized, MPEG-encoded videodiscs.

Fischetti, A., Y. Hemim, and J. Jouhaneau. "Differences between Headphones and Loudspeaker Listening in Spatial Properties of Sound Perception," *Applied Acoustics* 39, no. 4, 1993.

Fishman, Mark. "Time Alignment and Subwoofer Placement," *BAS Speaker*, July 1996. A review of Tom Nousaine's research.

Flindell, I. H., A. R. McKenzie, H. Negishi, M. Jewitt, and P. Ward. "Subjective Evaluations of Preferred Loudspeaker Directivity" (paper presented at the 90th convention of the AES, Paris, France, February 1991; available as a preprint from the Society).

Foster, Alvin. "To Fuse or Not to Fuse?" *BAS Speaker*, March-April 1981. A discussion of wire and fusing resistance effects on speaker sound.

Foster, Edward. "Home Movies: All about Home Theater Systems," *Stereo Review Buyer's Guide, 1995.*

———. "The Lowdown on Subwoofers," *Stereo Review*, December 1993.

Fostle, Don. "Digital Deliverance," *Audio*, April 1996. A discussion of HDCD technology.

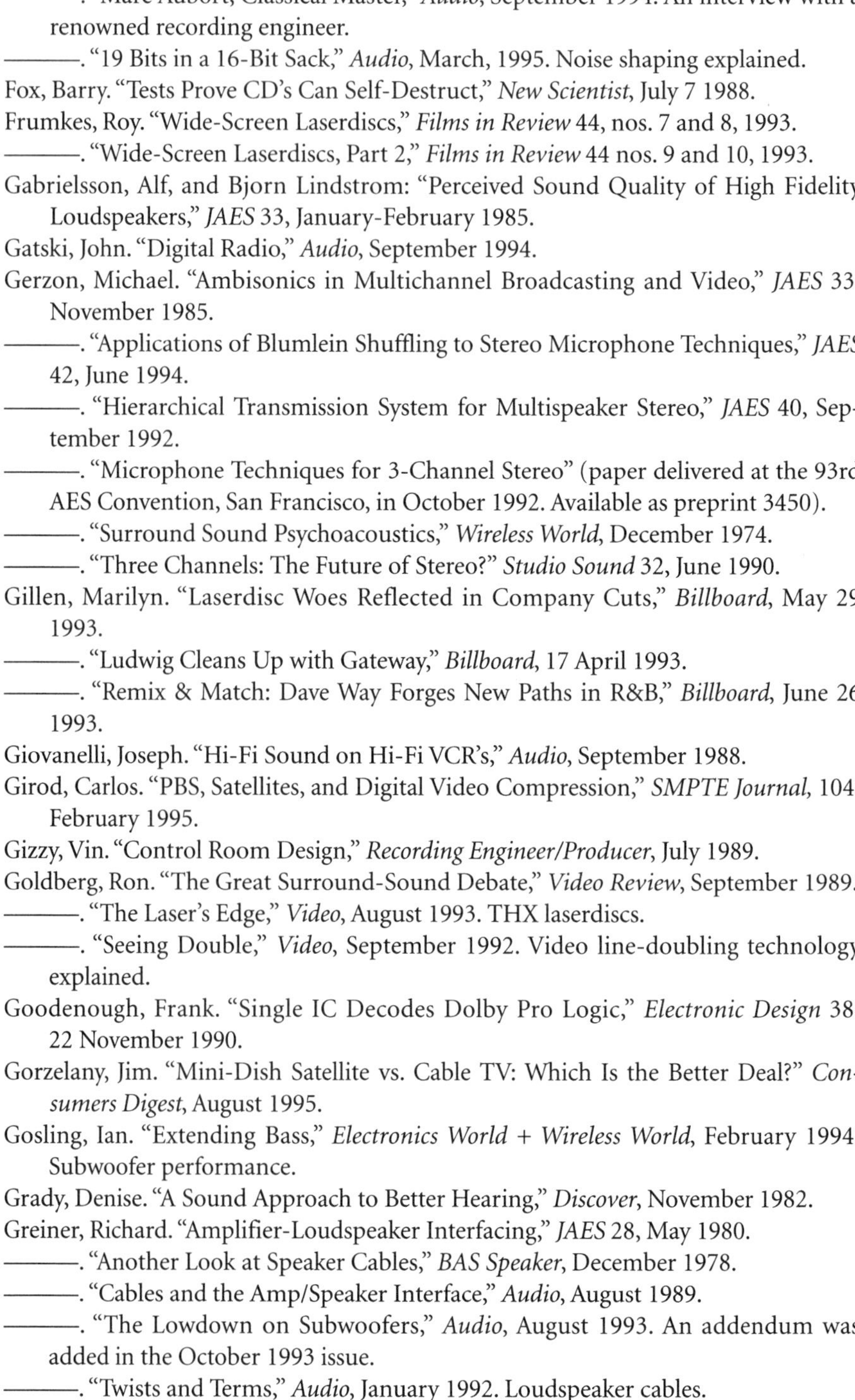

———. "Marc Aubort, Classical Master," *Audio*, September 1994. An interview with a renowned recording engineer.
———. "19 Bits in a 16-Bit Sack," *Audio*, March, 1995. Noise shaping explained.
Fox, Barry. "Tests Prove CD's Can Self-Destruct," *New Scientist*, July 7 1988.
Frumkes, Roy. "Wide-Screen Laserdiscs," *Films in Review* 44, nos. 7 and 8, 1993.
———. "Wide-Screen Laserdiscs, Part 2," *Films in Review* 44 nos. 9 and 10, 1993.
Gabrielsson, Alf, and Bjorn Lindstrom: "Perceived Sound Quality of High Fidelity Loudspeakers," *JAES* 33, January-February 1985.
Gatski, John. "Digital Radio," *Audio*, September 1994.
Gerzon, Michael. "Ambisonics in Multichannel Broadcasting and Video," *JAES* 33, November 1985.
———. "Applications of Blumlein Shuffling to Stereo Microphone Techniques," *JAES* 42, June 1994.
———. "Hierarchical Transmission System for Multispeaker Stereo," *JAES* 40, September 1992.
———. "Microphone Techniques for 3-Channel Stereo" (paper delivered at the 93rd AES Convention, San Francisco, in October 1992. Available as preprint 3450).
———. "Surround Sound Psychoacoustics," *Wireless World*, December 1974.
———. "Three Channels: The Future of Stereo?" *Studio Sound* 32, June 1990.
Gillen, Marilyn. "Laserdisc Woes Reflected in Company Cuts," *Billboard*, May 29 1993.
———. "Ludwig Cleans Up with Gateway," *Billboard*, 17 April 1993.
———. "Remix & Match: Dave Way Forges New Paths in R&B," *Billboard*, June 26 1993.
Giovanelli, Joseph. "Hi-Fi Sound on Hi-Fi VCR's," *Audio*, September 1988.
Girod, Carlos. "PBS, Satellites, and Digital Video Compression," *SMPTE Journal*, 104, February 1995.
Gizzy, Vin. "Control Room Design," *Recording Engineer/Producer*, July 1989.
Goldberg, Ron. "The Great Surround-Sound Debate," *Video Review*, September 1989.
———. "The Laser's Edge," *Video*, August 1993. THX laserdiscs.
———. "Seeing Double," *Video*, September 1992. Video line-doubling technology explained.
Goodenough, Frank. "Single IC Decodes Dolby Pro Logic," *Electronic Design* 38, 22 November 1990.
Gorzelany, Jim. "Mini-Dish Satellite vs. Cable TV: Which Is the Better Deal?" *Consumers Digest*, August 1995.
Gosling, Ian. "Extending Bass," *Electronics World + Wireless World*, February 1994. Subwoofer performance.
Grady, Denise. "A Sound Approach to Better Hearing," *Discover*, November 1982.
Greiner, Richard. "Amplifier-Loudspeaker Interfacing," *JAES* 28, May 1980.
———. "Another Look at Speaker Cables," *BAS Speaker*, December 1978.
———. "Cables and the Amp/Speaker Interface," *Audio*, August 1989.
———. "The Lowdown on Subwoofers," *Audio*, August 1993. An addendum was added in the October 1993 issue.
———. "Twists and Terms," *Audio*, January 1992. Loudspeaker cables.

Greiner, Richard, and Jeff Eggers: "The Spectral Amplitude Distribution of Selected Compact Discs," *JAES* 37, April 1989.

Greiner, Richard, and Douglas Melton: "Observations on the Audibility of Acoustic Polarity," *JAES* 42, April 1994. Further observations appeared in *JAES* 43, March 1995.

———. "A Quest for the Audibility of Polarity," *Audio*, December 1993.

Griesinger, David. "Improving Room Acoustics Through Time-Variant Synthetic Reverberation" (paper presented at the 90th convention of the AES, Paris, France, February 1991; available as a preprint from the Society).

———. "Spaciousness and Localization in Listening Rooms and Their Effects on the Recording Technician," *JAES* 34, 1986.

———. "Theory and Design of a Digital Audio Signal Processor for Home Use," *JAES* 37, January-February 1989. A useful analysis of surround sound.

Hall, David. "Phonorecord Preservation," *Special Libraries*, September 1971. Preserving LPs.

Harley, Robert. "Audio McCarthyism," *Stereophile*, January 1992.

———. "The Listener's Manifesto" *Stereophile*, January 1992. Entertaining, if bizarre and somewhat irrational, high-end audio attacks on objective audio testing by a noted advocate of golden-ear audio theory.

Hartmann, William. "Localization of Sound in Rooms: The Franssen Effect," *JASA* 86, October 1989.

———. "On the Minimum Audible Angle: A Decision Theory Approach," *JASA* 85, May 1989.

———. "Turning on a Tone," *JASA* 90, August 1991.

Herschelmann, Russ. "Please Touch That Dial," *Home Theater*, Spring 1996. A guide to aligning your TV with Reference Recordings' "A Video Standard."

———. "Video Dynamic Range and Room Color," *Widescreen Review*, July-August 1995.

Hibbing, Manfred. "XY and MS Microphone Techniques in Comparison," *JAES* 37, October 1989.

Hirsch, Julian. "Audio Cables: Fact and Fiction," *Stereo Review*, January 1994.

———. "*Consumer Reports*' Audio Tests," *Stereo Review*, September 1994.

———. "Hughes AK-100 Sound Retrieval System," *Stereo Review*, March 1992.

———. "Spatial Effects through Headphones," *Stereo Review*, July 1990.

Hirsch, Julian, and Michael Riggs: "Power: How Much Is Enough?" *Stereo Review*, August 1992.

Hodges, Ralph. "Ambisonics," *Stereo Review*, August 1992.

———. "Break-Ins," *Stereo Review*, March 1994. A nice analysis of the bizarre concept of breaking in an amplifier for better performance.

———. "An Equalizer," *Stereo Review*, January 1986. The importance of frequency response.

———. "Repercussions II," *Stereo Review*, November 1993. The use of conventional hi-fi speakers in home-theater applications.

———. "A Solid Center," *Stereo Review*, September 1990. Center-channel performance with surround-sound systems.

Hoffman, William. "Where Do You Begin Planning a Home Theater System?" *Audio*, November 1994.

Holman, Tomlinson. "Home THX," *Stereo Review*, April 1994.

———. "Loudspeaker Baffles Make an Impact: Eliminating Peaks and Dips in Auditorium Sound," *Film Journal*, September 1992.

———. "Motion-Picture Theater Sound System Performance: New Studies of the B-Chain," *SMPTE Journal* 103, March 1994.

———. "New Factors in Sound for Cinema and Television," *JAES* 39, July-August 1991. An excellent review of Dolby surround-sound techniques, recording practices, and playback system parameters.

———. "Stereo for the Movies: The Center Channel," *Audio*, April 1993.

———. "Surround Speakers," *Audio*, July 1995.

———. "10 Tips for TV Shopping," *Stereo Review*, November 1995.

———. "THX Sound System: Certified Hi-Fi for the Movies," *Audio*, September 1989.

Holl, Tim. "Room to Improve," *High Fidelity*, June 1986.

Honeycutt, R. A. "Will 'Beastie' Speaker Cables Improve Your Audio?" *Radio-Electronics*, February 1991.

Hoogendoorn, Abraham: "Digital Compact Cassette," *Proceedings of the IEEE* 82, October 1994.

Howard, Keith. "Beyond Stereo," *Gramophone*, June 1996.

———. "Beyond Stereo, Pt. 2," *Gramophone*, July 1996.

Howland, Rick. "Mechanics Hall, Meetinghouse for Music," *Audio*, Jan 1993. A discussion of concert-hall acoustics, with one particular hall used as a paradigm.

Jacobson, Robert, and J. Noel Gonzaga: "Hi-Fi: Countdown for CD's," *Opera News*, January 5 1985.

Jameson, Richard. "A View to a Laser," *Film Comment* 30, July-August 1994. Laser disc players and laser discs in general.

Johnson, Lawrence. "Making of a Classic," *CD Review*, May 1994. How the Chandos Records recording team functions.

Joly, Michael. "Plumbing the Secrets of Surround Sound: How to Get Full Bass Effects from Your System," *Film Journal* 94, October-November 1991. A piece on the controversial subject of bass response in the surround channels.

Julstrom, Stephen. "A High-Performance Surround Sound Process for Home Video," *JAES* 35, July-August 1987.

———. "An Intuitive View of Coincident Stereo Microphones," *JAES* 39, September 1991.

Kane, Joe. "Connectology," *Video*, April 1997. A discussion of video cable connections.

———. "A Lesson in Color," *Audio Video Interiors*, October 1995. The NTSC and HDTV color systems.

———. "Line Doublers," *Widescreen Review*, March-April 1996. A detailed analysis of IDTV performance.

———. "The Viewing Environment," *Widescreen Review*, July-August 1995.

Kantor, Ken. "DSP in Audio: The Promise, Reality, and Future of Digital Signal Processing," *Audio*, December, 1996.

———. "Plane Facts about Flat Speakers," *Audio*, August 1987. A discussion of flat-panel loudspeaker systems.

Kantrowitz, Philip. "Distortion Measurements of High-Frequency Loudspeakers," *JAES* 10, October 1962.

Kates, James. "Optimal Loudspeaker Directional Patterns," *JAES* 28, November 1980.

———. "A Perceptual Criterion for Loudspeaker Evaluation," *JAES* 32, December 1984.

Kaufman, Richard. "Build a Simple Surround Decoder," *Audio*, June 1991. A useful analysis of surround matrixing.

———. "Plain Wire/Fancy Reception," *Audio*, January 1991.

Keele, Donald. "Thunder in the Listening Room," *Audio*, November 1992. A test review of four interesting subwoofers, with an analysis of subwoofer design theory.

Killion, Mead, and Tom Tillman. "Evaluation of High-Fidelity Hearing Aids," *Journal of Speech and Hearing Research* 25, March 1982.

Klasco, Michael. "Acoustical Tune-Up: How to Get Your Listening Room into Shape," *Stereo Review*, January 1994.

Klayman, Arnold. "SRS: Surround Sound with Only Two Speakers," *Audio*, August 1992.

Klein, Larry. "Amplifier Damping Factor," *Radio-Electronics*, January 1989.

———. "Amplifier Transfer Functions: a Strange Audio Controversy," *Radio-Electronics*, December 1990.

———. "Transfer Functions, Part II," *Radio-Electronics*, March 1991.

———. "Audio Evaluations: A Non-Mystical Approach," *Electronics Now*, November 1992. Amplifiers and audible distortion.

———. "Audio Amplifiers: Do They Sound Different?" *Radio-Electronics*, April 1991.

———. "Cable Conflicts: The Mysteries of Getting Wired," *Electronics Now*, December 1993. Audio cable facts and myths.

———. "Cables (and Interconnects) Revisited," *Electronics Now*, January 1996.

———. "Getting the Noise Out of FM," *Radio-Electronics*, June 1988.

———. "Listening Tests," *Electronics Now*, August 1995. Blind vs. nonblind equipment comparison testing.

———. "Loudspeaker Power Ratings: Midrange and Woofer Problems," *Electronics Now*, July 1993.

———. "Loudspeaker Power Ratings: What Do They Mean and How Do You Avoid Problems?" *Electronics Now*, June 1993.

———. "Multi-Channel Made Easy," *Electronics Now*, February 1993. An explanation of the Hafler circuit.

———. "A Question of Power: What Is the Sound of One Amp Clipping?" *Electronics Now*, January 1994.

Krugman, Dean. "Video Movies at Home: Are They Viewed Like Film or Like Television?" *Journalism Quarterly* 68, Spring-Summer 1991.

Kumin, Daniel, "Getting the Most Out of FM," *Stereo Review*, August 1995.

———. "Making the Right Connections," *Stereo Review*, September 1993. Speaker-wire connections.

———. "Speakeasy: Everything You Need to Know about Shopping for Speakers," *Stereo Review,* March 1997. Well, maybe not quite everything you need to know.

———. "Stop! You're Surrounded," *CD Review,* July 1993. A useful analysis of surround sound.

———. "Surround Speakers: How to Choose Them and Use Them," *Stereo Review,* April 1995.

———. "Surrounded: Dolby Surround AC-3 Digital Surround Sound Lives Up to All the Hype," *Video,* September 1995.

———. "Surrounded: Intro to AC-3, Part II," *Video,* October 1995.

———. "Surrounded: Intro to AC-3, Part III," *Video,* November 1995.

———. "Three's Company," *Stereo Review,* December 1994. Three-piece speaker systems.

———. "THX 1993," *CD Review,* May 1993.

Lander, David: "Tom Jung: The Digital Music Man," *Audio,* August 1988. An interview with DMP's top recording engineer.

Lebegue, Xavier, and David McLaren: "Video Compression for the Grand Alliance: a Historical Perspective," *International Journal of Imaging Systems and Technology* 5, Winter 1994.

Lehnert, Hilmar, and Jens Blauert: "Principles of Binaural Room Simulation," *Applied Acoustics* 36, nos. 3–4, 1992.

Levine, Martin: "HDTV's Grand Alliance," *Video,* August 1993.

Levitin, Daniel: "Tom Stockham: Fidelity vs. Familiarity," *Audio,* November 1994.

Libbey, Ted: "Digital versus Analog," *Omni,* February 1995.

Linkwitz, Siegfried. "Active Crossover Networks for Noncoincident Drivers, *JAES* 24, January-February 1976.

———. "Passive Crossover Networks for Noncoincident Drivers, *JAES* 26, March 1978.

Lipshitz, Stanley. "Power Response of Loudspeakers with Noncoincident Drivers—The Influence of Crossover Design," *JAES* 34, April 1986.

———. "Stereo Microphone Techniques . . . Are the Purists Wrong?" *JAES* 34, September 1986. A seminal analysis of coincident and spaced-array techniques.

Lipshitz, Stanley, and John Vanderkooy. "The Acoustic Radiation of Line Sources of Finite Length." A pointed analysis of tall, line-source, and planar-type loudspeakers, presented at the 81st convention of the AES, November 1986. Available as preprint 2417 (D-4).

———. "The Great Debate: Subjective Evaluation," *JAES* 29, July-August 1981.

———. "More Straight Thinking on Line Sources," *BAS Speaker,* February-March 1985. Line-source and panel-type loudspeaker performance. This piece, essentially part of a literary discussion with Roy Allison, was a follow-up to two shorter entries in the August-September 1984 and December 1984–January 1985 issues.

———. "Uses and Abuses of the Energy-Time Curve," *JAES* 38, November 1990. Testing loudspeakers and sound measurement.

Lipshitz, Stanley, M. Pocook, and John Vanderkooy. "On the Audibility of Midrange Phase Distortion in Audio Systems," *JAES* 30, September 1982. See also D. Shanefield and J. Moir.

Long, Edward. "That Mysterious Source: The AC Power Line," *Audio,* June 1994.

Long, Robert. "How to Buy Your Last Turntable," *Stereo Review*, December 1992.

Lovelace, Eugene. "The Role of Vision in Sound Localization," *Perceptual and Motor Skills* 77, December 1993.

Lucas, Tim. "Who Framed Edward Scissorhands?" *Film Comment* 28, March-April 1992. Letterboxed video recordings evaluated.

Macaulay, Jeff. "Big Bass, Small Box," *Electronics World + Wireless World*, February 1994.

MacCabe, Culann, and Dermot Furlong. "Virtual Imaging Capabilities of Surround-Sound Systems," *JAES* 42, January-February 1994.

Maddox, Richard: "Psychoacoustics and the Home Theater Experience," *Home Theater*, April 1995. A look at the Arnold Klayman–designed, NuReality-built, SRS two-speaker surround-sound system.

Mannes, George: "Between the Lines," *Video Review*, May 1992. Evaluating video performance characteristics by observing the black, vertical blanking interval bar that exists between screens.

Markey, Edward (U.S. Representative, Massachusetts): "Digital Audio Broadcasting," Washington: Government Printing Office 1992. Hearing on the state of radio broadcasting in the USA.

Marshall, Mary. "Compact Disc's 'Indestructibility': Myth and Maybe," *OCLC Micro* 7, no. 1, February 1991.

Masters, Ian. "Ambisonics," *Sound and Vision*, March 1993.

———. "The Audibility of Distortion" *Stereo Review*, January 1989. A summary of a series of tests, done by Dave Clark, on various types of amplifier distortion.

———. "The Basics: Magnetics and Music—How the Sound Gets from There to Here," *Stereo Review*, December 1989.

———. "Do CD's Cost Too Much?" *Stereo Review*, November 1993.

———. "Fair Play: How to Coax the Last Smidgen of Fidelity from Your Soon-to-Be-Irreplaceable Analog Recordings," *Stereo Review*, July 1992. Deals mainly with how to preserve LP recordings.

———. "How to Get Better FM Reception," *Stereo Review*, April 1991.

———. "You Still Don't Get It?" *Stereo Review*, June 1994. Surround delay, center-channel localization, subwoofer placement, the decibel, oversampling, and tape bias.

Masters, Ian, and David Clark: "Do All CD Players Sound the Same?" *Stereo Review*, January 1986.

———. "Do All Amplifiers Sound the Same?" *Stereo Review*, January 1987.

Mastracco, Jim. "Recording the Sound of a Concert Hall," *Recording Engineer/Producer*, November 1989.

McComb, Gordon. "Antennas: A Ghostbuster's Guide," *Video*, October 1990.

———. "Macrovision under the Microscope," *Video*, October 1989.

———. "Super Pictures," *Popular Science*, January 1988. Super-VHS.

McKelvey, John. "CD Longevity," *American Record Guide*, September-October 1994.

Meares, D. J. "Sound Systems for High-Definition Television," *Applied Acoustics* 33.3 1991.

Merline, John. "CD's Forever?" *Consumer's Research Magazine*, May 1989.

Meyer, E. Brad. "The Amp/Speaker Interface," *Stereo Review*, June 1991.

———. "Audio for Video," *Stereo Review*, August 1990.
———. "Digital Film Sound: Rated S for Sound," *Audio*, June 1994. A comparison of SR-D and DTS theater sound systems.
———. "High-Performance Recording: What to Do When an Ordinary Cassette Deck Isn't Good Enough," *Stereo Review*, March 1995. Digital recording mediums.
———. "Magic Space," *Stereo Review*, August 1988. A comparison of the Yamaha and Lexicon surround-sound philosophies.
———. "A Music-Lovers Guide to Signal Processing," *Stereo Review*, May 1992. Equalizers, noise reducers, etc.
———. "Power: How Much Do You Really Need?" *The Boston Audio Society Speaker* 20 (6), March 1997.
———. "Romance of the Record: If CD Is So Good, Why Are Some People Still in Love with Vinyl?" *Stereo Review*, January 1996.
———. "Surround Sound for Music," *Stereo Review*, June 1994. Using a home-theater system for music-only listening.
Meyer, Jürgen. "The Sound of the Orchestra," *JAES* 41, April 1993.
Miller, Tom, and Tom Nousaine. "Down Time: How to Buy an Affordable Subwoofer," *Sound & Image*, April-May 1995.
Mitchell, Peter. "All about Subwoofers, Part 1," *Stereo Review*, June 1992.
———. "All about Subwoofers, Part 2," *Stereo Review*, July 1992.
———. "Amplifier versus Speaker," *Stereo Review*, December 1994.
———. "Choosing Speakers for Surround Sound," *Stereo Review*, September 1992.
———. "Digital on the Air," *Stereo Review*, November 1993. A review of digital radio options.
———. "Digital Sound: Myth and Reality," *Opus*, October 1985.
———. "EQ Review: A Practical Guide to Equalizers and Equalization," *Stereo Review*, December 1993.
———. "Exploring the Near Field," *Opus*, April 1987.
———. "The Ground Floor," *Stereophile*, March 1993. Subwoofers and deep bass.
———. "Imaging and How to Find It," *Opus* June 1987. Sound-stage localization and its relationship to speaker design.
———. "A Sensible Guide to Upgrading: From Two-Speaker Stereo to a Surround-Sound Home Theater," *Stereo Review*, February 1993.
———. "The Sound of AC-3," *Stereo Review*, July 1995.
———. "Surround Sound: Breaking Through to the Third Dimension," *Stereo Review*, April 1992.
Moir, James. "Comments on 'On the Audibility of Midrange Phase Distortion in Audio Systems,'" *JAES* 33, October 1985.
Moran, David. "Moran's Law of Imaging," *BAS Speaker*, February-March 1987. Speaker sound in real rooms.
———. "Speaker Imaging," *Stereo Review*, June 1987.
Nash, Michael. "Grateful Tapers: An Informal History of Recording the Dead," *Audio*, January 1988. A history and analysis of the techniques used to record the rock group the Grateful Dead.
Neuhaus, Mel. "Talk Video," *Sound & Image*, April-May 1995. A roundtable discus-

sion of video issues and video technology by several notables, including Yves Faroudja, Jay Cocks, and Sam Runco.

———. "Wide Scene: A Brief History of Widescreen Processes," *Video*, October 1995.

Nielsen, Soren. "Auditory Distance Perception in Different Rooms," *JAES* 41, October 1993.

Nousaine, Tom. "The Big Woof," *Car Stereo Review*, July-August 1989.

———. "Can You Trust Your Ears?" (paper presented at the 91st AES convention, October 4–8 1991; available from the Society as preprint 3177 [L-3]). A/B testing and the placebo effect.

———. "Do All Amplifiers Sound the Same?" *Sensible Sound* 10, no. 40.

———. "From Tweak to Geek," *The Audio Critic* 19, Spring 1993; "From Tweak to Weasel," *The Audio Critic* 20, Summer 1993. Two articles on audio equipment A/B testing and the lunatic fringe.

———. "Just Do It," *Car Stereo Review*, November-December 1995. An analysis of David Clark's DUMAX woofer-testing device.

———. "Listening Tests: Auto Sound vs. Home," *Car Stereo Review*, May 1990.

———. "Speaker Supports: Do They Make a Difference?" *Stereo Review*, January 1994.

———. "Subwoofer Secrets," *Stereo Review*, January 1995. The advantages of using a single, corner-positioned woofer.

———. "Surround Sanity," *Stereo Review*, April 1996. Placing surround speakers for optimum effect.

———. "THX vs. Non THX. Designed for Soundtracks, How Do Home THX Speakers Perform on Music?" *Stereo Review*, June 1995.

———. "Two-Channel Stereo Is As Dead As the LP." *Audio Critic* 24, Spring 1997.

O'Connell, Joseph. "The Fine-Tuning of a Golden Ear: High-End Audio and the Evolutionary Model of Technology," *Technology and Culture* 33, January 1992. Delusion and reaction in lunatic-fringe audio.

Olive, Sean, Peter Schuck, Sharon Sally, and Marc Bonneville. "The Effects of Loudspeaker Placement on Listener Preference Ratings," *JAES* 42, September 1994.

Oman, Ralph (Registrar of Copyrights, Library of Congress). "The Audio Home Recording Act of 1991," Washington, D.C.: Government Printing Office, 1992.

O'Neill, William. "All Screens Are Not Created Equal," *Home Theater*, February 1995. Projection television screens.

Otala, Matti, and Pertti Huttunen. "Peak Current Requirements of Commercial Loudspeaker Systems," *JAES* 35, 1987. A famous discussion of loudspeaker-amplifier interactions.

Peters, Bret. "The Buzz on Cable TV," *Stereo Review*, August 1994. Eliminating audio hum on cable.

Phillips, Wes. "Eyes on the Prize," *Video*, July-August 1995. A review of the THX laserdisc certification program.

Pinkwas, Stan. "Projection's New Direction," *Video*, August 1991. Liquid-crystal television.

Pisha, B. V., and Charles Bilello. "Designing a Home Listening Room," *Audio*, September 1987.

Pitts, Karen. "How Acceptable Is Letterboxing for Viewing Widescreen Pictures?" *IEEE Transactions on Consumer Electronics* 38, August 1992.

Pohlmann, Ken. "The Battle of the Balcony," *Stereo Review*, April 1995. Competing digital theater sound systems.

———. "The Big Chill," *Stereo Review*, May 1994. A well-thought-out dismissal of the CD cryogenic freezing fad.

———. "Bit Streams: MPEG Coding in Theory and Practice, from 23,000 Miles in Space," *Video*, September 1995.

———. "CD Magic," *Stereo Review*, July 1991. A clear-sighted analysis of CD rings, ink, magic fluids, and vibration-damping feet.

———. "Formal Fisticuffs: Mini Disc vs. Dolby S." *Stereo Review*, March 1997.

———. "Lawful Audio: How and Why the Serial Copy Management System Restricts Digital Recording," *Stereo Review*, March 1995.

———. "Science, Not Magic," *Stereo Review*, April 1991. Bogus CD treatments and how to make real improvements.

———. "Smoke and Mirrors," *Stereo Review*, August 1990. Bogus CD treatments.

———. "Squeeze Play," *Video*, April 1996. Audio data reduction.

Polk, Matthew. "Polk's SDA Speaker: Designed-In Stereo," *Audio*, June 1984. Crosstalk cancellation in two-channel loudspeaker installations.

Preis, Douglas. "Linear Distortion," *JAES* 24, June 1976.

———. "Phase Distortion and Phase Equalization in Audio Signal Processing—A Tutorial Review," *JAES* 30, November 1982.

Preis, Douglas, and P. J. Bloom: "Perception of Phase Distortion in Anti-Alias Filters," *JAES* 32, November 1984.

Prunty, Peter. "Delivery of TV Over Existing Phone Lines," *SMPTE Journal* 103, September 1994.

Quain, John. "Sound in the Round," *Video Review*, September 1990.

Queen, Daniel. "The Effect of Loudspeaker Radiation Patterns on Stereo Imaging and Clarity," *JAES* 27, May 1979.

Ranada, David. "Digital Chaos: Is Digital Audio Heading toward Disaster on the Superhighway?" *Stereo Review*, May 1994. Data compression/reduction.

———. "Digital vs. Analog Tapes," *Spectrum*, Winter 1991–1992.

———. "Hallmarks of Quality: Lexicon vs. Yamaha, *High Fidelity*, September 1988. DSP units—and philosophies—compared.

———. "How to Hook Up a Subwoofer," *Stereo Review*, September 1995.

———. "Inside MiniDisc," *Stereo Review*, March 1993.

———. "Interviewing the Best Interviewees in Audio, Part 1," *The Audio Critic* 18, Spring-Summer 1992. Conversations with John Eargle, Roy Allison, Kevin Voecks, and Floyd Toole, dealing with sound recording.

———. "Interviewing the Best Interviewees in Audio, Part 2," *The Audio Critic* 19, Spring 1993. Conversations with Mark Davis and Robert Carver, dealing with sound recording.

———. "Interviewing the Best Interviewees in Audio, Part 3," *The Audio Critic* 20, Summer 1993. A conversation with speaker designer Ken Kantor.

———. "Keep Your Eyes on the Size," *Video Review*, March 1993. How to choose a large-screen TV monitor.

———. "Picture Perfect," *Stereo Review*, April 1994. Adjusting the image on a TV screen.

———. "The Relapse of Chroma Phobia," *High Fidelity*, July 1989. Video picture artifacts.

———. "Sounding Off," *High Fidelity*, September 1988. A talk with recording engineer John Eargle.

———. "Surround-Sound Speaker Placement," *Stereo Review*, October 1994. Includes HDTV speaker placement.

Ranada, David, and Howard Ferstler. "Turntable Gremlins" *Stereo Review*, October 1984. Improving LP record player performance.

Reneau, Calix. "The Ear in Review," *EQ Magazine*, February 1993.

Rich, David. "Consumer and Designer Prejudices in High-End Audio: A New Way to Examine Them." *Audio Critic* 24, Spring 1997.

Riggs, Michael. "Amplifiers for Surround Sound," *Stereo Review*, November 1993.

———. "Digital Surround Comes Home," *Stereo Review*, May 1995. AC-3 surround, with additional information on DVD.

———. "Horizontal Resolution," *High Fidelity*, January 1985. Aspects of television monitor performance.

Riggs, Michael, and Ian Masters. "Speakers for Home Theater," *Stereo Review*, September 1993.

Riggs, Michael, and Richard Peterson. "Signals from the Sky," *Stereo Review*, November 1994. DBS satellite systems.

Roth, Cliff. "Citizen Kane," *Video*, June 1995. An interview with video guru Joe Kane.

Rothenberg, Jeff. "Ensuring the Longevity of Digital Documents," *Scientific American*, January 1995. Deals with magnetic digital recording.

Sabin, Rob. "Secrets of Youth: How to Keep Your CDs Spinning for a Lifetime," *Stereo Review*, December, 1996.

Sarver, Carleton. "Across the Lines," *High Fidelity*, December 1987. A discussion of video-monitor resolution capabilities by one of the more knowledgeable video writers.

———. "Clearing the Picture," *High Fidelity*, November 1988. An expert explanation of IDTV and line-doubling circuits.

Scheiber, Peter. "Four Channel and Compatibility," *JAES* 19, April 1971. A pioneering discussion of 4:2:4 surround-sound matrixing.

Schroeder, Manfred. "An Artificial Stereophonic Effect Obtained from a Single Audio Signal," *JAES* 6, April 1958.

Schroeder, Manfred, and F. K. Harvey. "Subjective Evaluation of Factors Affecting Two-Channel Stereophony," *JAES*, January 1961.

Schulein, Robert. "Does Surround Sound Add Dimension to Stereo?" *High Performance Review*, December 1987.

———. "Mixing Techniques for Stereosurround," *db* magazine, May-June 1990. The Shure surround-sound encoding process.

Schwartz, Steven. "The Blank Tape Rip-Off," *Video Review*, September 1988. Off-brand VHS tape quality.

Sehring, John. "One Thump or Two: The Advantages of Stereo Subwoofers," *Audio*, February 1994. See also Tom Nousaine, "Subwoofer Secrets," above.

———. "Roll Your Own Subwoofer," *Audio*, July 1995. A construction project that also has useful information on subwoofer performance in general.

———. "Search for a Budget Subwoofer," *Speaker Builder* 4, 1994.

Self, Douglas. "Ultra-Low-Noise Amplifiers and Granularity Distortion," *JAES* 35, November 1987.

Seredynski, Paul, and Lancelot Braithwaite. "Term Limits." *Video*, April 1997. A discussion of video performance terminology.

Shanefield, Daniel. "Comments on 'On the Audibility of Midrange Phase Distortion in Audio Systems,'" *JAES* 31, June 1983.

Shine, Jerry. "A Room for All Acoustics," *Popular Science*, September 1993. Electronic room simulation.

Small, Richard. "Closed Box Loudspeaker Systems Analysis, Part I," *JAES* 20, December 1972.

———. "Closed Box Loudspeaker Systems Analysis, Part II," *JAES* 21, January-February 1973.

———. "Direct-Radiator Loudspeaker Systems Analysis," *JAES* 20, June 1972.

———. "Vented-Box Loudspeaker Systems, Part I," *JAES* 21, June 1973.

———. "Vented-Box Loudspeaker Systems, Part II," *JAES* 21, July-August 1973.

———. "Vented-Box Loudspeaker Systems, Part III," *JAES* 21, September 1973.

———. "Vented-Box Loudspeaker Systems, Part IV," *JAES* 21, October 1973.

Smith, J. H. "The Soundfield Microphone," *db* magazine, July 1978.

Smith, Patrick. "Wired for Bel Canto," *Opera News* 57, August 1992. Microphone usage in live-music situations.

Snow, William. "Audible Frequency Ranges of Music, Speech and Noise," *JASA*, June 1931.

Stark, Craig. "Choosing the Right Tape," *Stereo Review*, March 1992. Analog audio tape.

Stover, Dawn: "Little Dish TV," *Popular Science*, January 1995. A good analysis of DSS.

Straley, Karl. "The Sound of Movies," *Stereo Review*, January 1995. A discussion of movie sound systems, including Dolby Surround.

Swedien, Bruce. "Grammy Recording Forum: Modern Engineering Production Techniques," *NARAS Journal*, Spring 1992.

Sweeney, Daniel. "Surround Sources: Programming Options for Home Theater," *Home Theater*, July 1995.

———. "Thrice Shy: Multichannel Music Formats Further Considered. " *Audio Critic* 24, Spring 1997.

———. "Twice Shy: On Reencountering Multichannel Music Formats." *The Audio Critic*, no. 23, Winter 1995–1996.

———. "Video Line Doublers: Plastic Surgery for NTSC," *Home Theater*, February 1995.

Terry, Kent, Michael Lodman, Gary Conti, and Pat O'Malley. "A Single Chip Stereo AC-3 Audio Decoder," *IEEE International Conference on Consumer Electronics, 13th*, IEEE Service Center, 1994.

Theile, Günther. "On the Naturalness of Two-Channel Stereo Sound," *JAES* 38, October 1991.

Theile, Günther, and Georg Plenge. "Localization of Lateral Phantom Sources," *JAES* 25, April 1977.

Theile, Günther, Martin Woehr, Hans-Juergen Goeres, and Alexander Persterer. "Room-Related Balancing Technique: A Method for Optimizing Recording Quality," *JAES* 39, September 1991.

Thiele, A. N. "Digital Audio for Digital Video," *Journal of Electrical and Electronics Engineering* (Australia), 13 September 1993.

———. "Loudspeakers in Vented Boxes, Part I," *JAES* 19, May 1971.

———. "Loudspeakers in Vented Boxes, Part II," *JAES* 19, June 1971.

———. "Optimum Passive Loudspeaker Dividing Networks," *Proceedings of the Institution of Radio and Electronics Engineers (Australia)* 36, July 1975.

Thompson, Robert. "Surround Sound Isn't Just for Movies Anymore," *Sensible Sound*, Winter 1995/96.

Thorpe, Laurence. "If Progressive Scanning Is So Good, How Bad Is Interlace?" *SMPTE Journal* 99, December 1990.

Toole, Floyd. "Audio Reviews and Reviewers," *Inside Audio Video* (Canada), May 1988. A pointed analysis of equipment reviewing foibles.

———. "Caged Sound," *High Fidelity*, March 1988. The relation between listeners, speakers, and the listening room.

———. "Good Sound: How to Evaluate Speaker Performance," *Stereo Review*, September 1992.

———. "Listening Tests—Turning Opinion into Fact," *JAES* 30, June 1982. The subjective performance of loudspeakers.

———. "Loudspeaker Measurements and Their Relationship to Listener Preferences: Part 1," *JAES* 34, April 1986.

———. "Loudspeaker Measurements and Their Relationship to Listener Preferences: Part 2," *JAES* 34, May 1986.

———. "Loudspeakers and Rooms for Stereophonic Sound Reproduction" (paper delivered at the 8th AES International Conference, May 1990).

———. "A Review of Measurement Options and Practices." An unpublished summary of the findings of a loudspeaker-measurement seminar at the 81st AES convention, Los Angeles 1986.

———. "Subjective Measurement of Loudspeaker Sound Quality and Listener Performance," *JAES* 33, January-February 1985.

———. "Stereo Hearing," *Stereo Review*, January 1973.

Toole, Floyd, and Sean Olive. "The Detection of Reflections in Typical Rooms," *JAES* 37, July-August 1989.

———. "Hearing is Believing vs. Believing Is Hearing: Blind vs. Sighted Listening Tests" (vailable from the Audio Engineering Society as preprint 3894 [H-6]). How failing to perform double-blind tests influences our opinions about products we are comparing.

———. "The Modification of Timbre by Resonances: Perception and Measurement," *JAES* 36, March 1988. How speaker-system resonances can influence system power response.

———. "The Perception of Sound Coloration Due to Resonances in Loudspeakers and Other Audio Components" (paper presented at the 81st AES convention, November 1986; available from the Society as preprint 2406 [A-6]).

Van Tuyl, Laura. "Now: The Best Seat in the House," *The Christian Science Monitor*, 21 January 1992. The impact of home-video rentals on American society and the movie-house business.

Vanderkooy, John, and Stanley Lipshitz. "Dither in Digital Audio," *JAES* 35, December 1987.

Villchur, Edgar. "A Method of Testing Loudspeakers with Random Noise Input," *JAES* 10 October 1962.

———. "Problems of Bass Reproduction in Loudspeakers," *JAES* 5, July 1957.

———. "Speaker Cables: Measurements vs. Psychoacoustic Data," *Audio*, July 1994.

Voelker, Ernst-Joachim. "Control Rooms for Music Monitoring," *JAES* 33, June 1985.

Wagenaars, W. M. "Localization of Sound in a Room with Reflecting Walls," *JAES* 38, March 1990.

Wallach, Hans. "The Precedence Effect in Sound Localization," *JAES* 21, December 1973.

Ward, Charles, James Thompson, and Mallory Harling. "Speaker Cables Compared," *BAS Speaker*, April 1980.

Warren, Rich. "DSS at Home." *Stereo Review*, January 1995.

Warren, Rich, and Michael Riggs: "Getting Wired: Today's Designer Cables Are a Far Cry from the Old '22 Gauge' Speaker Wire," *Stereo Review*, June 1990.

Watkinson, John. "Digital Audio Recorders," *JAES* 36, June 1988.

Welch, Walter. "Preservation and Restoration of Authenticity in Sound Recordings," *Library Trends*, July 1972. Deals only with LP recordings.

Westerink, Joyce. "Subjective Image Quality as a function of Viewing Distance," *SMPTE Journal* 98, February 1989. Television screen size and its effect on viewing habits.

Wetmore, R. Evans, Gerald Nash, and Glenn Berggren. "The Design of a Very High Performance Motion-Picture Screening Room," *SMPTE Journal* 100, August 1991.

Wielage, Marc. "Home-Theater Speakers," *Video Review* 15, 1994.

———. "Inside THX Home Theater," *Laserviews*, October 1993.

Wirtz, G. "PASC Data Coding for DCC," *IEE Colloquium* 62, March 9 1994.

Whyte, Bert. "Halls for All," *Audio*, July 1993. Concert-hall acoustics.

———. "Home Is Where the Hall Is," *Audio*, March 1992. Ambisonic surround sound.

———. "The Urge to Tweak," *Audio*, April 1991. Bogus CD treatments.

Wilson, Kim. "DTS: the Other Digital System," *Home Theater*, July 1995.

Woodcock, Roderick. "VHS vs. 8mm: Is Smaller Really Better?" *Video*, November 1991.

Woszczyk, Wieslaw. "Microphone Arrays Optimized for Music Recording," *JAES* 40, November 1992.

Wrightson, Jack. "Psychoacoustic Considerations in the Design of Studio Control Rooms," *JAES* 34, October 1986.

Wrightson, Jack, and Russ Berger. "Influence of Rear-Wall Reflection Patterns in Live End/Dead End Recording Type Recording Studio Control Rooms," *JAES* 34, October 1986.

Wylie, Fred. "Digital Audio Data Compression," *Electronics & Communication Engineering Journal* 7, February 1995.

Yoshida, Tadao. "The Rewritable MiniDisc System," *Proceedings of the IEEE* 82, October 1994.
Zumbahlen, Hank. "Zobels and All That," *Audio*, June 1995. Crossover design.
Zwicker, Eberhard. "Audio Engineering and Psychoacoustics: Matching Signals to the Final Receiver, the Human Auditory System," *JAES* 39, March 1991.

WORLD WIDE WEB SITES

More and more data on A/V are showing up on the Internet. Because most audio and video engineers like fooling with computers, some of that material is remarkably informative and is written by individuals with an intimate knowledge of hardware and software. Indeed, a fair number of top companies and company engineers have listings, and you can often learn a lot about products and philosophies by perusing them. (Be forewarned that this can often take a lot of time; it's not called the World Wide Web for nothing!) For example, both Dolby and LucasFilm have lengthy, informative, and frequently updated entries, as do individuals like Tomlinson Holman, Michael Riggs, and Roy Allison. Catalog shopping—either checking manufacturer's home pages for specs or looking at special mail-order catalogs online—is another useful pursuit on the Web. Remember that manufacturer's pages are usually filled with the same kind of hype that characterize their advertising brochures—in fact, many are little more than online ads. There are also classified ad listings if you're interested in buying used equipment—although some may be less reliable than others. And there are also the "lunatic fringe" home pages and Usenet links, posted by various individuals and companies with different audio and video axes to grind. Some of these can be fun reading, as long as you remember that the information may not be terribly reliable.

While it is impossible to list all Web sites here (the total number is huge), here are a few worth visiting.

Link Sites

Link sites offer "hyperlinks" to other parts of the Web and are excellent places to start your search. Obviously, there are many general Web index sites, the most famous being Yahoo, that cover the entire Web. The following are some of those dedicated to audio-video.

AVPointer (http://www.av.co.il): a good, easy-to-use site, indexing manufacturers, dealers, publications, and writers.

Audioweb (http://www.audioweb.com).
Boz's Home Page (http://www.engr.unic.ca/~wyung/audio/index.html): not as professional as some of the other indexes, but nicely organized, with Boz's own home-brewed philosophy along with good indices to selected audio-video manufacturers.
HiFi on WWW (http://www.unik.no/~robert/hifi/hifi.html).
Soundsite (http://www.soundsite.com): a basic, general information service.
Whoopie (http://www.Whoopie.com): similar to AVPointer.

Standard Setters

DBS Online (http://www.dbs-online.com/DBS/index.html).
Dolby Laboratories (http://www.dolby.com): informative with lots of specialized information, including their recommendations for designing your home theater.
THX/LucasFilm (http://www.thx.com): also informative, although colored by their own design philosophy (naturally).

Corporate Home Pages

Most major electronic corporations have home pages that will direct you to at least an online catalog. Some also offer more information. A few of those are listed below.

ACI/Audio Concepts Inc. (http://www.audioc.com).
Boston Acoustics (http://www.bostonacoustics.com/boston/).
Cambridge SoundWorks (http://www.hifi.com/).
Carver Corporation (http://www.carver.com): typical online brochure.
Crutchfield (http://www.crutchfield.com).
Definitive Technology (http://www.soundsite.com/definitive/): nicely designed site.
Digital Theater Systems (http://www.dtstech.com): interesting data on this company's 5.1-channel technology.
NAD (http://www.nad.com.au).
Panasonic (http://www.panasonic.com).
Philips (http://www.philips.com/sv): more than just product info; white papers, etc.
Roy Allison Labs (http://home1.gte.net/bobralab/index.html): to download Roy Allison's Bestplace speaker-placement software.
Snell Acoustics (http://www.primenet.com/mainpage).
Sony (http://www.sel.sony.com/SEL/consumer): nicely designed, with white papers giving detailed explanations of Sony's design philosophy.
TMH (http://www.lum.com:80/tmhlabs): Tomlinson Holman's company's home page.
Yamaha (http://www.yamaha.com).

Some Organizations and Online Magazines

Acoustical Society of America (http://www.asa.aip.org/index.html); Audio Engineering Society (http://www.aes.org): two organizations for the diehard techies in the crowd.

Audioshopper (http://cdrome.com:80/cdrome/): online shopping service for used high-end gear.

Cybertheater (http://www.cybetheater.com): electronic magazine written by audio enthusiasts that debuted in the spring of 1996.

Goodsound (http://www.goodsound.com/): an online guide to cheap audio gear.

Home Theater Online (http://www.hometheatermag.com): eye-popping graphics, with content drawn from the slick consumer magazine.

Besides the Web, various audio journals and company information are available through the proprietary services, such as America Online (where you'll find *Stereo Review*, for instance), Prodigy, and Compuserve/WOW!

Index